Introduction to Agricultural Engineering Technology

Introduction to Agricultural Engineering Technology

A Problem Solving Approach

Third Edition

Harry L. Field and John B. Solie
Oklahoma State University
Stillwater, OK, USA

Springer

Harry L. Field
Oklahoma State University, Stillwater, OK, USA
111 Agricultural Hall
Stillwater 74078
fharry@okstate.edu

John B. Solie
Oklahoma State University, Stillwater, OK, USA
111 Agricultural Hall
Stillwater 74078
jsolie@okstate.edu

Library of Congress Control Number: 2006930107

ISBN-10: 0-387-36913-9
ISBN-13: 978-0-387-36913-6
e-ISBN-10: 0-387-36915-5
e-ISBN-13: 978-0-387-36915-0

Printed on acid-free paper.

9 8 7 6 5 4 3 2 1

springer.com

Contents

1
Problem Solving

1.1. Objectives

1. Be able to define problem solving.
2. Be able to describe the common problem-solving methods.
3. Be able to select the appropriate method for solving a problem.
4. Understand the function and use of spreadsheets.
5. Understand the use of and common symbols for flow charts.

1.2. Introduction

Problem solving is a part of living. We are faced with a host of problems on a daily basis. Some of these problems involve people and human relations, whereas others require a mathematical solution. In this chapter we will deal with problems involving mathematical solutions, and several ways in which these problems can be approached.

1.3. Mathematical Problem Solving

Mathematical problem solving is the process by which an individual uses previously acquired knowledge, skills, and understanding to satisfy the demands of an unfamiliar situation. The essence of the process is the ability to use information and facts to arrive at a solution. There are two characteristics of problem solving that must be remembered when solving problems using mathematical processes.

1. The mathematical process does not always give you the answer—just more information so that you can make a more informed decision. Good decision-making requires good information.
2. Whenever perfection is not possible or expected, levels or intervals of acceptability must be established. When perfection is not possible someone must determine the amount of error that is acceptable. The acceptable level of error may be

determined by standards, manufacturers' recommendations, comparison to another situations or machines or personal experience.

Both of these characteristics will be utilized and explained in more detail in later chapters by using examples.

Problems can be solved in different ways. One of the objectives of this chapter is to increase the reader's knowledge of problem-solving methods. Seven different approaches to solving mathematical problems will be discussed: diagrams and sketches, patterns, equations and formulas, units cancellation, intuitive reasoning, spreadsheets, and flow charts.

1.3.1. Diagrams and Sketches

Some problems involve the determination of a quantity of items, such as the number of nails per sheet of plywood or the number of studs in a wall. In solving these types of problems it usually is helpful to draw a sketch or a diagram.

Problem: How many posts are needed to build a fence 100 ft long with posts 10 ft apart?

Solution: For many people the first response would be 10:

$$\text{Posts} = \frac{100 \text{ ft}}{10.0 \text{ ft/post}} = 10 \text{ posts}$$

but a diagram, Figure 1.1 shows that the correct number of posts is 11.

This is an example of a situation where a wrong answer is possible if you do not interpret the problem correctly. In this example, 10 is the number of spaces, not the number of posts.

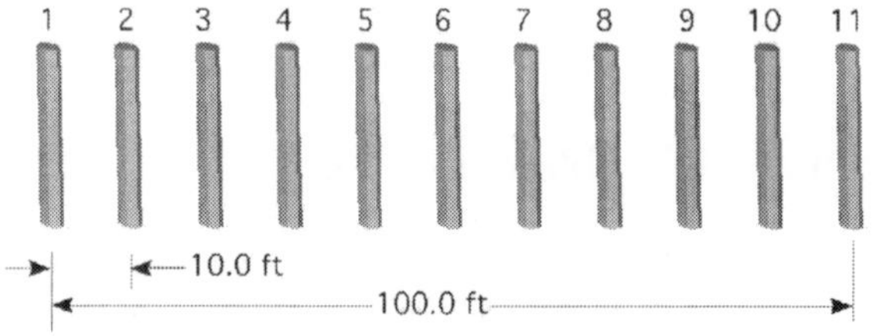

FIGURE 1.1. Number of posts.

1.3.2. Patterns

The solution to some problems may depend upon one's being able to discover a pattern in an array of numbers or values. Frequently, it is convenient to examine the patterns in a sample rather than the entire population. Once a pattern is discovered and shown to be consistent for the sample, it can be used to predict the solution for the entire population.

TABLE 1.1. Patterns in numbers, first sample.

		Cow number									
Ration	Child	1	2	3	4	5	6	7	8	9	10
Grain	1	N	N	N	N	N	N	N	N	N	N
Mineral	2		Y		Y		Y		Y		Y
Hay	3			Y			Y			Y	
Silage	4				Y				Y		
Water	5					Y					Y

Problem: A dairy farmer has five children. Each child is responsible for one part of the daily feed ration for the family's 100 dairy cows. The oldest is responsible for the grain, the second for the minerals, the third for the hay, the fourth for the silage, and the fifth for water. Instead of feeding each cow, the first child decides she will not feed the cows at all that day. The second child decides just to feed every other cow, the third child feeds every third cow, and so on. Dad soon discovers how the cows were fed, and needs to know which cows did not receive any feed or water.

Solution: When one is faced with this type of problem, it is usually helpful to set up a table. In this case, it would be very time-consuming to set up a table for all 100 cows. Instead, select a sample of the cows. If a pattern is true for the sample, there is a high probability that the pattern will be true for a large group. Determining the size of a sample is not always easy. Pick one, and if a clear pattern does not appear, increase the size until a pattern develops. We will start with the first 10 cows, Table 1.1.

In this sample, cows #1 and #7 did not receive any grain, mineral, hay, silage, or water. Is this enough information to establish a pattern? We will predict that the next cow that did not receive any feed or water is #11. Why? To test this prediction, the sample size must be extended to include a larger number of cows.

Table 1.2 shows that the prediction was right; cow #11, along with #13, #17, and #19, did not receive any grain, minerals, hay, silage, or water. It is now safe to consider that the prediction could be used to identify all of the animals within the herd that did not receive any grain, minerals, hay, silage, or water (those animals represented by prime numbers, that is, a number divisible only by itself and one).

TABLE 1.2. Patterns in numbers, second sample.

		Cow number									
Ration	Child	11	12	13	14	15	16	17	18	19	20
Grain	1	N	N	N	N	N	N	N	N	N	N
Mineral	2		Y		Y		Y		Y		Y
Hay	3		Y			Y			Y		
Silage	4		Y				Y				Y
Water	5					Y					Y

1.3.3. Equations and Formulas

Equations and formulas are very similar problem-solving tools. Some texts study them separately, but in this section they will be combined. Equations are a way of showing the relationship between different variables in a problem and are usually derived as needed for each problem. Formulas are equations that are used frequently enough or are some how unique enough that they are remembered and used without a derivation.

1.3.3.1. Equations

The solution to some problems requires the derivation of a mathematical equation based on a pattern or another type of relationship between the numbers. These equations will be unique for each problem.

Problem: How much wire is needed to build a single wire fence around a rectangular field measuring 450 ft long and 350 ft wide?

Solution: In this example there are three quantities: length, width, and perimeter. It should be obvious that the perimeter is a function of the other two. Begin by assigning the variables L to represent the length, W to represent the width, and Pr to represent the perimeter. Then, because a rectangle has two lengths and two widths, the perimeter can be found as follows:

$$\begin{aligned}\mathrm{Pr} &= (L + L) + (W + W)\\ \mathrm{Pr} &= (450\ \mathrm{ft} + 450\ \mathrm{ft}) + (350\ \mathrm{ft} + 350\ \mathrm{ft})\\ &= 900\ \mathrm{ft} + 700\ \mathrm{ft}\\ &= \mathbf{1{,}600\ ft}\end{aligned}$$

1.3.3.2. Formulas

For some problems the relationships of the variables are fixed and constant, so the equation for that problem is remembered and used. These equations are sometimes called formulas. Another characteristic of formulas is that they usually contain a constant. One example is the area of a circle: $A = \pi r^2$. The variable π is a constant.

There are at least two important considerations in using formulas.

1. You must enter the numbers with the correct units of measure. All formulas are designed with specific units for the numbers, especially if they have a constant. If the units are incorrect, the answer will be incorrect. An example is the equation used to determine the application rate of a boom type sprayer:

$$\text{Application rate } \frac{\text{gal}}{\text{ac}} = \frac{5{,}940 \times \text{Flow rate (gal/min)}}{\text{Speed (mi/hr)} \times \text{nozzle spacing (in)}}$$

It should be obvious that the units in the equation do not work (the units when combined do not result in the units for application rate, gal/min). This equation is an example of a situation in which units conversion values, that are always used each time the problem is worked, are combined into a units conversion constant (5,940). If any one of the values is entered in different units, the answer will be incorrect. When we solve for the application rate using units cancellation and conversion values, the source of the constant becomes apparent.

$$\begin{aligned}\frac{\text{gal}}{\text{ac}} &= \frac{\text{gal}}{\text{min}} \times \frac{60\text{ min}}{1\text{ hr}} \times \frac{1\text{ hr}}{1\text{ mi}} \times \frac{1\text{ mi}}{5,280\text{ ft}} \\ &\quad \times \frac{43{,}560\text{ ft}^2}{1\text{ ac}} \times \frac{12\text{ in}}{1\text{ ft}} \times \frac{1}{1\text{ in}} \\ &= \frac{31{,}363{,}200}{5{,}280} \\ &= 5{,}940\ \frac{\text{gal}}{\text{min}}\end{aligned}$$

2. You must be able to rearrange the formula to solve for the unknown value. For example, the application rate equation could be rearranged to solve for nozzle spacing in inches (nsi):

$$\text{Nozzle spacing (nsi)} = \frac{5{,}940 \times \text{Flow rate (gal/min)}}{\text{Speed (mi/hr)} \times \text{Application rate (gal/ac)}}$$

For the remainder of this text the terms equation and formula will be used as synonyms.

1.3.4. Units Cancellation

Some problems are more complex than the examples we have used, and many do not have patterns or previously developed equations. Equations can be developed for some of these problems, but an alternative approach is units cancellation. Problems of this type will usually involve several quantities. All of these quantities, except π, will have a unit such as feet, pounds, gallons, and so on. Units cancellation follows two mathematical principles: (1) the units of measure associated with the numbers (feet, gallons, minutes, etc.) follow the same mathematical rules as the numbers; (2) the units of the numbers behave according to the rules of fractions. For example:

$$2 \times 2 = 4 \text{ or } 2^2$$

With units of feet the same equation is:

$$2\text{ft} \times 2\text{ ft} = 2 \times 2 \text{ and ft} \times \text{ft or } 4\text{ ft}^2$$

To review the rules of fractions study the following example:

$$\frac{3}{4} \times \frac{4}{5} = \frac{3 \times 4}{4 \times 5} = \frac{3}{5}$$

In this example, the 4's in the numerator and denominator cancel out ($4/4 = 1$).

When the units of measure are included, they behave in the same way:

$$\frac{3\ \text{ton}}{4\ \text{hr}} \times \frac{4\ \text{hr}}{5\ \text{day}} = \frac{12\ \text{ton}}{20\ \text{day}} = 0.6\ \frac{\text{ton}}{\text{day}}$$

In this example, the 4's and the units associated with them cancel out. The uncancelled units become the units for the answer. The following example shows another variation of this principle (where gal = gallon and hr = hour):

$$5\ \frac{\text{gal}}{\text{hr}} \times 3\ \text{hr} = 15\ \text{gal}$$

In this example, the unit of hour in the numerator and denominator cancel each other, leaving the units of the answer in gallons.

Problem: What is the weight (lb) of one pint of water?

Solution: If a scale and a one-pint measure were available, it would be a simple task to weight one pint of water. An alternative is to use the conversion factors found in a table of weights and measures (Appendix I) and units cancellation.

Note, in this example two types of measure are used, volume and weight. The real nature of the problem is to find the conversion value(s) that will convert from volume (pints) to weight (pounds).

To begin, refer to Appendix I and identify conversion factors that use both volume and weight. You should find that 1 cubic foot contains 7.48 gallons, and 1 gallon contains 8 pints. This is a start, but you need something more. If you also know that water weighs 62.4 lb per cubic foot, the problem can be solved with (lb = pounds, gal = gallons, pt = pints, and ft^3 = cubic feet):

$$\frac{\text{lb}}{\text{pt}} = \frac{62.4\ \text{lb}}{1\ \text{ft}^3} \times \frac{1\ \text{ft}^3}{7.48\ \text{gal}} \times \frac{1\ \text{gal}}{8\ \text{pt}}$$
$$= 1.04\ \frac{\text{lb}}{\text{pt}}$$

The units of pints, gallons, and cubic feet all cancel each other leaving the answer in the desired units of pounds and pints. This example illustrates several principles of units cancellation.

- It is very important to begin by writing down the correct units for the answer.
- Write down the equal sign (=),
- Begin entering the values and their units. The first value entered should have one of the desired units in the correct position (numerator or denominator), even if it is a units conversion value from Appendix I or another source. Starting with one of the units of measure in the correct position will eliminate the possibility of having the problem inverted.
- Enter a value that will cancel out the unwanted units, if any, in the first value entered.
- Continue to add variables with the appropriate units until the only units that remain are the units of the answer.

- If all of the units cancel except those that are desired for the answer, and the units are in the correct position, then the only possible mistake is a math error.

The process of units cancellation is also useful for problems requiring the development of a new unit. For example, a very common quantity in agriculture is power. Power can have different units, depending on whether it is electrical or mechanical. The units of mechanical power are ft • lb per minute. The solution to a problem in which a 24-oz weight was moved 15 ft in 5 sec would look like this (with oz = ounces, sec = seconds, lb = pounds, ft = feet):

$$\text{Power}\left(\frac{\text{ft} \bullet \text{lb}}{\text{min}}\right) = \frac{15\ \text{ft}}{5\ \text{sec}} \times \frac{60\ \text{sec}}{1\ \text{min}} \times \frac{1\ \text{lb}}{16\ \text{oz}} \times \frac{24\ \text{oz}}{1}$$

$$= \frac{21{,}600\ \text{ft} \bullet \text{lb}}{80\ \text{min}}$$

$$= \frac{270\ \text{ft} \bullet \text{lb}}{\text{min}}$$

(Note that lb • ft is a compound unit, not feet minus pounds or feet times pounds.) This same process will work just as well for problems with units that are more complex and more variables.

1.3.5. Intuitive Reasoning

Intuitive reasoning is a process by which an individual arrives at a correct answer through insight or a hunch, usually without being able to explain the process used. The actual process depends on the individual and cannot be defined in progressive steps.

Problem: You ask your employees to determine how many vehicles in the parking lot need their seat covers replaced. They return with an answer of 150. Then you realize you need to know how many single seat pickups and how many cars with two seats. Your employees remember that there were the same number of cars in the lot needing seat covers as pickups; so how many car and how many pickup seat covers do you need?

Solution: Some people would solve this problem algebraically, but intuitive reasoning can be used to reason out a series of approximations. If there were 20 cars and 20 pickups, you would need 60 seat covers $[(20 \times 1) + (20 \times 2) = 60]$, if 40 cars and 40 trucks, 120 seat covers, if 60 of each then 180 seat covers. The answer must be less than 60 and more than 40. The correct number of vehicles is 50. Which means you would need 50 pickup seat covers and 100 car seat covers.

1.3.6. Spreadsheets

The development and adoption of computers has provided a very useful problem-solving method. This is the spreadsheet. A spreadsheet is a very powerful data

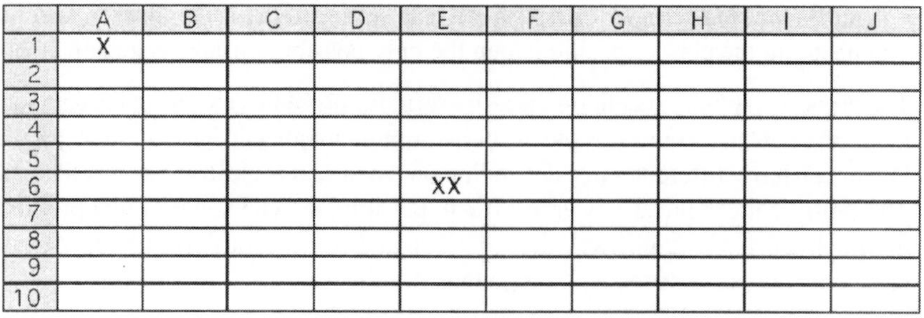

FIGURE 1.2. Columns and rows af a spreadsheet.

management tool that provides a means to enter, manipulate, and plot data and information. There is some variation in the features of different spreadsheets, but they all include several common features. Some of the common features and uses of spreadsheets will be presented in the following sections.

1.3.6.1. Data Entry

To be able to use data in a spreadsheet the computer must know where it is. This is accomplished by setting up a grid or array using columns and rows. It is common for the columns to be identified by letters and the rows to be identified by numbers. The junction of each row and column is called a cell. The labels used for the column and rows gives each cell a unique address or location, see Figure 1.2.

In the example spreadsheet shown in Figure 1.2, the cell that has the "XX" would be cell E6 located in column E and row 6. The number of columns and rows that are included varies with the producer of the spreadsheet. A popular one has columns that go through the alphabet several times and stops at I5. The rows are numbered to over 10,000. A grid with 256 columns and 10,000 rows would have 2,560,000 individual cells for information. The data are entered into the spreadsheet by typing the desired data into the appropriate cell.

Spreadsheets have many other features for recording and manipulating data. One feature is the ability to link information in different sheets. For example, one type of spreadsheet uses the term "workbook" to describe a spreadsheet file. Each

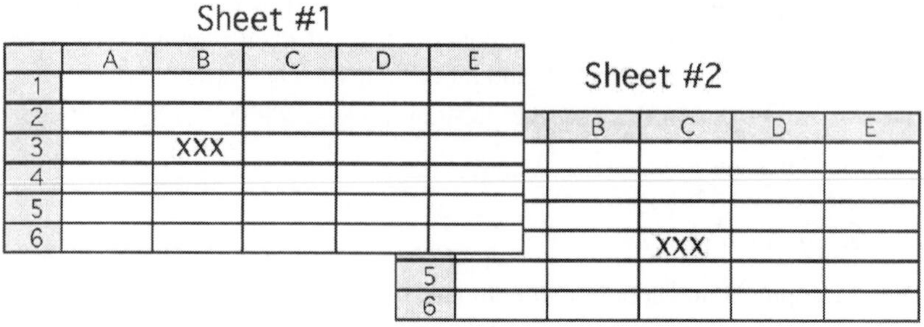

FIGURE 1.3. Linking sheets in a workbook.

FIGURE 1.4. Adding data in a spreadsheet.

	A	B	C	D	E
1					
2	3	4			
3					
4					
5					
6					

workbook can have several pages and the data in one page can be linked to another page. If cell B3 in sheet #1 contained the equation "= Sheet2!C4", that cell would display the contents of cell C4 in sheet #2, Figure 1.3.

Most spreadsheets also have the capability of importing the data from other sources. Experience with using spreadsheets teaches that the user should give some thought to how much data will be used and the best way to organize it before starting data entry. This will make the spreadsheet easier to use and the user will make fewer mistakes.

Although spreadsheets are primary for process data and calculate results, text can also be included in the cells and it can be used to process the data.

1.3.6.2. Data Processing and Calculation

Data in a spreadsheet can be processed with almost any mathematical operation or function. The math functions are used by entering the equation with the functions in the cell where the user wants the answer.

For example, if the desired operation is to add the two numbers in the example spreadsheet in Figure 1.4 and have the answer appear in cell D2, the equation to complete this operation would be inserted in cell D2, Figure 1.5. Note that the equation is started with the "=" sign. In most spreadsheets, starting with the "=" sign tells the computer that a mathematical operation or a text process will be performed on data from other cell(s) and the results displayed in the cell with the equal sign.

This is a simple example, but all mathematical operations are entered in the same way. To expedite the process, spreadsheets also include an option called function. Functions are preprogrammed mathematical operations, such as sum, square roots, and trigonometry functions. These functions are usually accessed through a separate menu option. Functions may be categorized as financial, date and time, math and trig, statistical, lookup and reference, database, text, logical, or information. Not all of these functions will be explained in this text. Two examples will be included to show how functions work. The addition example in Figure 1.5 can be completed using the sum function from the math and trig category.

	A	B	C	D	E
1					
2	3	4		=A2+B2	
3					
4					
5					
6					

FIGURE 1.5. Example of a mathematical operation in a spreadsheet.

	A	B	C	D	E
1					
2	3	4		=sum(A2:A3)	
3					
4					
5					
6					

FIGURE 1.6. Addition using the sum function.

In this example 7, the sum of 3 and 4 would appear in cell D2 (see Figure 1.5). In this example, there is no clear advantage for using the sum function instead of inserting the equation for adding two numbers. The advantage of the sum function is much more apparent when more numbers are added, Figure 1.7.

Using the sum function is much simpler than writing the equation to add several numbers. To write the equation for row 2 in Figure 14.7 using the "+" operator, each cell reference must be included with a "+" in between. The equation "= A2 + B2 + C2 + D2 + E2 + F2" would be entered into cell H2. This operation is easier using a function. The "sum" function is used by entering an = in the cell where the answer is wanted and then sum. After that enter the left parenthesis "("the beginning and ending cell to be added separated by a ":" and finish by entering the right parenthesis")", cell H2, Figure 1.7.

It cannot be demonstrated in this text, but the use of functions is enhanced because the cells can be entered in by highlighting the desired ones. The procedure is to enter "= SUM (then highlight the cells and end the equation with)" to complete the function. Also note that in Figure 1.7, cells can be summed both horizontally and vertically. All of the functions are used in a similar manner.

Another feature of spreadsheets is logic functions. Logic functions are expressions such as "IF", "AND", "FALSE", "NOT", "OR" and "TRUE". Logic functions are very useful because they can be used to compare or relate the values of different cells. The "if" function includes a logic statement so it can be used to compare cells and return a number or word. For example, a flour miller must constantly monitor the percent of flour milled from the wheat. A drop in yield rate would indicate a problem. The "IF" function can be used to trigger an error or warning message when the yield drops below the desired rate, see Figure 1.8.

	A	B	C	D	E	F	G	H	I	J
1										
2	2	4	8	21	45	234		=sum(A2:F2)		
3		5								
4		8								
5		10								
6		234								
7		782								
8										
9		=sum(B2:B7)								
10										

FIGURE 1.7. Using sum function with multiple numbers.

	A	B	C	D	E	F	G	H
1			Date					
2		10/10/04	10/11/04	10/12/04	10/13/04	10/14/04	10/15/04	10/16/04
3	Pounds wheat	10562	11584	11698	10962	11532	11846	10624
4	Pounds flour	8215	10634	10692	10854	10237	10687	9634
5	% Yield	77.78	91.80	91.40	99.01	88.77	90.22	90.68
6		Error				Error		
7								

FIGURE 1.8. Spreadsheet using the "IF" function.

In this example, an "IF" function was used to trigger the spreadsheet to insert the word "error" whenever the percent yield of flour was less than 90%. The function that was inserted into cell B6 written as: = IF(B5<90,"Error"," "). This function was then copied (filled) into the adjoining cells, C2 through H2. Spreadsheets are very powerful tools and allow a lot of flexibility in how they are set up. Through study and practice, the user can make them do very complex calculations and logic statements.

1.3.6.2.1. Fixed or Relational Reference

In the example in Figure 1.8, the reference cell "B5" was not fixed because the desired outcome was to use the same function to sum the columns B through H. This is an example of using a relational reference. Reference cells can also be fixed so they do not change when the function is copied. An example is if the miller in the flour example wanted the spreadsheet set up so that the yield percent could be changed without requiring the retyping of all of the equations. This could be accomplished by using a fixed reference cell for the yield percent, Figure 1.9.

In this example, the user can change the percent yield that will trigger the "Error" message by changing the value that is in cell B1. To accomplish this the function in cell B7 was changed and then copied (filled) to the adjoining cells. The function in cell B7 reads: = IF(B5 < B1,"Error"," "). In the spreadsheet used to develop these examples placing the "$" in front of the column and row label fixes these labels so they do not change when the function is copied into adjoining cells.

These examples are just a hint at the type of data manipulation that can be accomplished using spreadsheets. They are only limited by the amount of time and ingenuity of the user to figure out how to make them complete the desired tasks.

	A	B	C	D	E	F	G	H
1	Desired Yield	85	%					
2				Date				
3		10/10/04	10/11/04	10/12/04	10/13/04	10/14/04	10/15/04	10/16/04
4	Pounds Wheat	10562	11584	11698	10962	11532	11846	10624
5	Pounds Flour	8215	10634	10692	10854	10237	10687	9634
6	% Yeld	77.78	91.80	91.40	99.01	88.77	90.22	90.68
7		Error						
8								

FIGURE 1.9. Spreadsheet using fixed reference cell.

1.3.6.3. Graphing

Another powerful and useful feature of spreadsheets is the ability to produce graphs of the data in the spreadsheet. Some spreadsheets use the term chart instead of graph. Many people are visual learners and seeing a chart of the data will convey information faster and easier than studying the same data in a table. The charting function of most spreadsheets is not as powerful as a dedicated graphing or presentation program, but they will usually provide enough options to satisfy the needs of the average user.

Producing a graph with a spreadsheet has the same requirements as drawing one by hand. The computer must know which set of data should be plotted along the "X" axis and which along the "Y" and the "Z" if three-dimensional charts are used. The chart function should provide an opportunity to type in labels for the axis, the chart title and data legends and values if desired. To demonstrate the charting function a common problem of surveying will be used. A common survey is called a profile. A profile survey collects the data necessary for defining the topography of the earth's surface along a route. This is usually done for a utility, sidewalk, road, or retaining wall. In this example, an underground drainpipe will be used.

Table 1.3 contains the data for a profile survey. More information will be discussed about profile surveys in a later chapter, but to help understand the use of a chart, the numbers in the STA column are the distances from the starting point for each station and the numbers in the ELEV column are the elevations for each station along the profile. This information is combained with the calculations for the drain elevations at each station, Table 1.4, before the profile and the drain can be shown in a chart.

Assume the outlet of the desired drain is 2 ft below the elevation of station "0.0" and the drain will have a 1% slope. Table 1.4 shows the starting elevation, the surface elevation, and the elevation of the drain at each station. The drain elevation

TABLE 1.3. Data for profile survey.

STA	BS	HI	FS	IFS	ELEV
BM	3.56	103.56			100.00
0.0		103.56		4.89	98.67
27.3		103.56		4.67	98.89
35.6		103.56		5.10	98.46
41.2		103.56		5.89	97.67
56.9		103.56		4.68	98.88
63.4		103.56		3.61	99.95
75.9		103.56		4.01	99.95
80.7		103.56		4.65	98.91
93.5	4.04	103.92	3.68		99.88
BM			3.91		100.01
Sum	7.60		7.59		
Difference		0.01		0.01	
			0.01 = 0.01		
	AE=	0.02		.01<.02	

TABLE 1.4. Data and profile and drain.

	A	B	C
1	Starting		
2	Elevation		96.67 ft
3	% Slope		1%
4			
5	Station	Profile	Drain elevation
6	0.0	98.67	96.67
7	27.3	98.89	96.94
8	35.6	98.46	97.03
9	41.2	97.67	97.08
10	56.9	98.88	97.24
11	63.4	99.95	97.30
12	75.9	99.55	97.43
13	80.7	98.91	97.48
14	93.5	99.88	97.61

is determined by inserting an equation in cell C7 and copying it down through C14. The equation is: = \$C\$6 + (C7 × \$B\$3/100).

A person with experience in design and construction would be able to study the table and answer important questions about this drain. Such as, "What is the maximum depth?", or "What is the minimum depth?", and numerous others. The importance of these questions and their answers is easier to see if the data are plotted in a chart. In a chart it is easy to see the difference in elevation between the surface and the drain, the depth of the drain, etc. The chart or plot of the profile and the drain are shown in Figure 1.10.

Figure 1.10 shows that the maximum depth occurs at 62 ft from the start and the difference between the surface elevation and the elevation of the drain is about

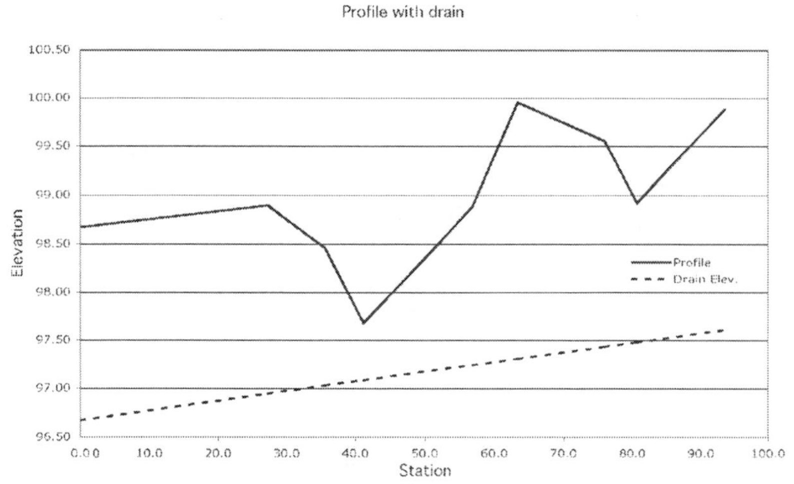

FIGURE 1.10. Chart of profile and drain.

TABLE 1.5. Drain data at 0.5% slope.

	A	B	C
1	Starting		
2	Elevation		96.67 ft
3	% Slope		0.5%
4			
5	Station	Profile	Drain elevation
6	0.0	98.67	96.67
7	27.3	98.89	96.81
8	35.6	98.46	96.85
9	41.2	97.67	96.88
10	56.9	98.88	96.95
11	63.4	99.95	96.99
12	75.9	99.55	97.05
13	80.7	98.91	97.07
14	93.5	99.88	97.14

100.0 − 97.4 or 2.6 ft. The minimum depth occurs at station 40 and it is 97.6 − 97.1 or 0.5 ft. This could be a problem because a drain only 6 inch depth is easily damaged and in cold climates could have problems with freezing.

The power and usefulness of spreadsheets can be shown using this example. In Table 1.4, the slope of the drain was in an individual cell and the equations used it with a fixed reference. This was done so it would be easy to do "What if?" scenarios. Such as, "What if the slope of the drain was changed to 0.5%?" The results are shown in Table 1.5 and in Figure 1.11.

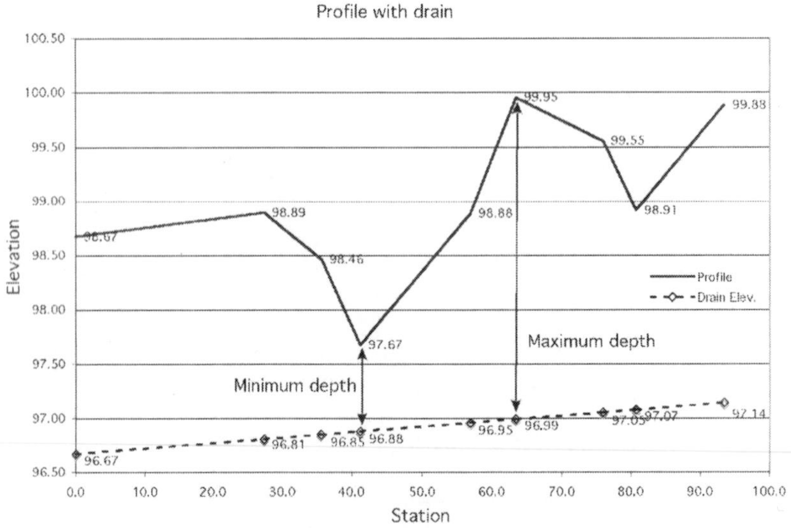

FIGURE 1.11. Drain with 0.5% slope.

Note that in this chart the values for each point were displayed. This is a common feature of spreadsheet charts. This shows that the maximum depth of the drain changes to 99.95 – 96.99 or 2.96 ft. The minimum depth changes to 97.67 – 96.85 or 0.82 ft. This "What if?" was accomplished by changing one value in the spreadsheet.

Spreadsheets are capable of much more complicated calculations and charts. In addition, they usually have the capabilities of recording "macro's" to automate actions that are frequently used and require multiple keystrokes. Another feature of some spreadsheets is visual basic programming language. This feature provides the tools for the user to write programs to manipulate or use data in some special way. This discussion of spreadsheets does not describe all of the capabilities and features of spreadsheets, but it attempts to explain their basic operation and show some of their useful capabilities. It is up to the reader to explore their capabilities in greater depth.

1.3.7. Flow Charts

Flow charts are used to graphically show the relationships of different parts or steps of a process. Two examples are the chain of command for an organization, the steps and alternatives in a manufacturing process. Different graphical symbols are used to represent different actions and lines and arrows can be used to show direction of flow. The complexity of the flow chart mirrors the complexity of the process being graphed. Flow charts can be categorized by type. Common types are sequential, loop, branch/decision making, and combination. Flow chart symbols represent different points along the process. An occupation or business may develop specific symbols for their specific needs, but several common symbols have been adopted by flow chart users. Figure 1.12 illustrates six of these.

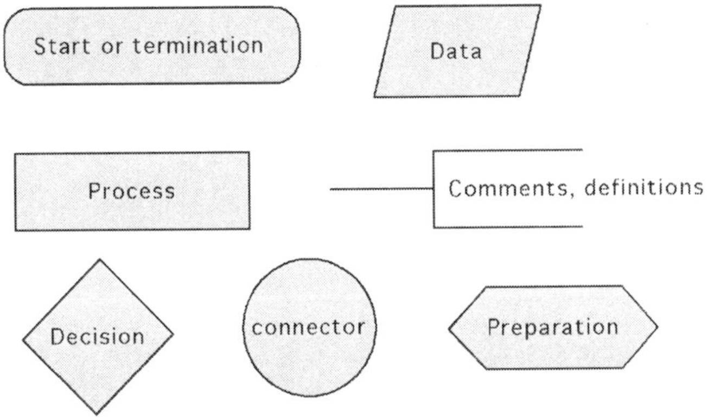

FIGURE 1.12. Common flow chart symbols.

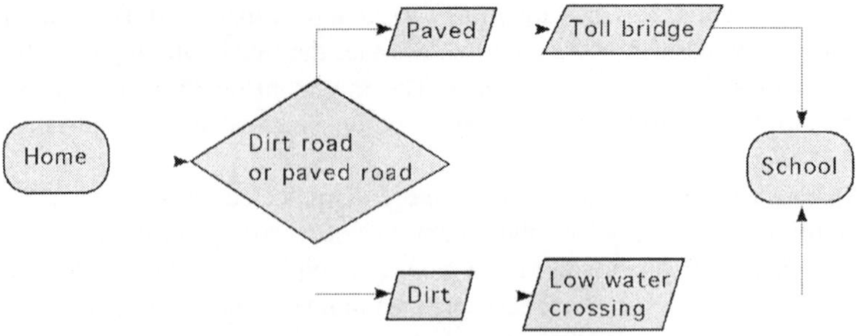

FIGURE 1.13. Example of flow chart.

In addition to different symbols representing different actions, different types of lines and arrows can be used to represent the relationship between the different steps or symbols in the process. Figure 1.13 is an example showing the alternatives and processes that might need to be completed to get from home to school.

2
Significant Figures and Standard Form

2.1. Objectives

1. Be able to define precision, accuracy, and uncertainty when working with numbers.
2. Understand the difference between exact and approximate numbers.
3. Be able to determine the number of significant figures.
4. Understand a technique for rounding numbers.
5. Understand the uses of scientific notation.

2.2. Introduction

There is no substitute for a good understanding of numbers and mathematical processes in solving modern agricultural problems. In this unit, we will discuss several features of numbers and techniques to use with numbers.

2.3. Precision, Accuracy, and Uncertainty

It is important to understand three characteristics of numbers in solving agricultural problems: precision, accuracy, and uncertainty. Precision refers to the size of the unit of measure used to obtain the number. For example, if a sack of feed is weighed on a scale measuring to the nearest 0.1 lb, the weight is not as precise as it would be if the smallest unit of measure was to the nearest 0.01 lb.

The accuracy of a number refers to the number of decimal places obtained in the answer. The greater the number of decimal places, the greater the accuracy. A measurement of 12.15 ft is more accurate than a measurement of 14.4 ft.

The uncertainty of a number is the amount it is expected to vary. If the uncertainty is not stated after the number, say, 14.5 ft $\pm$ 0.01, then you may assume it is $\pm$ half of the smallest unit, Table 2.1.

TABLE 2.1. Level of uncertainty.

Number	Uncertainty
25	± 0.5
15.7	± 0.05
2.567	± 0.0005

Source: *Theory and Problems of Technical Mathematics*, Schaum's Outline Series, McGraw-Hill Book Company, New York, 1979.

2.4. Exact and Approximate Numbers

Problems in agriculture use two different types of numbers, exact and approximate. The two common examples of exact numbers are those obtained by counting and ratios. For example, if you count the number of horses in a pen and arrive at 10, you have exactly 10 horses, not 10 and 1/2 or 9 and 3/4. Ratios are exact numbers because 3/4 of a circle is exactly 3/4 of a circle. One note of caution about ratios, ratios expressed as a decimal, say, $2/3 = 0.6666666\ldots$, are approximate numbers because some ratios expressed as a decimal contain repeating digits.

Any number obtained by a measurement is an approximate number. The actual value of an approximate number is uncertain because all measuring devices have a limit to their precision. If you have a ruler that is graduated in 1/16 of an inch, then the ruler is only precise to the nearest 1/16 of an inch. The following example illustrates this point.

In Figure 2.1, the length of the rectangle is not aligned with any of the marks on the scale (ruler). You must record the length of the box to the last shorter reading, 2–9/16's, or the next longer reading, 2–5/8. Regardless of the value you choose, the answer you record is only close to the actual length, and thus is an approximate number. If the desire is to reduce the amount of uncertainty in the measurement, then a ruler with higher precision, such as 1/32's of an inch should be used.

When approximate numbers are used, digits can be introduced into the problem, that are not accurate (significant). If these digits are included, the accuracy of the answer may decrease with every computation. The potential for error is especially great if a calculator is used because most calculators show eight or nine numbers, regardless of their significance to the problem. The task of the calculator operator is

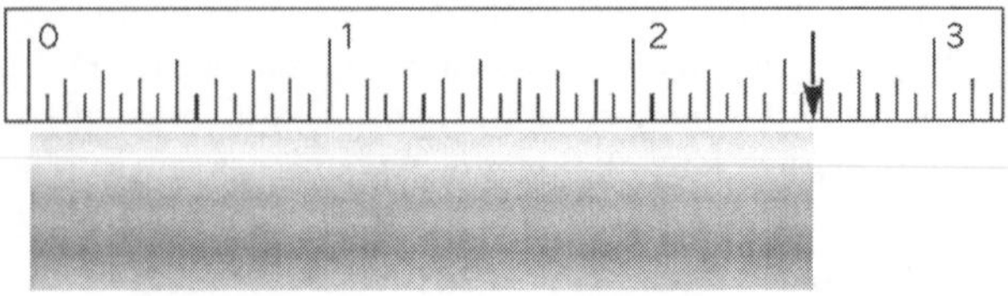

FIGURE 2.1. Example of an approximate number.

to determine how many digits are significant and to round accordingly. Determining significant figures is discussed in more detail in the following section.

2.5. Significant Figures

The principle of significant figures is important because the precision of a number should not be increased by mathematical computations. Calculators routinely carry 7 to 14 decimals places on screen and/or in memory even though the accuracy and precision of measurements are only two or three places. It is usually necessary to determine the number of digits that should be kept in a number after mathematical computations. The rules for determining significant digits are different for exact and for approximate numbers.

Because there is no uncertainty with exact numbers, all of the digits are considered to be significant. When rounding answers produced with exact numbers, assume that they have the same number of significant figures as the largest exact number.

Determining significant figures for approximate numbers is more complicated. The first issue is the significance of zeros. A common practice is to consider a zero significant if it is between another number and a decimal point or to the right of the decimal point. For example, 540.2 has four significant figures and 540.0 has four significant figures. A zero is not considered significant if it is to the left of the decimal point. For example 0.325 has three significant figures, but 0.0325 would have four. A zero is not considered significant when it is the last number and there is no decimal point. For example, 540 has two significant figures. An exception to this rule is when the zero is the result of rounding. The number 459.8 rounded to three significant figures is 460. In this case the zero is considered significant. The problem is when the reader doesn't know if the zero is the result of rounding. In this case the significance of the zero can be ambiguous.

For nonzero numbers the number of significant digits for an approximate number depends on the precision of the measuring instrument. If you are given the weight of a steer as 551 lb and know that the scale measured to the nearest 0.1 lb, then the actual weight could be less than 551.1 but more than 550.9 lb and be recorded as 551 lb. The number will have four significant figures. To be correct, if the weight of the steer was actually 551 lb, the number should have been written as 551.0.

When the precision of the measuring instrument is not known, determining significant figures is much more difficult. A common practice is to assume the precision is +/− one half of the smallest unit in the number. For example, a distance of 347 ft would have four significant figures because the measurement could be between 347.4 and 346.6 ft and be recorded as 347 ft.

Problem: You are helping measure the weight of a calf on scales that measure to the nearest 0.5 lb. You are told the calf weighs 102 lb. How many significant figures does the weight have?

Solution: The number of significant figures is four. The scale can read 101.5, 102.0, or 102.5. If the pointer on the scale is between 101.5 and 102.0 but closer to 102.0, then the reading is recorded as 102.0. Similarly, if the pointer is between 102.0 and 102.5 but closer to 102.0, then reading is recorded as 102.0. If the weight is exactly 102 lb, the weight of the calf should be recorded as 102.0 lb. Then the weight has four significant digits.

Consider the following situation:

Problem: What is the area (ft^2) of a room if the width is 12 ft 3 in and the length is 22 ft 3/16 in?

Solution: The first step is to convert the dimensions to decimal form, 12 ft 3 in = 12.25 ft, and 22 ft 3/16 in = 22.1875 ft. Completing the multiplication gives a value of 271.79687 ft^2. How many digits are significant?

Two rules have been developed to help determine the number of significant digits during mathematical computations.

- Adding or subtracting: the answer should be rounded to the number of decimal places in the least precise number.
- Multiplication and division: the answer should be rounded to the number of significant figures in the least accurate number.

For this problem, rule number two applies. The product, 271.79687, is reduced to four significant figures. The correct area for the room is 271.8 ft^2.

Problem: You need to know the perimeter of the room to calculate the amount of paint needed to paint the walls. The dimensions of the floor are 12.25 ft by 12.1875 ft.

Solution: In this case the perimeter is the sum of the lengths of the four walls in the room, $P = 12.25 + 12.25 + 22.1875 + 22.1875$. For this problem, rule number one applies. The sum, 68.1750, is reduced to four significant figures. The answer is 68.18 ft.

2.6. Rounding Numbers

Rounding is used to eliminate figures that are not significant. When the digits to be rounded are to the left of the decimal point, the digits are replaced by zeros. For example, if you need to reduce the area of an acre, 43,560 ft^2, to two significant digits, round to 44,000 ft^2. When the digit being dropped is greater than 5, add one to the next digit remaining. When the digit being dropped is less than 5, the first remaining digit is unchanged. If the digit being dropped is exactly 5, leave the remaining digit even. When the remaining digit is odd, add one, if it is even, leave it as is. For example, if the number 43,560 is rounded to one significant digit, the answer is 40,000; rounded to two it is 44,000; and rounded to three it is 43,600.

2.7. Scientific Notation and Standard Form

Scientific notation and standard form were developed to express large or small numbers in a more convenient form. They both use powers of 10 to replace the nonsignificant digits of a large or small number. Standard form uses a whole digit, then a decimal point, and two significant figures. It was primarily used with slide rules because three significant digits was the practical limit. Scientific notation uses the same format. The difference is that in scientific notation the number of significant figures is not limited to three. For example, if the number 1,000,000 has one significant figure, it is expressed as 1.0×10^6. This number is read as "one point zero times ten to the sixth power." The number 43,560,000 expressed in scientific notation with four significant figures would be 4.356×10^7. To use scientific notation effectively you must understand the powers of numbers and how they can be manipulated during mathematical computations.

First a review of the powers of 10:

$$10^0 = 1$$

$$10^1 = 10 \qquad 10^{-1} = 0.1 \text{ or } \frac{1}{10}$$

$$10^2 = 100 \qquad 10^{-2} = 0.01 \text{ or } \frac{1}{100}$$

$$10^3 = 1{,}000 \qquad 10^{-3} = 0.001 \text{ or } \frac{1}{1{,}000}$$

$$10^4 = 10{,}000 \qquad 10^{-4} = 0.0001 \text{ or } \frac{1}{10{,}000}$$

There are several helpful rules to use when working with powers of 10:

- When a number 10 and its exponent are moved from the denominator of a fraction to the numerator, or from the numerator to the denominator, the sign of the exponent is changed. Thus:

$$\frac{1}{10^{-3}} = \frac{10^3}{1} \text{ and } \frac{10^{-4}}{1} = \frac{1}{10^4}$$

- When two or more numbers in standard form are multiplied together, the powers of 10 can be added. Thus:

$$10^3 \times 10^5 = 10^{(3+5)} = 10^8$$
$$10^5 \times 10^{-2} = 10^{(5+(-2))} = 10^3$$
$$10^{-4} \times 10^{-3} \times 10^4 = 10^{((-4)+(-3)+4)} = 10^{-3}$$

It is important to note that different methods are used to express scientific notation. One method replaces the 10 and exponent with an "E" and the power of 10.

$$4.0\,E3 \times 5.0\,E5 = 20\,E^{3+5} = 20.0\,E8$$

Another method reduces the confusion of using an "×" to represent both multiplication and powers of 10 by replacing the "×" for multiplication with a •.

$$4.0 \bullet 10^3 \times 5.0 \bullet 10^5 = 20 \bullet 10^{3+5} = 20 \bullet 10^8$$

Other methods are also used.

- When two or more numbers in standard form are divided, the powers of 10 are subtracted. Thus:

$$\frac{10^4}{10^3} = 10^1 \text{ or } 10 \qquad \frac{4 \bullet 10^4}{2 \bullet 10^2} = 2 \bullet 10^2$$

$$\frac{10^{-4}}{10^2} = 10^{-6} \qquad \frac{4 \bullet 10^{-4}}{2 \bullet 10^2} = 2 \bullet 10^{-6}$$

- When two or more numbers in standard form are added or subtracted, the numbers must be converted to the same power of 10, and the power of 10 is not affected during the addition or subtraction. Thus:

$$4 \bullet 10^4 - 3 \bullet 10^3 \text{ becomes } 40 \bullet 10^3 - 3 \bullet 10^3 = 37 \bullet 10^3$$
$$\text{reduced to } 3.7 \bullet 10^3$$

and

$$4 \bullet 10^4 + 3 \bullet 10^3 \text{ becomes } 40 \bullet 10^3 + 3 \bullet 10^3 = 43 \bullet 10^3$$
$$\text{reduced to } 4.3 \bullet 10^3$$

3
Common Units of Measure

3.1. Objectives

1. Be able to explain the different type customary units of measure.
2. Be able to convert from one unit of measure to another.
3. Be able to use the units associated with each measurement.
4. Be able to calculate horsepower.
5. Be able to explain the SI measuring system.
6. Be able to explain differences between weight, force, and mass.

3.2. Introduction

All measurements require a unit to define the precision of the measurement. The preferred unit depends on the magnitude of the measured variable. For example in the customary system, commonly used in the United States, when the distance is large the mile is the preferred unit. When the distance is very small the unit of measure could be in decimal inches (0.01 in) or fractions of an inch (3/64 in). The units of measure can be divided into several categories. The common categories used in agriculture are:

1. Distance
2. Area
3. Temperature
4. Volume
5. Weight or Force
6. Pressure
7. Time
8. Velocity
9. Torque
10. Power.

3.3. Systems of Units

In modern agriculture we deal with two systems of units, the U.S. Customary system (sometimes called the English or gravimetric system) and the Metric or SI system. Both of these systems will be discussed in this chapter.

The bases for many of the units of measure in the customary system are lost in history. The Weights and Measures Division of the National Institute of Standards and Technology has the responsibility of storing and maintaining the standards for all of the customary units of measures. They have the standard foot, the standard yard, the standard pound, etc. All of the measuring devices used in the United States are based on one of these standards.

3.3.1. Distance

The dimension of distance has two common meanings, displacement and length. Displacement is the movement from one point to another. If you walk one mile, you have displaced a distance of 5,280 ft. Length refers to the physical size of an object. For example, the lengths of a standard piece of paper are 8.5 and 11 inches. The common units for distance are inches (in), feet (ft), yards (yd), and miles (mi).

3.3.2. Area

The unit of area is defined as the number of unit squares equal in measure to the surface. This indicates that an area is a surface, and the size of the surface is measured in terms of units squared. The common units for area are the units for distance squared, in^2, ft^2, mi^2, etc. Agriculture uses the acre (ac) as the unit for large land areas. There are 43,560 ft^2 in one acre.

3.3.3. Temperature

Temperature is the degree of hotness or coldness of an object. Five different measures of temperature have been used. The customary unit of temperature is the degree Fahrenheit.

3.3.4. Volume

Volume is defined as the amount of space occupied by a three-dimensional figure. The basis for a volume is a distance cubed; in^3, ft^3, etc. In the current system of weights and measures additional units of volume are also used, such as gallon, quart, pint, and cup. The Bureau of Weights and Measures has standardized the quantity of each one of these units in terms of distance cubed. For example there are 231 in^3 in one liquid gallon, Appendix I.

3.3.5. Weight

Weight can be defined as the force of gravity acting on a body. The common units for weight are ounces, pounds, and tons.

3.3.6. Force

Force is that action which causes or tends to cause motion or a change of motion of an object. To describe a force completely, its direction of action, magnitude, and point of application must be known. What is commonly referred to as a "force" is really two forces, as forces are never present singly, but always in pairs. The two parts are called action and reaction. They are always of equal magnitude, but in opposite directions. In this text, the weight of an object will be considered a force. Forces are commonly measured in units of ounces (oz), pounds (lb), and tons (ton).

3.3.7. Pressure

Pressure is the amount of force or thrust exerted over a given area. Pressure is the combination of two units, force and area. Therefore the common units for pressure will be a combination of these two, lb/in^2, oz/in^2, lb/ft^2, etc.

3.3.8. Time

The concept of time has its root in the natural cycles of the earth. One very visible cycle is the ocean tides. The words time and tide both come from the same root. The current idea of time is as a measure of an interval of duration. Time may be better described as an accounting technique for relating events. The common units for time are seconds (sec), minutes (min), and hours (hr).

3.3.9. Velocity

Velocity, speed, is the time rate of movement. Velocity is also a combined unit. It is the combination of distance and time. The common units of velocity are ft/min, mi/hr, etc.

3.3.10. Power

Power is the rate of doing work. Work (W) is the result of a force acting (or moving) through a distance. Written as an equation:

$$\text{Work } (W) = \text{Distance } (D) \times \text{Force } (F)$$

A numerical value for work may be obtained by multiplying the value of a force by the displacement.

Problem: If a force of 100.0 lb displaces 12.0 ft, how much work has been performed?

Solution:

$$\begin{aligned} W &= D \times F \\ &= 12.0 \text{ ft} \times 100.0 \text{ lb} \\ &= 1{,}200.0 \text{ ft-lb} \end{aligned}$$

In this situation 1,200 ft-lb of work was completed. Notice that according to this definition, unless both distance and force are present, no work is being accomplished.

Problem: A loaded wagon weighing 10,000.0 lb requires 400.0 lb of force to pull it along a horizontal surface. How much work is done if the wagon is pulled for 100.0 ft?

Solution:

$$\begin{aligned} W &= D \times F \\ &= 100.0\text{ ft} \times 400.0\text{ lb} \\ &= 40{,}000.0\text{ ft-lb} \end{aligned}$$

In this problem, the 400.0 lb of force is not related to the weight of the wagon. It is the force required to pull it.

Written as an equation Power $= \frac{\text{Work}}{\text{Time}}$ because work equals distance times force. Then:

$$\text{Power} = \frac{\text{Distance} \times \text{Force}}{\text{Time}} \text{ or } P = \frac{W}{T} \text{ which also means } P = F \times \frac{D}{T}.$$

Because D/T equals velocity (speed), power is the force times the velocity. This demonstrates that power is a combination of distance, force, and time.

Problem: How much power is developed when a force of 100.0 lb moves through a distance of 12.0 ft in 2.0 min?

Solution:

$$P = \frac{D \times F}{T} = \frac{12.0\text{ ft} \times 100.0\text{ lb}}{2.0\text{ min}} = \frac{1200.0\text{ ft-lb}}{2.0\text{ min}} = 600.0\,\frac{\text{ft-lb}}{\text{min}}$$

Notice that the unit associated with power is a combination of the individual units for the variables. In this case, the answer is read as "600 foot-pounds per minute." This is the "time-rate" at which work is being done. Remember; always write down the units that are associated with a number.

Problem: A person loads a 60.0-lb bale onto a truck platform 4.0 ft high in 0.50 min. How much power is being developed?

Solution:

$$P = \frac{D \times F}{T} = \frac{4.0\text{ ft} \times 60.0\text{ lb}}{0.50\text{ min}} = \frac{240.0\text{ lb}}{0.50\text{ min}} = 480\,\frac{\text{ft-lb}}{\text{min}}$$

Up to this point we have used easy-to-understand values with units of feet for distance, pounds for force, and minutes for time. Suppose that in the previous problem the individual could load three 60.0-lb bales in 0.50 min. In this example it is easy to make a mistake in determining a value for the force. The solution to this problem is:

$$P = \frac{4.0\text{ ft} \times \frac{3\text{ bales}}{1} \times \frac{60.0\text{ lb}}{\text{bale}}}{0.50\text{ min}} = \frac{720\text{ ft-lb}}{0.50\text{ min}} = 1{,}400\,\frac{\text{ft-lb}}{\text{min}}$$

Here the average power produced is 1,400 ft-lb/min because the weight moved in 0.50 min is the weight of all three bales (3 × 60 lb).

This problem illustrates a principle of power. If three times the amount of work is done in the same amount of time, the power will be increased three times. What is the impact on the power produced if the distance changes, or if the time changes?

Problem: If a person could load three 60.0-lb bales onto the 4.0 ft platform in 10.0 sec, instead of 0.50 min, how would this change the power produced?

Solution:

$$P = \frac{D \times F}{T} = \frac{4.0 \text{ ft} \times 180.0 \text{ lb}}{10.0 \text{ sec}} = 720 \frac{\text{ft-lb}}{\text{sec}} = 72 \frac{\text{ft-lb}}{\text{sec}}$$

The amount of power changed, but this answer cannot be compared to the previous one because the units are different. You might ask, is foot-pounds per second an acceptable unit of measure for power? Yes, but to compare this value for power to the previous one the units must be converted. This can be accomplished in more than one way. When the desired units are ft-lb/min, a conversion value can be added to the equation. To change the unit of time from seconds to minutes:

$$P = \frac{4.0 \text{ ft} \times 180.0 \text{ lb}}{10.0 \text{ sec} \times \dfrac{1 \text{ min}}{60.0 \text{ sec}}} = \frac{720 \text{ ft-lb}}{0.166\ldots \text{min}} = 4{,}320 \text{ or } 4{,}300 \frac{\text{ft-lb}}{\text{min}}$$

Now the two values can be compared. It should be obvious that it takes a greater amount of power to complete the same amount of work in less time. A similar relationship is true for the distance moved. The power requirement will change as the distance moved changes, assuming that the force and the time remain the same.

In summary, power is directly proportional to distance and force, and is inversely proportional to time.

In working with agricultural machinery, speed is usually measured in miles per hour (mph). When this is the case, the units must be changed. Otherwise the answer will be incorrect. Study the following statements:

If Power is equal to work divided by time, then:

$$P = \frac{F \times D}{T}.$$

This can be changed to:

$$P = F \times \frac{D}{T}$$

and because D/T (distance/time) is speed, if D/T is in miles/hour, it must be converted to feet/minute. The common conversion factor for speed is: 1 mph = 88 ft/min. This factor is obtained as follows:

$$88 \frac{\text{ft}}{\text{min}} = \frac{5{,}280 \text{ ft}}{1 \text{ mi}} \times \frac{1 \text{ hr}}{60 \text{ min}} \times \frac{1 \text{ mi}}{1 \text{ hr}}$$

Therefore power can also be found by:

$$P = F \times S \times 88$$

where F = force (lb); S = Speed (mi/hr); 88 = Units conversion value.

3.3.11. *Torque*

Torque is the application of a force through a lever arm. It is a force that causes or tends to cause a twisting or rotary movement. In equation form:

$$\text{Torque} = \text{Force} \times \text{Lever arm length}$$

or:

$$\text{To} = F \times LA$$

where To = Torque (lb-ft or lb-in); F = Force (lb); LA = Lever arm length (ft or in). The lever arm is the distance from the point the force is applied to the center of revolution.

Because force is measured in pounds and length in feet or inches, the common units of torque are pound-feet (lb-ft) or pound-inches (lb-in).

Problem: How much torque is developed when a 50.0-lb force is applied at the end of a wrench that is 1.0 ft long?

Solution:

$$\text{To} = F \times LA = 50.0 \text{ lb} \times 1.0 \text{ ft} = 50 \text{ lb-ft}$$

Notice that the answer has been written as "50 pound-feet." To distinguish torque from work, torque is written with units with the force unit first, "pound-feet," and the units for work is written with the distance unit first, "foot-pounds."

There is one additional difference between torque and work. We stated that unless there is movement, there is no work. Because torque is a force working through a lever arm, torque can exist without movement.

Problem: Which of the following will cause a greater torque to be exerted on a shaft: (1) 50.0 lb of force applied at the end of a 6.0-in (1/2-ft) wrench, or (2) 15.0 lb of force applied at the end of a 24.0-in (2-ft) wrench?

Solution:
(1) $\text{To}_1 = 50.0 \text{ lb} \times 0.5 \text{ ft} = 25 \text{ ft-lb}$
(2) $\text{To}_2 = 15.0 \text{ lb} \times 2.0 \text{ ft} = 30 \text{ ft-lb}$

Situation (2) will cause greater torque (twisting effect) on the shaft.

3.3.12. *Horsepower*

Although power is a basic unit, in agriculture the more common unit is horsepower. Different measures of horsepower are used. For example, when discussing tractors, you may use engine horsepower, brake horsepower, drawbar horsepower, or power take-off horsepower. In the following section we will investigate the principles of horsepower. Horsepower is an arbitrary unit that was developed by James Watt to promote his early steam engines. He watched horses pulling loads of water out of mine shafts and concluded that one horsepower was equal to performing work at the rate of 33,000 foot-pounds per minute. Expressed algebraically the conversion from power to horsepower is:

$$1\text{ hp} = \frac{\text{Power}\ \dfrac{\text{lb-ft}}{\text{min}}}{33{,}000\ \dfrac{\text{lb-ft}}{\text{min}}} \text{ or} 1\text{ hp} = \text{Power}\ \frac{\text{lb-ft}}{\text{min}} \times \frac{1\text{ min}}{33{,}000\text{ lb-ft}}$$

Note that unless power is in the units of ft-lb/min, the use of the conversion factor 1 hp = 33,000 ft-lb/min will not produce correct results. The common horsepower equation is:

$$\text{hp} = \frac{F \times D}{T \times 33{,}000}$$

This is an example of an equation with a conversion constant. This means all of the variables must have the correct units. This equation requires distance (D) expressed in feet, force (F) in pounds, and time (T) in minutes.

Problem: How many horsepower are developed if a person loads six 60-lb bales onto a truck platform 4.0 ft high in 1.5 min?

Solution:

$$\text{hp} = \frac{F \times D}{T \times 33{,}000} = \frac{4.0\text{ ft} \times \left(6\text{ bales} \times \dfrac{60.0\text{ lb}}{\text{bale}}\right)}{1.5\text{ min} \times 33{,}000}$$

$$= \frac{1{,}440}{49{,}500} = 0.02909\ldots \text{ or } 0.029\text{ hp}$$

Now consider what would happen if the time required to load the hay is measured as 90.0 sec instead of 1.5 min. Obviously 90.0 sec equals 1.5 min, but if the same equation is used either the seconds must be converted to minutes or a different conversion factor from power to horsepower must be used. If the 33,000 conversion unit is used with time measured as 90.0 sec, the answer will be incorrect. The problem can be solved in two different ways. The first way is to add a conversion unit for time to the equation. This would be the preferred method for just a few calculations.

$$\text{hp} = \frac{F \times D}{T \times 33{,}000} = \frac{4.0\text{ ft} \times \left(6\text{ bales} \times \dfrac{60.0\text{ lb}}{\text{bale}}\right)}{\left(90.0\text{ sec} \times \dfrac{1\text{ min}}{60\text{ sec}}\right) \times 33{,}000}$$

$$= \frac{1{,}440\text{ ft-lb}}{49{,}500} = 0.02909\ldots \text{ or } 0.029\text{ hp}$$

The second way is to determine the appropriate conversion value for the units being used. In a situation where horsepower will be calculated several times and time is measured in seconds, it would be more efficient to use a conversion value from power to horsepower appropriate for when time is measured in seconds instead of minutes. This can be accomplished using units cancellation:

$$\frac{\text{ft-lb}}{\text{sec}} = \frac{33{,}000 \text{ ft-lb}}{1 \text{ min}} \times \frac{1 \text{ min}}{60 \text{ sec}} = 550 \frac{\text{ft-lb}}{\text{sec}}$$

Using this conversion value, the previous problem can be solved by:

$$\text{hp} = \frac{F \times D}{T \times 550} = \frac{1{,}440 \text{ ft-lb}}{90.0 \text{ sec} \times 550} = \frac{1{,}440}{49{,}500} = 0.02909 \ldots \text{or } 0.029 \text{ hp}$$

Horsepower can also be calculated using torque and the speed of the shaft. This is called shaft horsepower. This equation was used to evaluate the horsepower being produced by early engines using a Prony Brake, Figure 3.1.

The early Prony Brake used the friction between a rotating flywheel and a stationary block of wood to produce a force on the lever. Once the force on the lever and the speed of rotation of the flywheel is known, the brake horsepower can be determined. Mathematically brake horsepower is:

$$\text{Bhp} = \frac{\text{FLN}}{5252}$$

where F = Force produced (lb); L = Length of the lever arm (ft); N = Rotary speed of the Prony brake shaft (rpm); 5252 = Units conversion constant.
This equation is derived from the horsepower equation:

$$1 \text{ hp} = \frac{\frac{2\pi}{\text{rev}} \times F \times D \times N}{\frac{1 \text{ hp}}{33{,}000 \frac{\text{ft-lb}}{\text{min}}}} = \frac{F \times D \times N}{5252}$$

where the length of the lever arm equals the radius of the circle and the distance the force is working through.

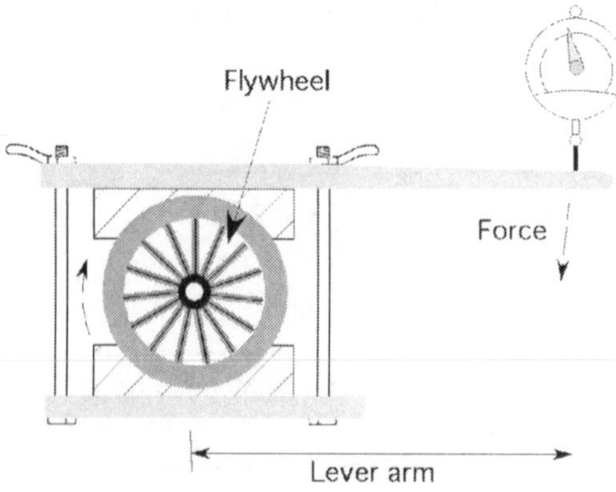

FIGURE 3.1. Prony brake.

Problem: How many horsepower is an engine producing when 8 lb of force are measured at the end of an 18.0 inch Prony brake arm when rotating at 1,700 revolutions per minute?

Solution:

$$\begin{aligned}\text{Bhp} &= \frac{\text{FLN}}{5252} = \frac{8\ \text{lb} \times \left(18.0\ \text{in} \times \dfrac{1\ \text{ft.}}{12\ \text{in.}}\right) \times 1{,}700\ \dfrac{\text{rev}}{\text{min}}}{5252} \\ &= \frac{244{,}800}{5252} \\ &= 3.8842\ldots \text{ or } 3.9\ \text{hp}\end{aligned}$$

[Note: the answer was rounded to two significant figures because the significance of the zero's in 1,700 are ambiguous, but at least two significant figures should be used because 8 lb is a measurement and the uncertainty is +/− 0.5 lb.]

Brake horsepower contains a length and a force. We know from earlier discussions that if we have a force times a length, we are dealing with torque. Therefore, the brake horsepower equation can be rewritten to use torque:

$$\text{Bhp} = \frac{\text{To} \times N}{5252}$$

A Prony brake can still be used to measure the horsepower of engines, but electrical and hydraulic dynamometers are more accurate than the Prony brake. The rotary power of modern engines is measured by connecting directly to the flywheel or by using the power take off (pto) shaft.

3.4. Using Units in the Metric (SI) System

A major problem with the customary units is that there is no logical relationship between units. For instance, there is no obvious reason why the standard mile is 5,280 ft. The metric system was created in the late 18th century by the French to address this problem.

One of the advantages of the SI system is that all of the units are based on natural phenomena. For example, a meter is the distance light travels, in a vacuum, in 1/299,792,458th of a second. The U.S. agricultural industry is based on the customary units of measure, however many manufactures are converting to SI units and agriculture has become more international in buying and marketing products. It is critical to understand the differences between the two systems.

In the previous section the several customary units of measure were discussed. The following section will explain the comparable units in the SI system. The SI system is a decimal-based system. To use the different units requires knowing the prefixes used for the different powers of 10, Table 3.1.

Although the recommended practice in the metric system is to use the standard units or 1,000 unit multiples of the units, Table 3.1, there are nonstandard units

TABLE 3.1. Prefixes for SI units.

yotta [Y]	$= 10^{24}$	yocto [y]	$= 10^{-24}$
zetta [Z]	$= 10^{21}$	zepto [z]	$= 10^{-21}$
exa [E]	$= 10^{18}$	atto [a]	$= 10^{-18}$
peta [P]	$= 10^{15}$	femto [f]	$= 10^{-15}$
tera [T]	$= 10^{12}$	pico [p]	$= 10^{-12}$
giga [G]	$= 10^{9}$	nano [n]	$= 10^{-9}$
mega [M]	$= 10^{6}$	micro [μ]	$= 10^{-6}$
Kilo [k]	$= 10^{3}$	milli [m]	$= 10^{-3}$
hecto [h]	$= 10^{2}$	centi [c]	$= 10^{-2}$
deca [da]	$= 10$	deci [d]	$= 0.1$

frequently used in the SI system. It is critical to understand and correctly apply the differences between the U.S. Customary and the SI system, Appendix II.

3.4.1. Distance

The Standard unit of distance in the SI system is the meter. A meter is about 39 inches long. The other lengths that are commonly used are the millimeter (mm), and the kilometer (km). A nonstandard commonly used length is the centimeter.

3.4.2. Area

The principle of area is the same—it is two distances multiplied together. The units become millimeters squared, etc. In the customary system the standard unit of land measurement is the acre. In the SI units it is the hectare. An acre is about 0.4047 hectares.

3.4.3. Temperature

In the SI system the preferred unit for measuring temperature is Celsius. The Celsius scale was established so that the boiling point of water is at 100°C and the freezing point is at 0°C. The equation for converting from the Fahrenheit scale to the Celsius scale is:

$$T_{\mathrm{C}} = \frac{5}{9} \times (T_{\mathrm{F}} - 32)$$

and the conversion from Celsius to Fahrenheit is:

$$T_{\mathrm{F}} = \frac{9}{5} \times (T_{\mathrm{C}} + 32)$$

3.4.4. Volume

The standard unit of volume for the SI system is the liter. A liter is defined as 1,000 cubic centimeters. A liter is slightly larger than a quart.

3.4.5. Weight and Force (Mass)

The U.S. Customary system (sometimes referred to as a gravitational system) is a force-based system and the SI is mass-based system. The key to understanding how to solve problems in these two systems is to understand the relationship between mass and force. The definition of mass was proposed by Sir Isaac Newton.

One useful definition of mass (Hibbler, R.C. *Engineering Mechanics*, 6th edition. Macmillan, New York) is "Mass is a property of matter by which we can compare the action of one body with that of another. This property manifests itself as a gravitational attraction between two bodies and provides a quantitative measure of the resistance of matter to a change in velocity." This implies several things. Gravity is a form of acceleration. Thus, mass is independent of gravity. This means that your mass is the same on the earth, moon, or in rocket ship leaving the earth. Your mass is also the same in space orbiting the earth.

Mass can be determined by comparing an object of known mass to your object whose mass is not known. One way to do this is to place the object with the unknown mass in one pan of a balance scale and add objects of known mass in the other pan until the scale is balanced. Gravity is the acceleration acting on both pans of the balance scale. The balance scale will give the same results in the rocket ship, on the earth or on the moon.

Force or weight changes with gravity or acceleration. You will weight about 1/6th as much on the moon as on the earth, because the force attracting you to the moon, gravity is about 1/6th that of the earth's gravity. In orbit about the earth, where the net gravitational force is zero, your weight is zero.

In this book we are primarily interested in force rather than mass. Sir Isaac Newton's second law defines the relationship among mass (m), acceleration (a), and force (F). This equation is $F = ma$.

The SI unit of force is the Newton (N), the unit of mass is the kilogram, and the unit of acceleration is meter/second^2 (m/s^2). In the SI system we specify the mass of the object and calculated the weight. Since gravity is a form of acceleration, we can rewrite Newton's second law as:

$$W = mg$$

where W is the weight in Newtons (N) and g is the acceleration by gravity at the earth's surface, 9.81 m/s^2.

In the customary system we measure the force in pounds and calculate the mass. The unit of mass in the customary systems is the slug. Mass is calculated by:

$$m = \frac{W}{g} \text{ where } W \text{ is the weight in pounds (lb) and } g = 32.3 \text{ ft/s}^2.$$

The relationship of mass and weight can be explained using an example of a class 1 lever. Note: class one levers are explained in chapter 4. You are using a class 1 lever to pry a large rock from the ground. You push down on the end of the lever with enough force to raise your feet off the ground and the rock doesn't move. You have recently "weighed" yourself on a metric scale and your mass is 81 kg. The force arm length is 1.5 m and the resultant arm length is 0.3 m. What is the weight of the rock?

The first step in determining the weight of the rock is to convert your body mass to weight.

$$W = m \times g$$

$$W = 81\text{ kg} \times 9.81\ \frac{\text{m}}{\text{s}^2} = 794.61\ \frac{\text{kg} \bullet \text{m}}{\text{s}^2} = 795\text{ N}$$

The second step is to calculate the weight of the rock using the formula for a class 1 lever and solving for F_r.

$$F_a \times A_a = F_r \times A_r$$

The equation for a class 1 lever is written in terms of a force, but weight is a force so $F_A = W$

$$F_r = W \times \frac{A_a}{A_r}$$

$$F_r = 795\text{ N} \times \frac{1.5\text{ m}}{0.3\text{ m}} = \frac{1192.5}{0.3} = 3975 \text{ or } 4{,}000\text{ N}$$

The weight of the rock is greater than 4,000 N or 4 kN.

3.4.6. Pressure

Pressure is force per unit area. In the SI system a combination of force and area can be used, but the preferred unit is Newtons per meter squared (N/m^2) or commonly called Pascals (Pa). Therefore,

$$1\frac{\text{N}}{\text{m}^2} = 1\text{ Pa.}$$

3.4.7. Time

The units used to measure time are the same for both the customary and SI units.

3.4.8. Velocity, Speed, and Acceleration

Velocity and speed are measured by determining a distance and the time it takes to travel a distance. For vehicles the standard unit for distance in the SI system is kilometers and the standard unit for time is hours, therefore the SI units for velocity are meters per hour (km/hr). Acceleration is the rate at which velocity changes. The units in the SI system are m/s/s or m/s^2. The units of acceleration in the U.S. Customary system are ft/s/s or ft/s^2.

3.4.9. Power

Power is the rate of doing work and work is force times a distance. In the SI system force is measured in units of Newtons and distance in meters, except for vehicles where the distance is in kilometers. This means work is measured in units of Newton · meters. This is an acceptable unit of measure for work, but for many calculations the preferred unit is the joule. One joule is the amount of work done when an applied force of 1 Newton moves through a distance of 1 m in the direction of the force.

The units for time are the same, therefore in the SI system power could be measured in units of joules per hour. This measurement of power is not usually used, instead the preferred unit of measure for power in the SI system is the watt. One watt is equal to 1 joule per second.

3.4.10. Torque

Torque is a measure of a force working through a lever. In the SI system force is measured in Newtons and distance is measured in meters. The units for torque are Newton · meters (N · m).

3.4.11. Horsepower

The SI unit system does not have a comparable conversion from power to horsepower. In the SI system Watts are used to measure power and the appropriate prefix is used. For many agricultural problems using the unit of Watt results in a large number, therefore it is customary to use the unit of kilowatt (kW).

4
Simple Machines

4.1. Objectives

1. Be able to explain the common simple machines.
2. Be able to give an example of each type of simple machine.
3. Be able to use the principles of simple machines to solve problems.

4.2. Introduction

A machine is any device that either increases or regulates the effect of a force or produces motion. All agricultural machines are composed of combinations and modifications of two basic machines, the lever and the inclined plane. We will study the basic principles surrounding these two machines and illustrate some of their common modifications and uses. (*Note*: In the following discussion of simple machines two assumptions are made: losses due to friction are ignored, and the strength of the materials is not considered.)

4.3. Lever

A lever is a rigid bar, straight or curved, capable of being rotated around a fixed point (fulcrum). When a fulcrum and a bar are used, two different forces exist, the applied force (F_a) and the resultant force (F_r). The forces, bar, and fulcrum can be used in three ways, called classes, Figure 4.1.

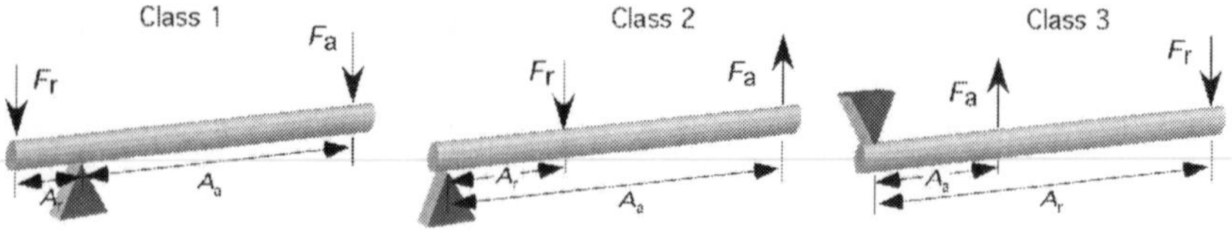

FIGURE 4.1. Three classes of levers.

The principle of levers can be expressed mathematically as:

$$\begin{pmatrix}\text{Force}\\\text{Applied}\end{pmatrix}\times\begin{pmatrix}\text{Arm}\\\text{Applied}\end{pmatrix}=\begin{pmatrix}\text{Force}\\\text{Resultant}\end{pmatrix}\times\begin{pmatrix}\text{Arm}\\\text{Resultant}\end{pmatrix}$$

$$F_a \times A_a = F_r \times A_r.$$

4.3.1. Class One Lever

Class one levers are used primarily for their mechanical advantage. The mechanical advantage for a first-class lever is the ratio of the lengths of the two arms. In our discussion of simple machines, mechanical advantage will be defined as the increase of force that occurs through the use of a lever. Expressed mathematically:

$$\text{Mechanical advantage} = \frac{\text{Force arm length}}{\text{Resultant arm length}}$$

The principles of a class one lever are illustrated by the problem in Figure 4.2.

Problem: How much weight can a 140.0-lb person lift with a class one lever if the force arm is 4.0 ft long, and the resultant arm is 1.0 ft long?

Solution: In this problem three of the variables are known: $F_a = 140$ lb, $A_a = 4$ ft, and $A_r = 1$ ft. To solve the problem, we must use one of the techniques of problem solving—rearranging an equation. In this example we need to rearrange the equation to solve for F_r and then insert the values.

$$F_a \times A_a = F_r \times A_r$$

$$F_r = \frac{F_a \times A_a}{A_r} = \frac{140.0 \text{ lb} \times 4.0 \text{ ft}}{1.0 \text{ ft}} = 560 \text{ lb}$$

[*Note*: In this example two significant figures were used. This is an example where the significance of the zero in the number would be ambiguous unless the reader had excess to the entire problem. To remove this ambiguity the answer should be written as 5.5 E2.]

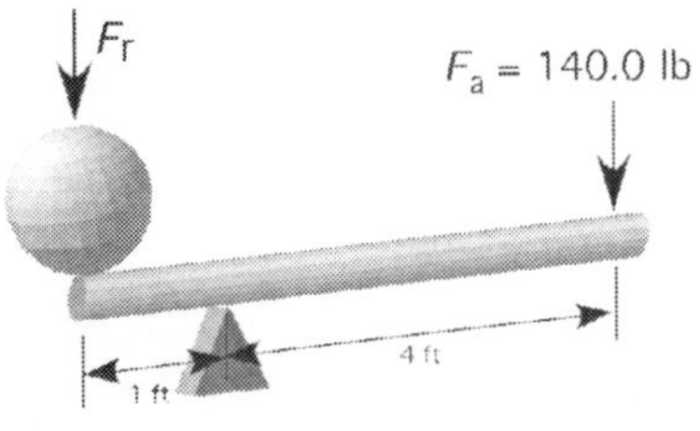

FIGURE 4.2. Example of class one lever.

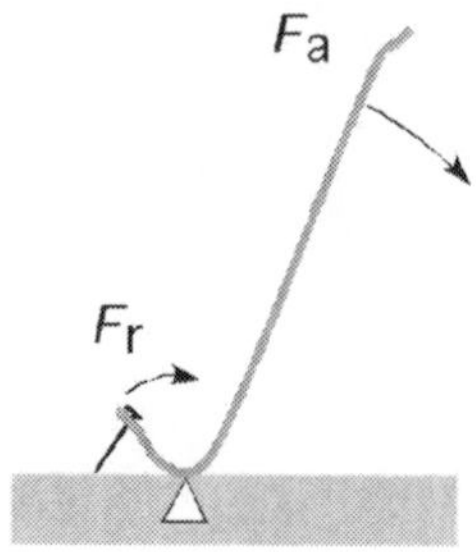

FIGURE 4.3. Example of the use of a class one lever.

With this lever, 140 lb of applied force is capable of lifting 560 lb. This demonstrates the mechanical advantage of the class one lever. The source of the mechanical advantage is clearer if the equation is written as:

$$F_r = 140.0 \text{ lb} \times \frac{4.0 \text{ ft}}{1.0 \text{ ft}}$$

When the equation is arranged in this manner, it is easy to see that the increase in force or mechanical advantage is the ratio of the lengths of the two arms. In this example, the amount of force the person could produce (mechanical advantage) was increased 4/1 or 4 times.

In addition to the mechanical advantage, the distance moved and the speed of movement for the two ends of the bar also can be determined. Calculations will show that the distance moved is proportional to the ratio of the length of the resultant arm to the length of the force arm. The speed of movement is proportional to the ratio of the length of the force arm to the length of the resultant arm.

An example of the use of a class one lever is a wrecking bar pulling a nail, Figure 4.3.

4.3.2. Class Two Lever

The second class of lever also produces a mechanical advantage. In this lever the mechanical advantage is the ratio of the distance between the fulcrum and the applied force and the distance from the fulcrum to the resultant force, Figure 4.4. The same applied force is used to illustrate this lever.

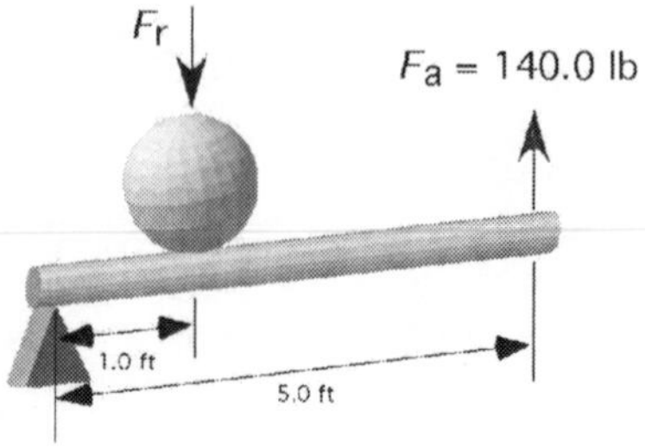

FIGURE 4.4. Example of class two lever.

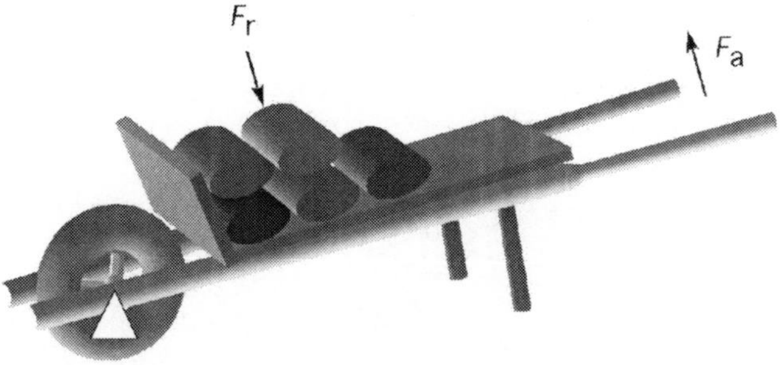

FIGURE 4.5. A two-wheeled dolly as a class two lever.

Using the lever equation, rearranged for the unknown value, determines the resultant force for the applied force:

$$F_a \times A_a = F_r \times A_r$$

$$F_r = \frac{F_a \times A_a}{A_r} = \frac{140.0 \text{ lb} \times 5.0 \text{ ft}}{1.0 \text{ ft}} = 700 \text{ lb or } 7.0 \text{ E2 lb}$$

For the class two lever the mechanical advantage will always be greater than one. In this case the mechanical advantage is 5 divided by 1 or 5. This is why a 700-lb load can be moved with only a force of 140 lb supporting the load at the end of the fulcrum.

In the class two lever the distance moved and speed of movement also are proportional to the ratio of the lengths of the two arms.

A common use of a class two lever is the wheelbarrow, Figure 4.5.

It is important to remember that the determination of the class of lever is based on the location of the fulcrum, the length of the applied force arm and the length of the resultant force arm. If the relationship of these three components changes, the class of lever will change. A simple machine, like a two-wheeled dolly, can act as a class one or class two lever. Figure 4.6 shows that when the center of the mass is to the right of the fulcrum (axle) the dolly acts as a class one lever.

When the center of mass moves to the left of the fulcrum (axle) the dolly will behave as a class two lever, Figure 4.7.

Also note that the direction of the force on the handle changes. When used as a class one lever, the operator must provide a constant pull on the handles. When used as a class two lever, the operator must lift up on the handles.

4.3.3. Class Three Lever

The class three lever does not have a mechanical advantage. It is primarily used to increase speed and movement. The same applied force and distances are used to illustrate a class three lever, Figure 4.8.

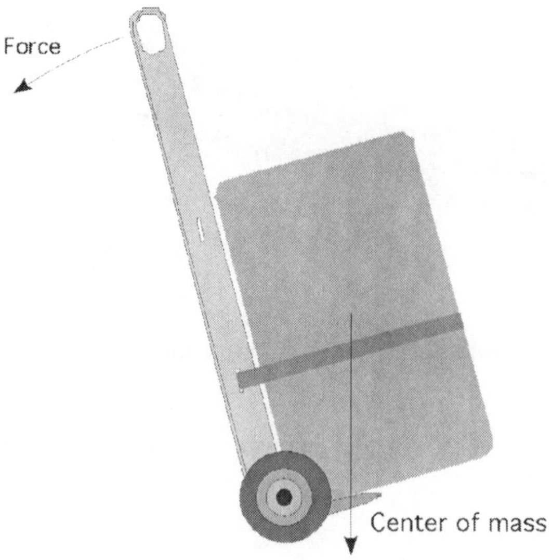

FIGURE 4.6. Example of class two lever in use.

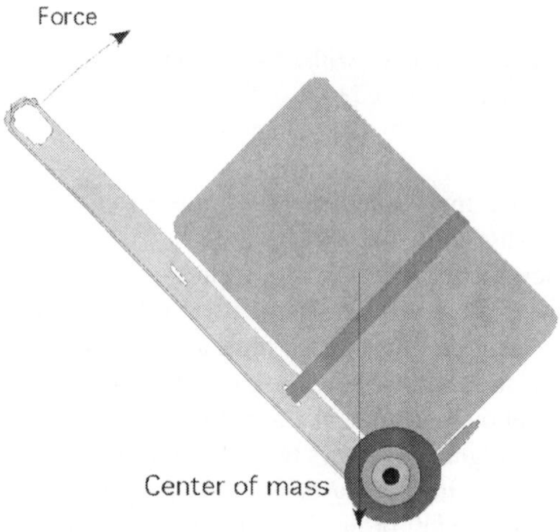

FIGURE 4.7. A two-wheeled dolly as a class two lever.

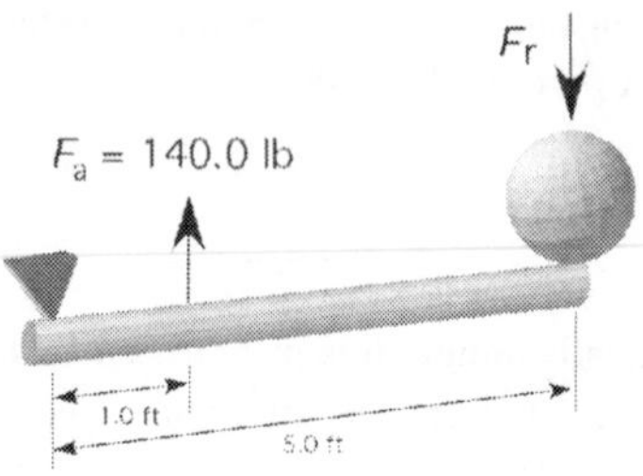

FIGURE 4.8. Example of class three lever.

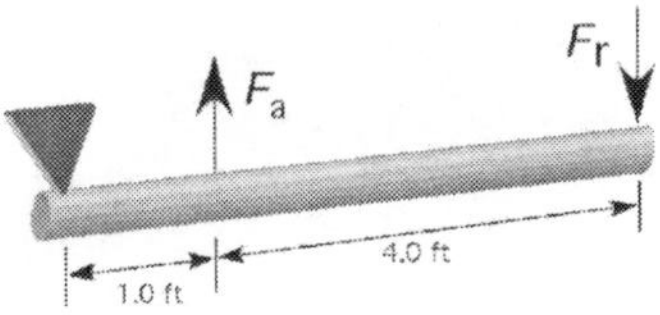

FIGURE 4.9. Class three lever with split measurement of length.

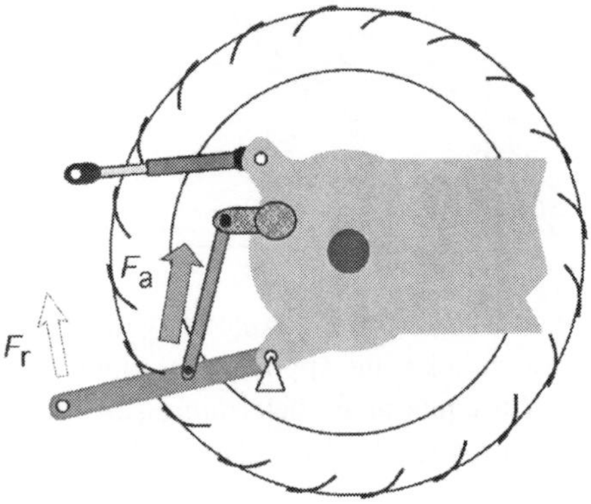

FIGURE 4.10. Example of the use of class three lever.

The resultant force for a class three lever is determined by rearranging the lever equation and inserting the values:

$$F_a \times A_a = F_r \times A_r$$

$$F_r = \frac{F_a \times A_a}{A_r} = \frac{140.0 \text{ lb} \times 1.0 \text{ ft}}{5.0 \text{ ft}} = 28 \text{ lb}$$

As this example illustrates, the mechanical advantage for the third-class lever always will be less than one. Here 140 lb of applied force only can lift a weight of 28 lb. The distance moved and the speed of movement of the resultant point compared to the applied point is increased proportionally.

It is important to remember that when all three levers are used, the two lengths, applied and resultant, are measured from the point of application to the fulcrum. In both the class two and three levers, the length of the applied arm or the resultant arm may be broken into two measurements, see Figure 4.9.

In Figure 4.9, the length of the resultant arm is 5 ft because the distance is measured from the point of application to the fulcrum.

A common use of the class three lever is the 3-point hitch on a tractor, Figure 4.10.

4.4. Wheel and Axle

A wheel and axle behave as a continuous lever. The center of the axle corresponds to the fulcrum. A wheel and axle has a mechanical advantage if the radius of the wheel is the applied arm and the radius of the axle is the resultant arm. It does not

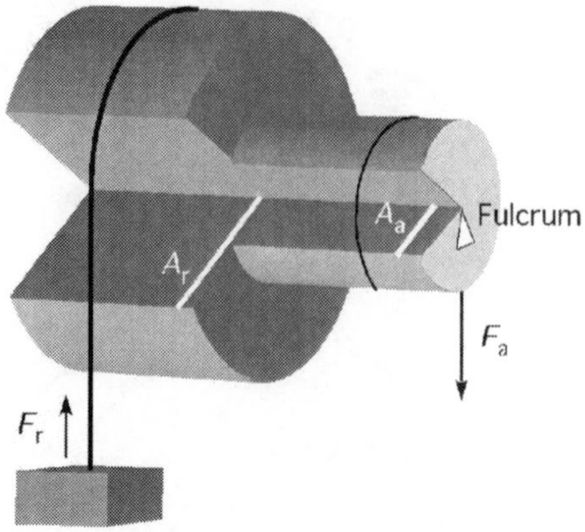

FIGURE 4.11. Illustration of wheel and axle.

have a mechanical advantage when the wheel radius is the resultant arm and the axle radius is the applied arm, Figure 4.11.

The equation for determining the mechanical advantage is the same as the equation used for the three classes of levers.

Problem: How much force will it take to lift a 10.0-lb weight with a wheel and axle, used as in Figure 4.11, when the axle is 2.0 inches in diameter, and the wheel is 10.0 inches in diameter?

Solution: The first step is to rearrange the equation to solve for the applied force. Remember that the length of the force arm is the radius of the wheel and the length of the resultant arm is the radius of the axle. This gives us:

$$F_a \times A_a = F_r \times A_r$$

$$F_a = \frac{F_r \times A_r}{A_a} = \frac{10.0 \text{ lb} \times 5.0 \text{ in}}{2.0 \text{ in}} = 25.0 \text{ lb}$$

The wheel and axle is not normally used alone. It is usually used in conjunction with a gear reduction system to form a hand operated or electric motor winch found on boat trailers and other applications, Figure 4.12.

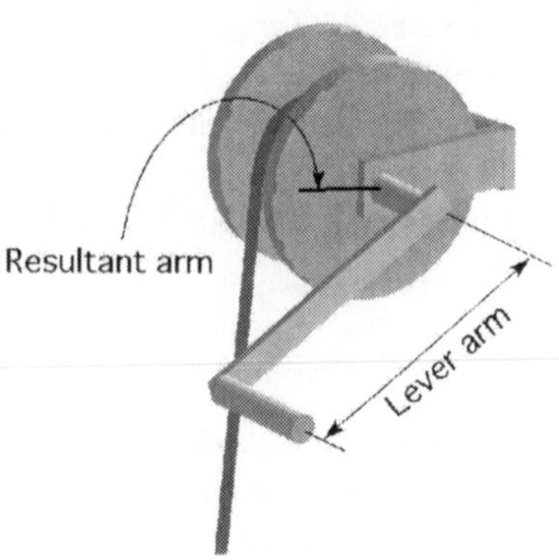

FIGURE 4.12. Wheel and axle used as a winch without a gear reducer.

FIGURE 4.13. A single pulley.

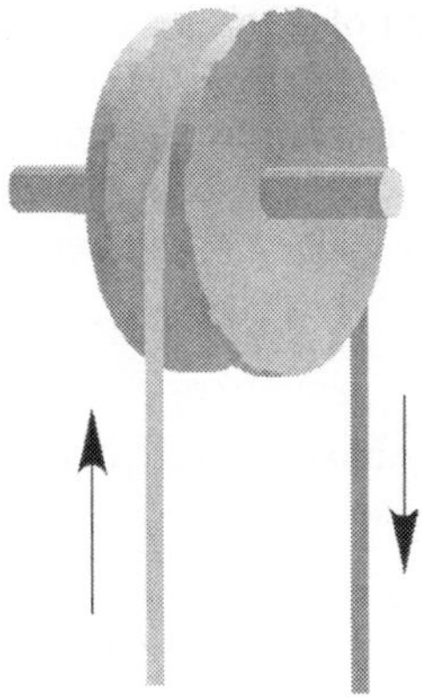

4.5. Pulley

A pulley is a modification of a first or second class lever. A single pulley does not produce any mechanical advantage, just a change of direction in force, Figure 4.13.

If the power loss to friction is not ignored, then a single pulley has a mechanical advantage of less than one. When pulleys are combined in pairs, a mechanical advantage is produced. This use of pulleys is commonly called a block and tackle, Figure 4.14.

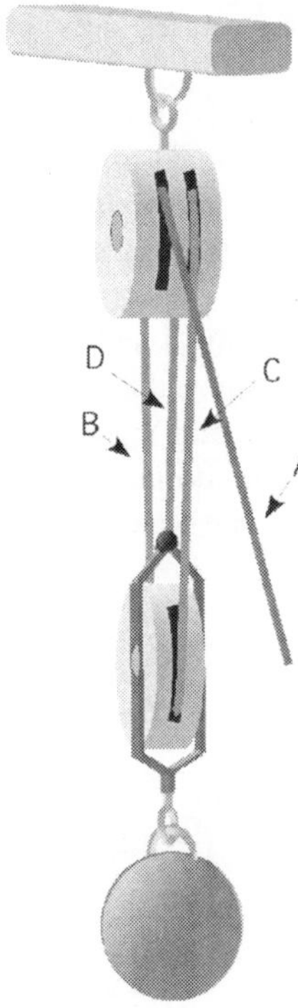

FIGURE 4.14. Block and tackle.

The amount of mechanical advantage produced by a block and tackle is proportional to the number of ropes that support the weight. The block and tackle in Figure 4.14 has four ropes, but only three, R_B, R_C, and R_D, support the weight. In this arrangement the amount of weight that can be lifted is three times the amount of force being applied. Expressed mathematically:

$$R_n = \frac{W}{F}$$

where W = Amount of weight to be lifted; F = Amount of force applied to the block and tackle; R_n = Number of ropes.

Problem: How much pull (pounds of force) would it take on the block and tackle rope in Figure 4.14 to lift a 545-lb ball?

Solution: From Figure 4.14, the number of ropes supporting the load is three. The pull can be found by rearranging the pulley equation to solve for F:

$$F = \frac{W}{R_n} = \frac{545 \text{ lb.}}{3} = 181.666\ldots \text{ or } 182 \text{ lb.}$$

With a three rope, block and tackle, 182 lb of force will lift a 545-lb load, but the rope where the force is applied will move a three times the distance the load moves.

4.6. Inclined Plane

An inclined plane is an even surface sloping at any angle between vertical and horizontal. An inclined plane produces a mechanical advantage. The amount is determined by the ratio of the length of the inclined plane to the change in elevation. Instead of lifting the entire weight vertically, part of the weight is supported by the inclined plane.

Compare drawings I and II in Figure 4.15. Intuitive reasoning suggests that if the weight being moved and the distance AC are the same in both cases, then less force (ignoring friction) will be required to move the wagon up the inclined plane in the situation represented by drawing I because the change in height is less in I than it is in II for the same length of inclined plane. If we need to know the pounds of force required to pull the wagon, then we must use an equation based on the

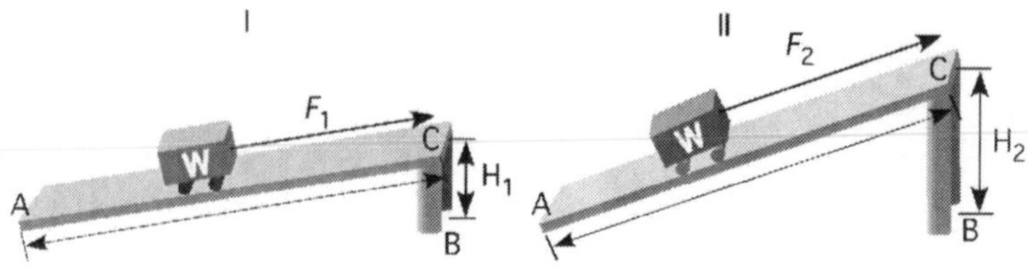

FIGURE 4.15. Two inclined planes.

principles of an inclined plane. Expressed mathematically:

$$F \times AC = W \times BC$$

where F = Amount of force to pull the wagon (ignoring friction); AC = Length of the inclined plane; W = Weight of the wagon; BC = Height of the inclined plane.

If we analyze drawing I, Figure 4.15, first and assume that the total weight is 100.0 lb, the height (BC) is 2.0 ft, and the length of the inclined plane (AC) is 12.0 ft, then the amount of force that would be required to pull the wagon up the inclined plane is:

$$F = \frac{W \times BC}{AC}$$

Substituting the values gives:

$$F = \frac{100.0 \text{ lb} \times 2.0 \text{ ft}}{12.0 \text{ ft}} = 16.666\ldots \text{ or } 17 \text{ lb}$$

Now we can see if the conclusion was right about the situation in drawing II. We will use the same equation to calculate the force in this situation. If we assume the length of the plane is the same (AC), then:

$$F = \frac{100.0 \text{ lb} \times 4.0 \text{ ft}}{12.0 \text{ ft}} = 33.33\ldots \text{ or } 33 \text{ lb}$$

It thus is obvious that an inclined plane with a steeper angle will require more force for the same weight.

4.7. Screw

The screw is a modification of the inclined plane combined with a lever. The threads of a screw or a bolt are an inclined plane that has been rolled into the shape of a cylinder. A lever is used to turn the threads, which causes the load to move along the cylinder. An example of this principle is the screw type jack.

Figure 4.16 illustrates this principle as it is used in a jack, where the jack handle is the force arm. The same principle applies to a bolt and nut. In the case of a

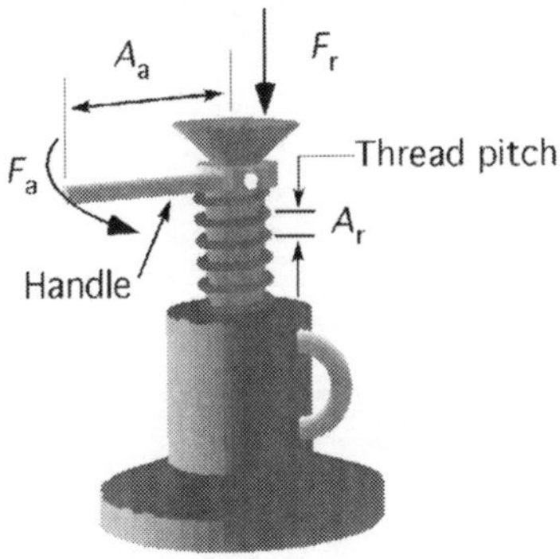

FIGURE 4.16. Lever equation applied to a screw jack.

bolt and nut the lever arm is the wrench, and the resultant force is the clamping pressure.

The distance between any two threads, called pitch, determines the amount of movement per revolution. The mechanical advantage is determined by the ratio of the radius of the lever and the pitch distance. The lever equation can be used to express this relationship mathematically:

$$F_a \times A_a = F_r \times A_r$$

where F_a = Forced applied at the end of the lever arm; F_r = Amount of weight the jack will lift; A_a = Length of the lever arm; A_r = Pitch of the threads.

Problem: If the lever arm in Figure 4.16 is 18 in long and the pitch of the threads is 0.125 in, how much weight will the jack lift when 50.0 lb of force is applied (ignoring friction)?

Solution: Rearranging the equation to solve for F_r gives:

$$F_r = \frac{F_a \times A_a}{A_r} = \frac{50.0 \text{ lb} \times 18.0 \text{ in}}{0.125 \text{ in}} = 7{,}200 \text{ lb or } 7.20 \text{ E3 lb}$$

This problem illustrates that when using a screw jack a small amount of force will lift a large load. The disadvantage is that the load will only be lifted 0.125 inches (1/8 of an inch) for every revolution of the handle.

Friction will affect the performance of this machine more than the others. With the proper lubrication it can be kept to a manageable level.

4.8. Combining Machines

All agricultural machines are a combination of these five simple machines. The analysis of the forces on a complex machine exceeds the objectives of this text, but Figure 4.17 is an example of how two or more of these machines can be used together.

This example is also useful for discussing the problem solving process. The first inclination is to try and solve for the amount of weight first. If this approach is

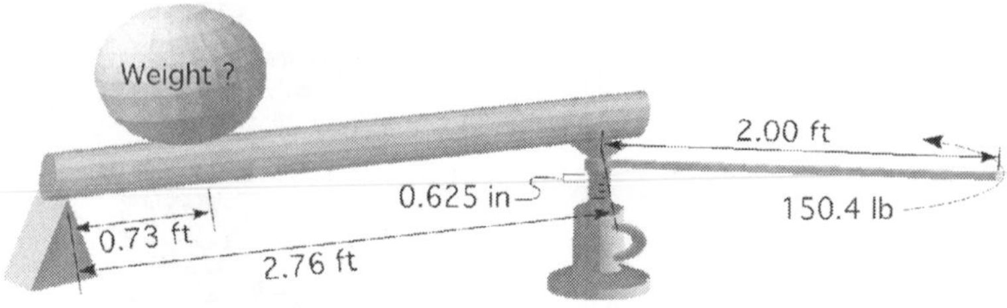

FIGURE 4.17. Two machines working in combination.

used it will soon become evident that an important piece of data is missing. What is the value for the applied force? To solve this problem it is important to realize that the first step is solving the amount of force produced by the jack.

$$F_a \times A_a = F_r \times A_r$$

$$F_r = \frac{F_a \times A_a}{A_r} = \frac{150.4 \text{ lb} \times 2.00 \text{ ft}}{0.625 \text{ in} \times \frac{1 \text{ ft}}{12 \text{ in}}} = \frac{300.8}{0.052 \ldots} = 5775.36 \text{ lb}$$

[*Note*: in this example the answer was not rounded to the appropriate number of significant figures because it is just the first step in a multiple step problem.]

Ignoring friction and the strength of the materials, the jack will be able to produce 5,775.36 lb of force. The force produced by the jack is the applied force for the lever. The amount of weight that can be lifted by the lever is determined by:

$$F_a \times A_a = F_r \times A_r$$

$$F_r = \frac{F_a \times A_a}{A_r} = \frac{5775.36 \text{ lb} \times 2.76 \text{ ft}}{0.73 \text{ ft}} = \frac{15939.9936}{0.73} = 21835.607 \ldots$$

or 22, 000 lb

The machines in the illustration are capable of lifting 22,000 lb with an application of 150.4 lb on the jack handle.

4.9. Metric Problems

The procedure for solving lever and problems involving forces can be illustrated with the following example. You are using the 3-point hitch on a tractor to lift a large bale of hay. You have recently "weighted" the bale on a metric scale and its

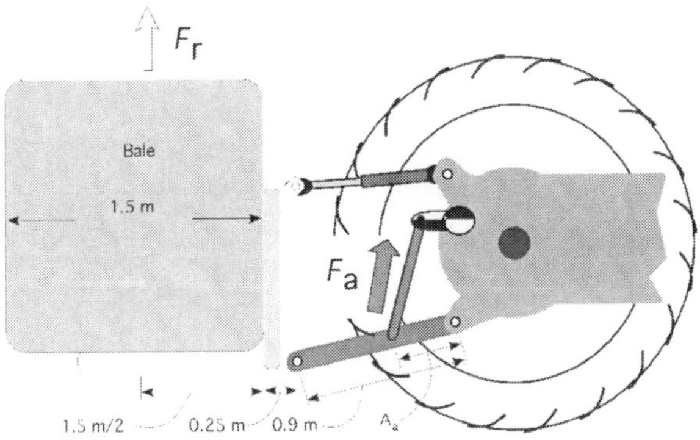

FIGURE 4.18. Illustration of bale problem.

mass is 840 kg. The lift arm is attached to the lower arm of the hitch 0.5 m from the pivot point and the length of the lower hitch arm is 0.95 m. How much force must be produced by the lift arms of the 3-point hitch to raise the bale?

The first step in determining the required amount of force to raise the bale is to convert its mass to weight.

$$\begin{aligned} \text{Weight } (W) &= \text{mass } (m) \times \text{gravity } (g) \\ &= 840 \text{ kg} \times 9.81 \ \frac{\text{m}}{\text{s}^2} \\ &= 8240.4 \text{ N} \end{aligned}$$

The second step is to calculate the required force using the equation for a class three lever. The equation for a class three lever is written in terms of a force or weight of an object, $F_r = W$. This is an example where a serious error can occur because the actual length of the resultant arm is not the length of the lower hitch arm, but the distance from the pivot point to the center of the mass of the bale, Figure 4.18.

Assuming a 1.5 m bale and an additional distance of 25.0 cm from the end of the lower lift arm to the front edge of the bale, the length of the resultant arm is:

$$m = 0.95 \text{ m} + 0.25 \text{ m} + \frac{1.5 \text{ m}}{2} = 1.95 \text{ m}$$

The lift arm force required is:

$$F_a = W \times \frac{A_r}{A_a} = 8240.4 \text{ N} \times \frac{1.95 \text{ m}}{0.50 \text{ m}} = 32137.56 \text{ or } 3.2 \text{ E4 N}$$

The 840 kg bale will require the lift arms on the 3-point hitch to produce 3.2 E4 N of force to lift the bale.

The same procedure can be used to solve any of the other problems in this chapter.

5
Internal Combustion Engines

5.1. Objectives

1. Be able to list and describe the events that occur in an internal combustion engine.
2. Be able to describe how a spark ignition (Otto-cycle) engine differs in operation from a compression ignition (Diesel-cycle) engine.
3. Be able to diagram and describe the events that occur in sequence during each stroke of a four-stroke cycle engine.
4. Be able to diagram and describe the events that occur in a two-stroke cycle engine.
5. Given the bore, stroke, number of cylinders, and clearance volume, be able to calculate the piston displacement, engine displacement, and compression ratio in both customary and SI units.

5.2. Introduction

There are six primary sources of power in agriculture: human labor, domestic animals, wind, flowing water, electricity, and heat engines. The primary sources of power for modern agriculture are the internal combustion engine and electric motor. Some day, the primary source of power may change to fuel cells, solar energy, or atomic energy, but in the immediate future the primary sources of power for agriculture will continue to be internal combustion heat engines and electric motors.

Internal combustion engines used for agricultural applications range from those used for one-horsepower garden tools to the hundreds of horsepower required for very large tractors. Because engines are very common, knowledge of how and why they work is essential to a successful agricultural manager. In this chapter, the basic engine types and functions are discussed, as are some basic calculations concerning engine size (displacement) and compression ratio.

5.3. Theory of Operation

The function of all internal combustion engines is to convert fuel (chemical energy) to power. This is accomplished by burning a fuel in a closed chamber and using the increase in temperature within the closed chamber to cause a rise in pressure. The pressure produces a force on the head of the piston causing it to move. The linear movement of the piston is converted to rotary motion (at the crankshaft), Figure 5.1. Rotary motion is more useful than linear movement.

All internal combustion engines have eight requirements for operation:

1) Air (oxygen) is drawn into the engine cylinder.
2) A quantity of fuel is introduced into the engine.
3) The air and the fuel are mixed.
4) The fuel–air mixture is compressed.
5) The fuel–air mixture is ignited by the spark plug in gasoline engines or by the heat of compression in diesel engines.
6) The burning of the fuel–air mixture causes a rapid pressure increase in the cylinder, which acts against the piston, producing a force on the piston.
7) The use of a connecting rod and a crankshaft converts the linear movement of the piston to rotary motion. The force on the piston is converted to torque on the crankshaft.
8) The products of combustion are expelled from the engine.

The next sections illustrate two ways in which these events are arranged in—four-stroke cycle and two-stroke cycle engines.

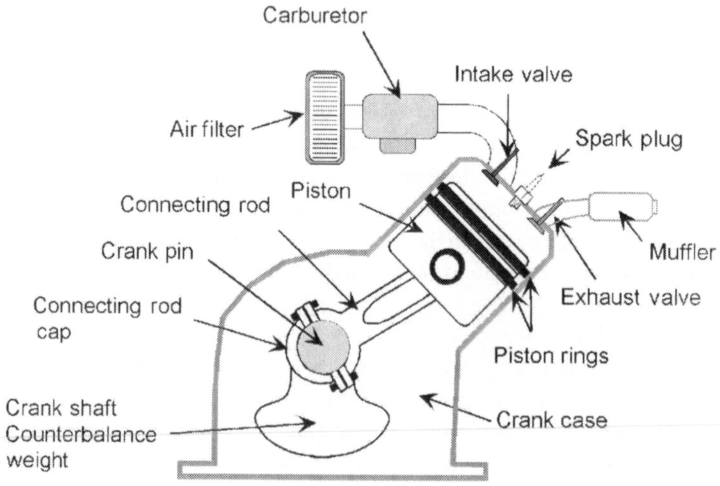

FIGURE 5.1. Parts of an internal combustion gasoline engine.

5.3.1. Four-Stroke Cycle

The four-stroke cycle engine, commonly called 4 cycle engine, is one of the two common types of engine cycles used for both spark and compression ignition engines. In 4 cycle engines, the eight events occur during four strokes of the piston or two revolutions of the crankshaft. The four strokes are called:

1. Intake
2. Compression
3. Power
4. Exhaust

Study Figure 5.2 and the following description of the events. During the intake stroke, the intake valve opens, and the piston travels from the top of the cylinder (TDC) to the bottom (BDC). In both the spark and the compression ignition engines, during the intake stroke, the movement of the piston reduces the pressure inside the cylinder. Higher atmospheric pressure outside the engine causes air to flow through the intake system and into the cylinder as the intake valve opens. In the carbureted engines fuel is introduced into the air stream as it flows through the carburetor. Shortly after the piston reaches the bottom of the intake stroke, the intake valve closes, thus trapping the air–fuel mixture inside the cylinder. Any restrictions in the flow of air into the engine reduces its volumetric efficiency. Volumetric efficiency is the comparison of how much air flows into the cylinder compared to the amount of air the cylinder can hold when it is totally full at atmospheric pressure. Reduction in volumetric efficiency reduces the horsepower produced by the engine. One common cause is a dirty air filter.

The compression stroke follows. Because the air–fuel mixture is trapped in the cylinder, as the piston returns to the top of the cylinder during this stroke, the air is compressed, and as it is compressed, the temperature rises. As the piston nears the top of the compression stroke, ignition occurs. Pressure is the key for converting the energy of the fuel to power. If any of the combustion pressure escapes from the combustion chamber, less power will be produced. On engines with a lot of hours and/or poor maintenance, the rings become less capable of

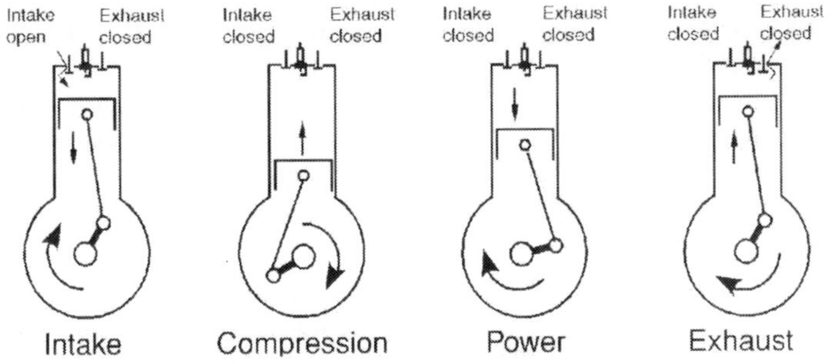

FIGURE 5.2. The four strokes of the four-stroke cycle.

sealing the combustion chamber and some of the compressed gases blow by the rings and into the crankcase.

The rapid expansion of the burning mixture causes the pressure to rise very quickly. The rapid rise in pressure (lb/in^2) causes a force (lb) on the face of the piston,

$$F(\text{lb}) = P\left(\frac{\text{lb}}{\text{in}^2}\right) \times A\left(\text{in}^2\right).$$

This pressure forces the piston away from the cylinder head. This is called the power stroke. It is during this stroke that the chemical energy of the fuel is converted to power.

Before the piston reaches the bottom of the cylinder on the power stroke, the exhaust valve opens, and the exhaust gases start to flow out of the engine. As the piston returns toward the top of the cylinder, the remaining byproducts of combustion are expelled from the cylinder through the open exhaust valve. This is called the exhaust stroke.

At this point, the intake valve opens, and the process repeats. Four strokes of the piston and two revolutions of the crankshaft have been completed.

5.3.2. Two-Stroke Cycle

In a two-stroke cycle engine (commonly called the 2 cycle engine) the eight requirements for operation occur during two strokes of the piston and one revolution of the crankshaft. Two very noticeable differences exist between a 2 and a 4 cycle engine. Some variability exists in the ways 2 cycle engines are constructed, but Figure 5.4 will be used to illustrate the primary functions of the common designs. In the 2 cycle, the carburetor is attached to the crankcase, and there are no intake and exhaust valves. Instead gases flow into and out of the cylinder through ports in the cylinder wall as they are exposed and covered by the movement of the piston, Figure 5.3.

Study Figure 5.4 and the following description of events to understand the 2 cycle engine. In a 2 cycle engine, intake and exhaust occur at almost the same time. The description of the cycle starts after combustion. As the piston moves away from the cylinder head, it first exposes the exhaust port. The combustion pressure starts the process of expelling the gases. As soon as the piston moves away from the cylinder head, the reed valve closes. This causes the crankcase to become pressurized. As the piston continues to travel it exposes the intake port. The pressurized air–fuel–oil mixture in the crankcase flows into the cylinder. This flow delivers the next fuel–air charge for combustion and helps expel the exhaust gases.

As the piston continues toward the cylinder head, the intake port is closed and then the exhaust port is closed. As soon as the intake port is covered, the continuing movement of the piston lowers the pressure in the crankcase, and the air–fuel–oil mixture flows from the carburetor, through the reed valve, and into the crankcase. As soon as the exhaust port is closed and the piston continues to move toward the cylinder head, compression occurs.

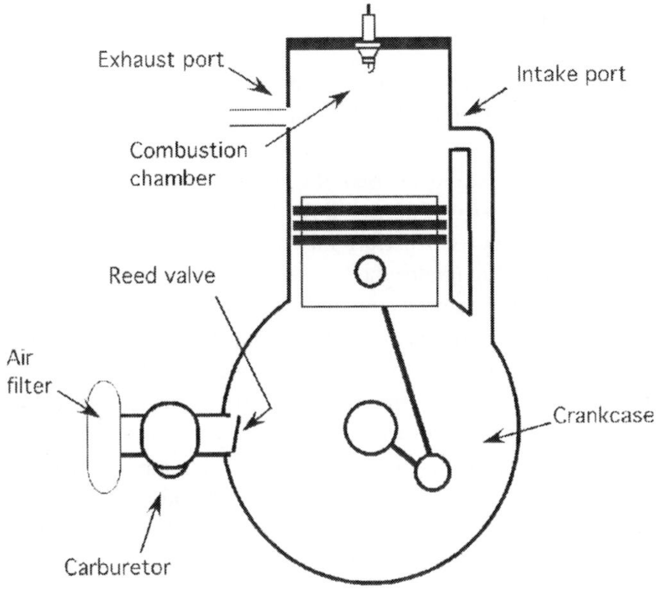

FIGURE 5.3. Parts of a two-stroke cycle engine.

The spark plug fires, the piston reaches top dead center (TDC), burning occurs, and pressure builds rapidly in the cylinder. This is called the ignition and power event. As soon as the piston moves away from the head far enough to expose the exhaust port, the cycle begins again.

If you count the movements of the piston, you will find that it has traveled two strokes or one revolution of the crankshaft. Because both the exhaust and the intake processes are not as efficient as in the 4 cycle engine, the 2 cycle engine will produce less power per stroke. However, because it has a power event each crankshaft revolution, the total power produced by the engine is comparable to that of a 4 cycle engine.

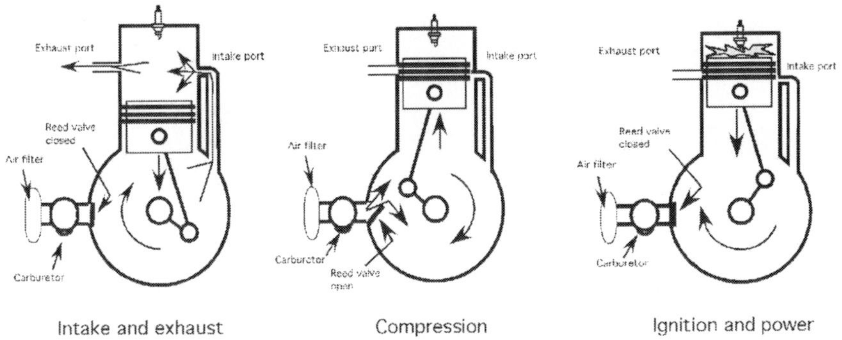

FIGURE 5.4. Two stroke cycles.

5.4. Types of Engines

The two primary categories of engines are the spark ignition (Otto-cycle) and the compression ignition (Diesel-cycle). In the spark ignition engine, the fuel is metered and introduced into the engine by either a carburetor or an injection system, Figure 5.5. When a carburetor is used, the fuel is metered by jets and orifices in the carburetor and added to the air as it flows through the venturi of the carburetor. Mixing occurs as the fuel–air mixture moves through the intake system and into the cylinder. A spark plug ignites the fuel–air mixture at the proper time.

Fuel injection systems improve the fuel efficiency of a gasoline engine because the multiple sensors and computer can better match the fuel needs of the engine. Historically the additional cost of the system over a carburetor fuel system has limited its use to larger, multicylinder engines, but rising fuel costs and emission standards have been the impetus for increasing the use of fuel injection systems on smaller engines. An engine that uses injection does not have a carburetor; the fuel is usually injected into the air stream at the throttle body or intake port. In some engines, a single injector is located at the throttle body (throttle body injection) that replaces the carburetor. This single injector meters the fuel for all of the cylinders. In the port injection system, an injector is positioned just outside the intake valve (port) leading into each cylinder. The injector in conjunction with the injector pump meters and delivers the required amount of fuel to the cylinder. The fuel and air are mixed as they flow into the cylinder. Engineers are continuing to work on direct inject for gasoline engines and the technology is starting to be used in automobile engines.

In a compression ignition engine (Diesel) the fuel is injected directly into the combustion chamber, Figure 5.6. The injection is timed to occur just before TDC on the compression stroke. Combustion occurs almost instantaneously because compression causes the air temperature within the combustion chamber to rise to 1,000°F and above. Thus, there is no need for a spark plug to ignite the fuel. The injector pump controls the amount of fuel delivered to each cylinder.

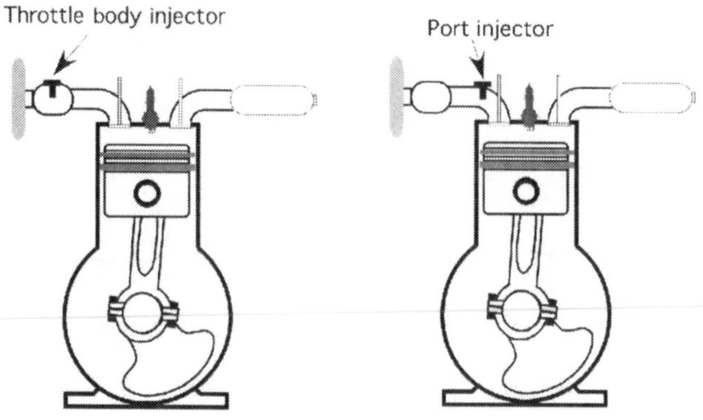

FIGURE 5.5. Throttle body and port injection.

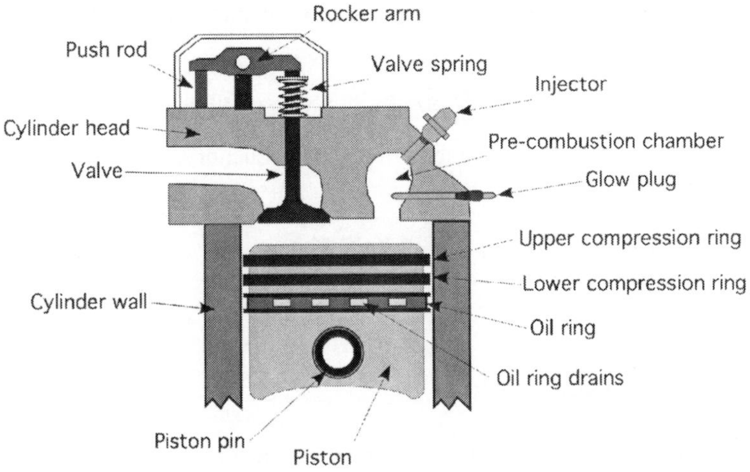

FIGURE 5.6. Parts of diesel engine.

5.5. Displacement

Displacement is the cylindrical volume that a piston displaces as it moves through one stroke. It is equal to the area of the piston multiplied by the length of the stroke. Displacement is one of the factors that determine the amount of horsepower an engine will produce, the greater the displacement, the greater the volume of air and fuel that is burned during combustion, which means more power. Expressed mathematically, piston displacement is:

$$\mathrm{PD} = \frac{\pi \times B^2}{4} \times S$$

where PD = Piston displacement in cubic inches (in^3); B = Bore of cylinder (diameter), in inches (in); S = Stroke of piston in inches (in).

The bore and the stroke of an engine has been traditionally expressed as $B \times S$ with the dimensions in inches. Thus, if $B \times S = 3.50 \times 4.00$, the piston displacement is:

$$\mathrm{PD} = \frac{\pi \times B^2}{4} \times S = \frac{\pi \times (3.50\ \mathrm{in})^2}{4} \times 4.00\ \mathrm{in} = 38.484\ldots \text{ or } 38.5\ \mathrm{in}^3$$

Larger engines are constructed with more than one cylinder. For multicylinder engines, the term engine displacement (ED) is used. Engine displacement is the product of the cylinder displacement times the number of cylinders. Expressed in equation form:

$$\mathrm{ED} = \mathrm{PD} \times n$$

where ED = Engine displacement (in^3); PD = Piston displacement (in^3); n = Number of cylinders.

If the engine in the previous example is a four-cylinder engine, then the engine displacement is:

$$\text{ED} = \text{PD} \times n = 38.5 \times 4 = 154\ \text{in}^3$$

It is a common practice to combine these two equations. For a single cylinder engine the value for $n = 1$.

$$\text{ED} = \frac{\pi \times B^2 \times S}{4} \times n$$

5.6. Compression Ratio

The compression ratio is the ratio of the total volume in a cylinder to the clearance volume. The clearance volume is the volume of the combustion chamber when the piston is at TDC, Figure 5.7. The total volume is the clearance volume plus the displacement.

The compression ratio is an engine characteristic related to engine efficiency, that is, the ability of the engine to convert energy in the fuel to useful mechanical energy. The greater the compression ratio, the greater the potential efficiency of the engine. The maximum compression ratio that can be obtained is a function of the type of fuel and the physical strength of the engine components. A compression ratio of 7 to 8:1 is common for gasoline engines and a ratio of 15 to 22:1 is common for diesel engines.

Expressed mathematically compression ratio is:

$$\text{CR} = \frac{\text{PD} + \text{CV}}{\text{CV}}$$

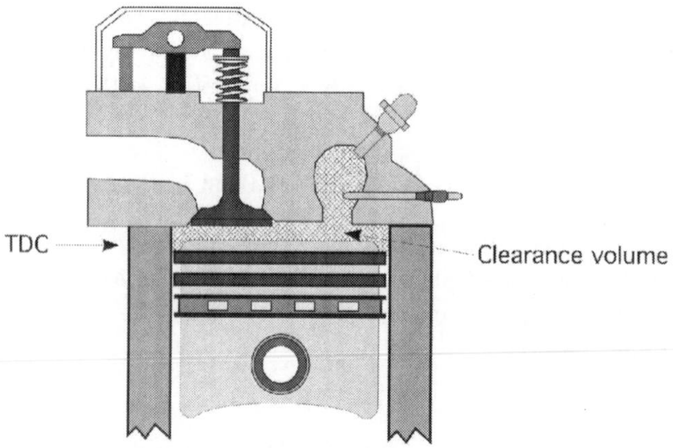

FIGURE 5.7. Cylinder clearance volume.

where CR = Compression ratio; PD = Piston displacement (in^3); CV = Clearance volume (in^3).

Problem: What is the compression ratio for a single cylinder engine with a $B \times S$ of 3.25×3.75 and a clearance volume of 6.20 in^3?

Solution: The first step is to calculate the displacement:

$$PD = \frac{\pi \times B^2}{4} \times S \times n = \frac{3.14 \times (3.25\ \text{in})^2}{4} \times 3.75 \times 1 = 31.093\ldots \text{ or } 31.1\ \text{in}^3$$

Then the compression ratio is:

$$CR = \frac{31.1\ \text{in}^3 + 6.2\ \text{in}^3}{6.2\ \text{in}^3} = 6.0 \text{ as a ratio } 6.0{:}1$$

Notice that the answer has no units. To have a ratio the units in the numerator and the denominator must be the same; thus the units cancel each other. A compression ratio of 6.0 would be expressed as 6.0:1.

5.7. Theoretical Power

In the previous chapter, theoretical power was defined as the calculated horsepower based on the bore, cylinder pressure, and engine speed. Figure 5.8 shows a typical cylinder pressure curve for a naturally aspirated engine. During the intake stroke the pressure is below zero, atmospheric. The pressure starts to rise as the compression stroke occurs and jumps rapidly at combustion. During the power stroke and exhaust stroke, the cylinder pressure gradually drops back to zero. The average pressure for the complete cycle is used to calculate engine power.

The theoretical or indicated engine horsepower can be determined using the mean effective cylinder pressure (MEP). The indicated or theoretical power is the power produced in the engine by combustion of the fuel. It does not account for the power lost to friction and other losses. The MEP can be determined from the graph, but these calculations are beyond the math level for this text. The equation for calculating indicated engine power in units of horsepower for 4 cycle engines is:

$$E_{\text{hp}} = \frac{P \times S \times A \times N}{33{,}000 \times 2} \times n$$

The equation is changed for 2 cycle engines:

$$E_{\text{hp}} = \frac{P \times S \times A \times N}{33{,}000} \times n$$

where P = Mean effective pressure (psi); S = Stroke (ft); A = Cylinder area (in^2); N = Engine speed (rpm); n = Number of cylinders. Note: the dimension for stroke in this equation is feet to cancel out the units of feet in the 33,000 constant.

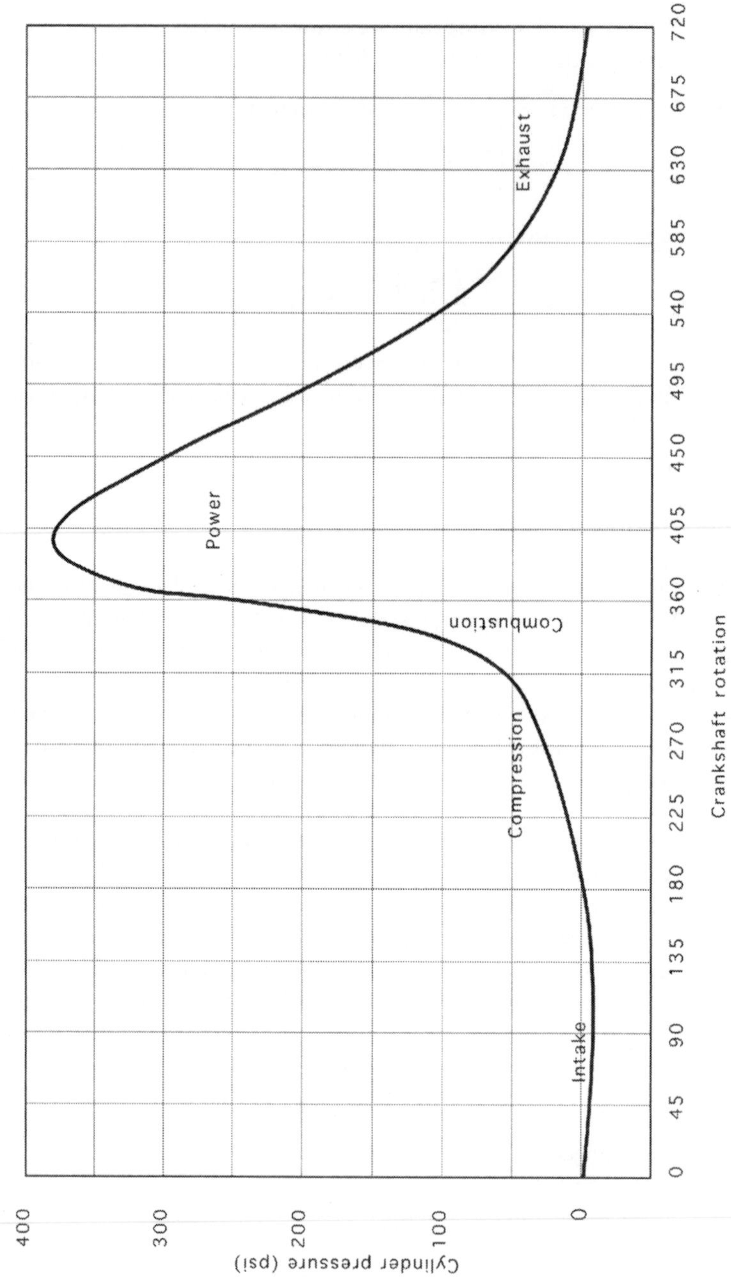

FIGURE 5.8. Cylinder pressure example.

Problem: Determine the indicated horsepower for a four cycle single cylinder engine, that has a mean effective pressure of 152.0 psi. The bore is 2.75 in, the stroke is 2.65 in, and the engine speed is 3,000 rpm.

Solution: The engine power for 4 cycle engines can be calculated using the following equation:

$$E_{\text{hp}} = \frac{P \times S \times A \times N}{33{,}000 \times 2} \times n$$

$$= \frac{152.0\,\frac{\text{lb}}{\text{in}^2} \times \left(2.65\text{ in} \times \frac{1\text{ ft}}{12\text{ in}}\right) \times \left(\pi \times \left(\frac{2.75\text{ in}}{2}\right)^2\right) \times 3{,}000\text{ rpm}}{33{,}000 \times 2} \times 1$$

$$= \frac{152 \times 0.22083\ldots \times 5.939\ldots \times 3{,}000}{66{,}000}$$

$$= \frac{598115.0626}{66{,}000} = 9.0623\ldots \text{ or } 9.06\text{ Ehp}$$

5.8. Metric Problems

The unit of engine power in the SI system is the kiloWatt. Piston bore and stroke are measured in millimeters. Displacement is reported in liters. One liter is equal to 1,000 cm^3 or 1,000,000 mm^3. Pressure is measured in kiloPascals. The equation for engine power in kW is:

$$E_{\text{kw}} = \frac{P \times S \times A \times N}{60{,}000 \times 2} \times n$$

P = pressure (Pa), S = stroke (m), A = piston area (m^2), N = speed (rpm).

Problem: Determine the indicated engine power for an engine in units of kW, that has a mean effective pressure of 1,050 kPa. The engine is a four-stroke cycle engine and has three cylinders. The bore (B) is 109 mm, the stroke (S) is 115 mm, and the speed is 3,000 rpm.

First convert the units to base SI units

$$P = 1{,}050\text{ kPa} = 1{,}050\text{ kPa} \times \frac{1{,}000\text{ Pa}}{1\text{ kPa}} = 1{,}050{,}000\text{ Pa}$$

$$B = 109\text{ mm} = 109\text{ mm} \times \frac{1\text{ m}}{1,000\text{ mm}} = 0.109\text{ m}$$

$$S = 115\text{ mm} = 115\text{ mm} \times \frac{1\text{ m}}{1,000\text{ mm}} = 0.115\text{ m}$$

Next calculate the area of the piston:

$$A(\text{m}^2) = \pi \times \frac{B^2}{4} = \pi \times \frac{(0.109\text{ m})^2}{4} = 0.0093313\ldots \text{ or } 0.00933\text{ m}^2$$

Finally determine the power (KW):

$$E_{\text{kw}} = \frac{P \times S \times A \times N}{60{,}000 \times 2} \times n$$

$$= \frac{1.050\ \text{E6 Pa} \times 0.115\ \text{m} \times 0.009331 \ldots \text{m}^2 \times 3{,}000\ \frac{\text{rev}}{\text{min}}}{120{,}000} \times 3$$

$$= \frac{3380269.069}{120{,}000}$$

$$= 28.163 \ldots \text{ or } 28.2\ \text{kW}$$

Problem: Calculate the compression ratio for the engine in the previous problem if the clearance volume is 0.063124 L.

Solution: Since we are calculating a ratio any consistent set of units can be used. Engine displacement is usually expressed in liters, therefore is this example the displacement is converted to liters because the clearance volume is in units of liters.

Start by calculating the piston displacement:

$$\text{PD(L)} = S \times \pi \times \frac{B^2}{4} \times \frac{1\ \text{L}}{1.0\ \text{E} - 3\ \text{m}^3}$$

$$= 0.115\ \text{m} \times 3.1416 \times \frac{(0.109\ \text{m})^2}{4} \times \frac{1\ \text{L}}{1.0\ \text{E} - 3\ \text{m}^3}$$

$$= \frac{0.00429240516}{4.0\ \text{E} - 3}$$

$$= 1.0731 \ldots \quad \text{or} \quad 1.07\ \text{L}$$

Next calculate the compression ratio:

$$\text{CR} = \frac{\text{PD} + \text{CV}}{\text{CV}}$$

$$= \frac{1.07\ \text{L} + 0.063124\ \text{L}}{0.063124\ \text{L}}$$

$$= \frac{1.133124}{0.063124}$$

$$= 17.950 \ldots \quad \text{or} \quad 18 \text{ to } 1.$$

6
Power Trains

6.1. Objectives

1. Be able to describe mechanical, pneumatic, and hydraulic power trains.
2. Be able to determine the appropriate size for a pulley, sprocket, or gear.
3. Be able to use speed ratios to determine pulley, sprocket, and gear sizes.
4. Be able to determine the speed and direction of rotation at any point in a power train.
5. Understand the relationship between speed and torque.
6. Be able to determine the torque and power at any point in a power train.
7. Be able to calculate the movement and changes in force for a hydraulic power train.

6.2. Introduction

In some applications, such as a push lawn mower with a vertical crankshaft engine, the power is used at the location that it is produced. The blade is attached directly to the end of the crankshaft. Most machines require a more complex system for transporting the power from the source to the load. In addition, many machines require that the power produced by the motor or engine be modified and/or transported to another part of the machine. These modifications could include changing the speed of rotation, direction of rotation, and location of the power. The collection of machine components that are used to accomplish this are called the transmission, power train, or drive train. In complex agricultural machines, such as a combine, the power must also be modified several times to meet the needs of the different machine components.

The transportation of power can be accomplished using mechanical, pneumatic, hydraulic, or electrical systems. Mechanical systems are very different so they will be discussed separately. Hydraulic and pneumatic systems have enough similarities that they will be discussed together. Electrical systems will not be discussed in this text. Each of these systems use a different means for completing the modifications and delivery of the power and each has advantages and disadvantages. The remaining sections of this chapter will explain how these power transfer systems work. *Note*: none of the systems are 100% efficient in the transfer of power, but energy losses will be ignored during the following sections to make it easier to understand how each system functions.

6.3. Mechanical Systems

Mechanical power systems are composed of pulleys, sprockets, gears, bearings, shafts, and numerous other components that are used to deliver power to the individual components of a machine and to change it to meet each component's requirement. Changes in speed, direction of rotation, or torque may be needed. The next sections illustrate some of the principles of mechanical power trains and provide examples of how these principles are used.

6.3.1. Pulleys

V-pulleys or sheaves are used in conjunction with V-belts because the V shape of the belt wedges in the V shape of the pulley. This increases friction between the belt and the pulley, which results in greater allowable power transmission. A belt drive will always have at least two pulleys, the driver and the driven. Consider Figure 6.1, which represents two pulleys mounted on shafts and connected by a belt.

When we apply torque to the driver shaft and cause it to turn or rotate as indicated by the arrow on the driver pulley in Figure 6.1, a tension or force will be created in the belt. If the tension exceeds the load (torque) of the driven pulley and the friction of the drive train, it will cause the driven pulley and shaft to turn. Drive trains using belts and pulleys may contain more than two pulleys. If more than two are used, one will be the driver and the others will be driven.

If the driver pulley in Figure 6.1 has a diameter of 10.0 in and rotates at 100.0 revolutions per minute (rpm), and the driven pulley has a diameter of 5.0 in, we can determine the speed of the driven pulley (and the shaft it is attached to) because

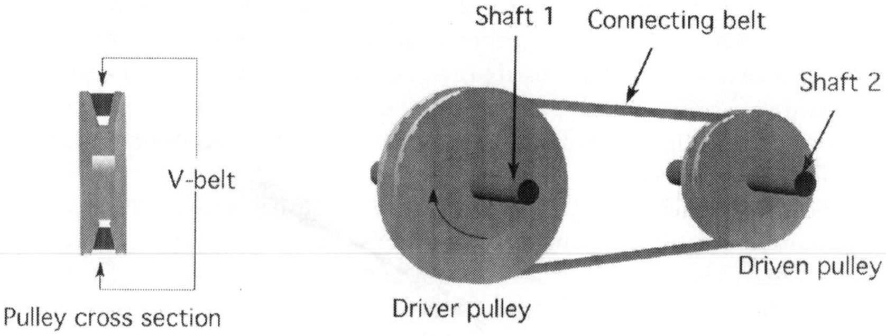

FIGURE 6.1. Parts of a V-belt system.

the diameter of the driver pulley times the speed of the driver pulley is equal to the diameter of the driven pulley times the speed of the driven pulley. Expressed mathematically the pulley equation is:

$$D_1 \times N_1 = D_2 \times N_2$$

where D = Diameter of a pulley; N = Pulley speed in revolutions per minute.

This statement is true because the linear speed (feet per minute) of the belt remains constant. If we identify D_1 as the diameter of the driver pulley and D_2 as the diameter of the driven pulley, we can determine the speed of the driven pulley by rearranging the pulley equation.

$$N_2 = \frac{D_1 \times N_1}{D_2} \text{ or } \frac{D_1}{D_2} \times N_1$$

Using the example problem:

$$N_2 = \frac{10.0 \text{ in} \times 100 \text{ rpm}}{5.0 \text{ in}} = 200 \text{ rpm}$$

When you understand the pulley equation, you should be able to state with confidence that when the driven pulley is smaller than the driver, the speed will increase, and when the driven pulley is larger than the driver is, the speed will decrease. Thus by visual inspection and intuitive reasoning you should be able to tell if a pulley drive train will increase or decrease the speed. To determine the actual amount of change, use the pulley equation.

One application of this principle is represented in the fan drive in Figure 6.2. Large ventilation fans of this type are commonly used in greenhouses, livestock buildings, and other agricultural applications.

Problem: Assume that you have a fan and an electric motor, but no pulleys. The fan is designed to operate at 500 rpm, and the electric motor operates at 1725 rpm. What sizes of pulleys will be needed to operate the fan?

Solution: First, intuitive reasoning tells us that a large change in speed will require a large difference in pulley diameters. Second, we know that the pulley equation

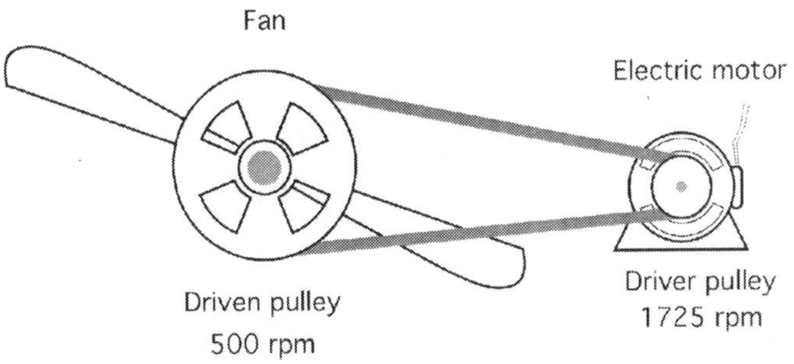

FIGURE 6.2. Determining unknown pulley size.

includes four variables, and at this point we only know two, the pulley speeds. To find a solution we must select one of the pulley sizes and then determine the other.

We could begin by selecting a pulley for the fan, but because the fan speed (driven) is less than the motor speed (driver), we know that the pulley on the motor will be smaller than the pulley on the fan. If we selected a pulley for the fan that is too small, the calculated pulley size for the electric motor may be smaller than what is physically possible. Therefore, begin by selecting the pulley size for the electric motor, and then calculate the required pulley size for the fan. If we select a 2.5-in pulley for the motor, then the size of the fan pulley can be determined by rearranging the pulley equation:

$$D_2(\text{fan}) = \frac{D_1(\text{motor}) \times N_1}{N_2} = \frac{2.50 \text{ in} \times 1725 \text{ rpm}}{500 \text{ rpm}} = 8.62 \text{ in or } 8\,\frac{5}{8} \text{ in}$$

Because pulleys are manufactured with diameters in increments of fractions of an inch, an 8 and 5/8-in pulley probably would be used.

Note that the ratios of the pulley diameters, D_2/D_1, will be equal to the ratios of the pulley speeds, $N_1/N_2 = 3.45$. Thus, if for any reason the 2.5-in pulley for the motor is not available, any combination of pulley diameters with a ratio of 3.45 (or approximately 3.5) will provide the correct speed. For example, pulleys with diameters of 8.625 and 2.5 or 17.25 and 5.0 or 34.5 and 10.0 will produce the same change in speed. This ratio of two pulleys is called the speed ratio and it will be discussed in more detail later in this chapter.

6.3.2. Sprocket Sizes

Roller chains and sprockets, Figure 6.3, have two advantages over belts and pulleys. They are capable of transmitting greater amounts of power, and because it is impossible for the chain to slip, the sprockets stay in time.

The equation used with pulleys can be used to determine sprocket sizes by replacing the diameter of the pulley with the teeth for each sprocket. For sprockets,

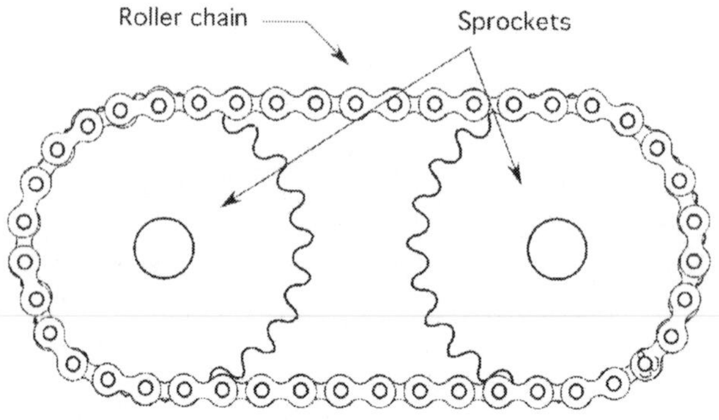

FIGURE 6.3. Roller chain and sprockets.

the pulley equation becomes:

$$T_1 \times N_1 = T_2 \times N_2$$

Problem: A hydraulic pump will be powered by a tractor power take-off (PTO). The pump must turn at 2100 rpm, and the PTO operates at 540 rpm. What sizes of sprockets are needed?

Solution: This problem is the same as the previous one where we determined the size of pulleys needed to power the fan. The first step is to select the size of one of the sprockets. In this example the driven sprocket (pump) will have fewer teeth than the driver sprocket (PTO). For this example, we will start by selecting an 18-tooth sprocket for the pump. We need to determine the size of the PTO sprocket.

Rearranging the sprocket equation for the number of teeth in the driver sprocket give us:

$$T_1 = \frac{T_2 \times N_2}{N_1} = \frac{18 \text{ teeth} \times 2{,}100 \text{ rpm}}{540 \text{ rpm}} = 70 \text{ teeth}$$

The pump will turn at the correct speed if the PTO sprocket of the tractor has 70 teeth.

6.3.3. *Gear Sizes*

It is common to use gears in power trains when the shafts are very close together, when a large amount of power is being transmitted or in transmissions where selectable speed ratios are needed. The sizes of gears are determined in the same way as those of chains and sprockets. The sprocket equation can be used without modification.

6.4. Speed Ratios

In some situations when the speeds of the driver and driven shafts are known, pulley, sprocket, and gear sizes can be determined by using speed ratios instead of the pulley or sprocket equations.

In the previous problem we determined that a 70-tooth sprocket was needed to power the pump. A 70-tooth sprocket will have a large diameter and may be too large to fit on a small tractor. Does this mean that the pump cannot be used?

The solution is to use a smaller sprocket on the pump and then determine the size of sprocket needed for the PTO. In the original problem we saw that a 18 and 70-tooth sprockets would provide the correct speed. Thus we know that the ratio of the two sprockets is 70/18 or 3.9:1. In other words, the sprocket on the PTO must have approximately four times as many teeth as the sprocket on the pump.

With the speed ratio between the two shafts known, different sprocket combinations could be used:

$$\frac{T_1}{T_2} = \frac{70}{18} \text{ or } \frac{35}{9} \text{ or } \frac{43}{11} = 3.9$$

In these situations the ratio does not usually need to be exact for the system to operate. A hydraulic pump that runs slightly faster or slower, because the exact speed ratio cannot be used, usually performs adequately.

Knowing the speed ratio makes it easier to select sprocket combinations that will operate the pump at the correct speed and still fit within the physical limitations of the machine.

6.5. Direction of Rotation

In the previous section the examples illustrated that one use of power trains is to change the speed between two or more shafts or components. Another function of mechanical power trains is to change the direction of rotation. In many applications, a shaft or component may need to turn in the opposite direction as the driver. The two directions of rotation are labeled clockwise (CW) and counterclockwise (CCW) when viewed from the end of the shaft. The direction of rotation can be changed through the use of belts and pulleys, chains and sprockets, and gears. Each of these methods will be illustrated.

6.5.1. *Belts and Pulleys*

Two different techniques can be used to change the direction of rotation, Figure 6.4.

Notice that twisting the belt changes the direction of rotation. This technique is acceptable if the pulleys are some distance apart, the belt is operating at slow speed and it is not transmitting a large load. The belts will rub at the center and these three conditions will increase the amount of friction at that point. The friction produced from rubbing will shorten the life of the belt.

The addition of a third pulley also will reverse the direction of rotation. In this situation, a six-sided, or double-V, belt will be used if the third pulley requires a lot of torque, if not, a flat pulley will be used that runs on the back side of the belt. This is very common in serpentine drives used on modern automobiles.

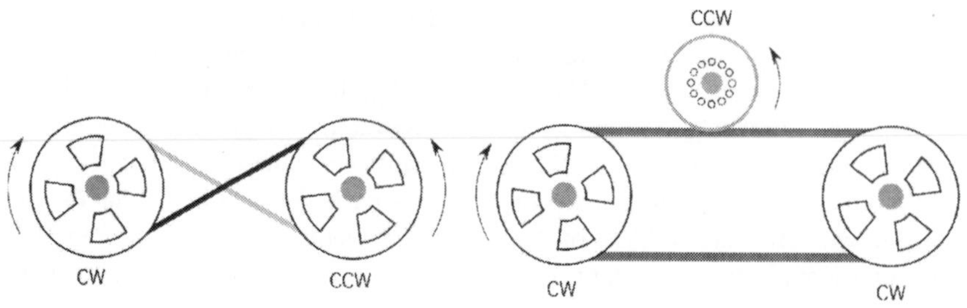

FIGURE 6.4. Changing the direction of rotation using belts.

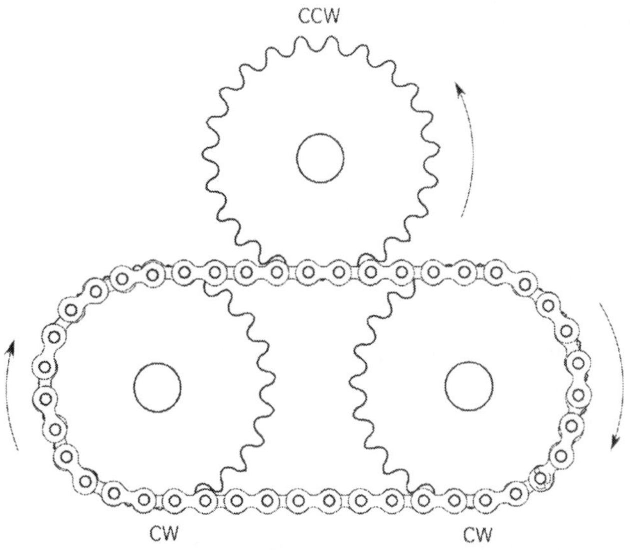

FIGURE 6.5. Changing the direction of rotation with a roller chain and sprockets.

The use of the third pulley also provides an opportunity to change the speed of rotation. A difference in diameter between the third pulley and the driver will cause a change in speed.

6.5.2. Roller Chains and Sprockets

Figure 6.5 shows that chains and sprockets can be used to reverse the direction of rotation. Because roller chains cannot be twisted, changing the direction of rotation requires the use of a third sprocket. Changing the number of teeth in the third sprocket also allows the third sprocket to turn at a different speed.

6.5.3. Gears

Gears are unique in that each pair will change the direction of rotation, Figure 6.6.

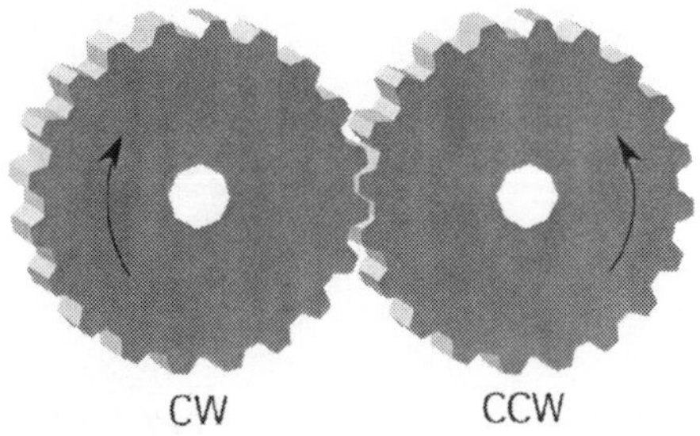

FIGURE 6.6. Changing the direction of rotation with gears.

6.6. Complex Power Trains

Many machines use power trains that are more complex than what has been discussed up to this point. Figure 6.7 represents a power train designed to allow an engine or motor to supply the power for two different components. These components are attached to output shafts A and B. This figure illustrates the application of direction and speed of rotation for more complex power trains.

Problem: What are the speed and the direction of rotation of shaft A and shaft B in Figure 6.7?

Solution: In this power train the output shaft of a 50-horsepower motor powers gear 1. Gear 2 is powered by gear 1, and because gear 3 is attached to the same shaft, it is also powered by gear 1. Gear 4 is powered by gear 3, and gear 5 by gear 4. Shaft A will turn at the same speed and direction of rotation as gear 5, and shaft B will turn at the same speed and direction of rotation as gears 2 and 3.

More than one approach can be used to determine the speed of shaft A and shaft B. The first one that we will show uses the gear equation. Because there is more than one pair of gears, the equation must be used more than once. The speed of gear 2 is:

$$N_2 = \frac{T_1 \times N_1}{T_2} = \frac{24 \times 2{,}200}{12} = 4{,}400 \text{ rpm}$$

Because gear 2 is fixed to shaft B, shaft B and gear 3 turn at 4,400 rpm.

To determine the speed of shaft A, the gear equation is used two more times, first to find the speed of gear 4:

$$N_4 = \frac{16 \times 4{,}400}{8} = 8{,}800 \text{ rpm}$$

and then the speed of gear 5:

$$N_5 = \frac{8 \times 8{,}800}{12} = 5866.66\ldots \text{ or } 5{,}900 \text{ rpm}$$

Therefore, shaft A turns at 5,900 rpm.

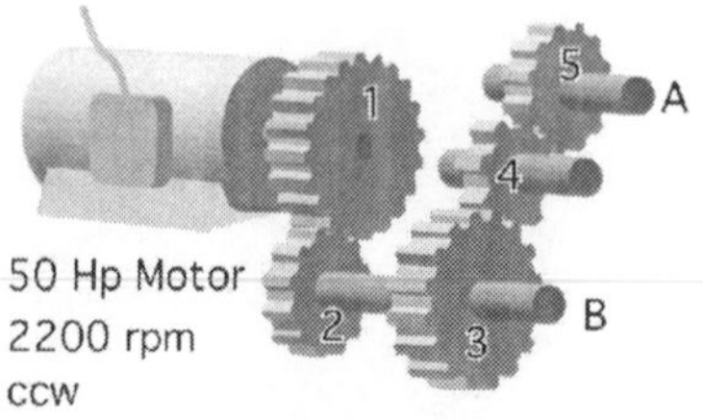

Gear	Number of teeth
1	24
2	12
3	16
4	8
5	12

FIGURE 6.7. Calculating shaft speed and direction of rotation in multiple shaft drives.

Remembering the discussion of speed ratios, we can develop an alternative approach for solving this problem. If the speed of shaft B is determined by the speed of the driver gear times the gear ratio of the two gears, then the speed of shaft B can be determined by:

$$N_B = N_1 \times \frac{T_1}{T_2} = 2{,}200 \times \frac{24}{12} = 4{,}400 \text{ rpm}$$

and the speed of shaft A can be determined by:

$$\begin{aligned} N_A &= N_1 \times \frac{T_1}{T_2} \times \frac{T_3}{T_4} \times \frac{T_4}{T_5} \\ &= 2{,}200 \times \frac{24}{12} \times \frac{16}{8} \times \frac{8}{12} \\ &= \frac{6{,}758{,}400}{1{,}152} \\ &= 5866.66\ldots \text{ or } 5{,}900 \text{ rpm} \end{aligned}$$

When the speed ratio approach is used, it is important to place the number of teeth of the driver in the numerator and the number of teeth of the driven in the denominator.

6.6.1. Direction of Rotation

The best way to determine the direction of rotation of shafts A and B is to use intuitive reasoning.

We know, from our discussion of gears, that the direction of rotation changes with every pair of gears in the power train. Thus, shaft B is driven by one pair of gears, and changes direction of rotation one time. Shaft B turns clockwise. Shaft A is powered by three pairs of gears, and changes the direction of rotation three times. Shaft A turns clockwise also. Study Figure 6.8 to check these answers.

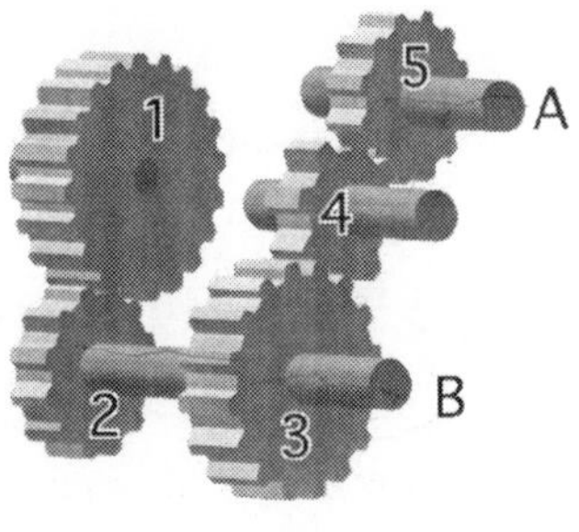

FIGURE 6.8. Determining the direction of rotation for shafts A and B.

6.7. Speed and Torque

Torque is a rotating force applied to a lever arm. In the case of a belt and pulley, the force is the tension on the belt and the length of the lever arm is the radius of the pulley. Suppose that a set of scales is attached to the belt, as in Figure 6.9, and assume that the tension on the belt is 100.0 lb.

Because the tension or force in the belt is constant along its length, there is 100.0 lb of force pulling at the edge of both the driver (A) and the driven (B) pulley.

When the driver pulley has a diameter of 10.0 in and the driven pulley has a diameter of 5.0 in, then the torque on the driver pulley in Figure 6.9 is:

$$\text{To} = F \times R$$

$$\text{To} = F \times R = 100.0\text{ lb} \times 5.0\text{ in} = 500\text{ lb-ft}$$

and the torque on the driven shaft is:

$$\text{To} = F \times R = 100.0\text{ lb} \times 2.5\text{ in} = 250\text{ lb-ft}$$

The torque is different on the two pulleys because pulleys behave as class one levers; the radius (one-half the diameter) is the length of the lever arm. A larger diameter pulley will have a longer lever arm. This example also illustrates that because the belt speed (in inches per minute) stays the same; the torque on the driven shaft is inversely proportional to the change in speed between the driver and driven pulleys. When the driven shaft turns at a higher speed, its torque is decreased relative to the driver shaft, and vice versa.

For every speed change, there is a change in the torque. Reviewing the PTO power equation is one way of showing this relationship:

$$P_{\text{Hp}} = \frac{\text{To} \times N}{5252}$$

When the horsepower stays the same, assuming that there are no drive train power losses, then as torque increases, the rpm must decrease, and vice versa. This relationship also can be expressed as:

$$\text{To}_1 \times N_1 = \text{To}_2 \times N_2$$

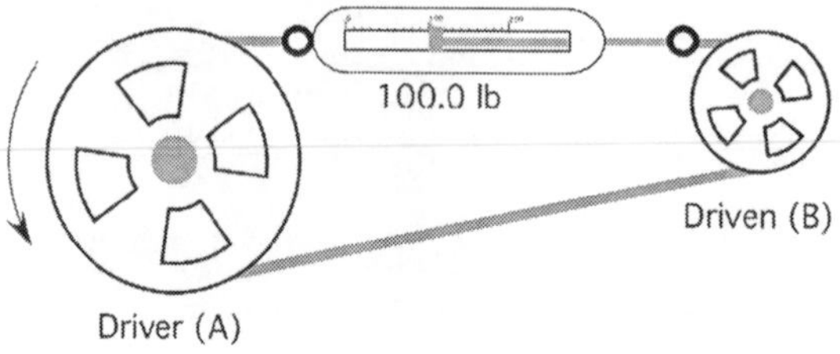

FIGURE 6.9. Speed and torque.

Problem: The driver shaft of a belt power train turns at 300 rpm and applies 20.0 lb-ft of torque, and the driven pulley turns at 50 rpm, how much torque will be developed at the driven pulley?

Solution: Rearranging the torque equation to solve for To_2 gives us:

$$\text{To}_2 = \text{To}_1 \times \frac{N_1}{N_2} = 20.0\ \text{lb-ft} \times \frac{300\ \text{rpm}}{50\ \text{rpm}} = 120\ \text{lb-ft}$$

Notice that the speed of the driven shaft is one-sixth that of the driver shaft, but the torque on the driven shaft is six times that on the driver shaft. This relationship suggests the reason for a tractor transmission. The engine turns at high speed with low torque, and the drive axle turns at low speed with high torque. The transmission uses different gear pairs to produce different speed–torque combinations at the drive axle.

6.8. Transmission of Power and Torque

In managing agricultural machinery, it is sometimes useful to know the amount of torque being transmitted by a component of a power train. For example, it would be useful to know the amount of torque and the horsepower available at shafts A and B in Figure 6.7.

Using the previous discussion of the relationship between torque and speed and the torque equation the maximum torque available at each shaft can be determined:

$$\text{To}_\text{B} = \text{To}_1 \times \frac{N_1}{N_2}$$

To solve for To_B the values for the three variables must be known. The two rpm's were calculated in the previous section. The remaining variable is the torque supplied by the motor. Horsepower will be discussed in greater detail in a later chapter, but rotary horsepower can be determined by the horsepower equation used in the previous section.

$$P_\text{hp} = \frac{\text{To} \times N}{5252}$$

Because we know the horsepower and the rpm being produced by the engine, we can rearrange this equation to solve for the engine torque:

$$\text{To}_\text{eng} = \frac{P_\text{Hp} \times 5252}{N} = \frac{50\ \text{hp} \times 5252}{2{,}200\ \text{rpm}} = 119.36\ldots\ \text{or } 120\ \text{lb-ft}$$

Once the torque produced by the motor is known, the torque on shaft B can be calculated.

$$\text{To}_\text{B} = \text{To}_1 \times \frac{N_1}{N_2} = 120\ \text{lb-ft} \times \frac{2{,}200\ \text{rpm}}{4{,}400\ \text{rpm}} = 60\ \text{lb-ft}$$

The torque at shaft A can also be calculated:

$$To_A = To_{motor} \times \frac{N_{motor}}{N_A} = 120 \text{ lb-ft} \times \frac{2{,}200 \text{ rpm}}{6{,}000 \text{ rpm}} = 44 \text{ lb-ft}$$

Note that because we already knew the speed of shaft A, we used the speed ratio to solve for the torque. If we did not know the speed of shaft A, we would solve for it first.

6.9. Power Transmission Through Power Trains

Power is transmitted through a power train. Study Figure 6.8 again. Can you predict how much power is available at either shaft A or shaft B? The answer 50 horsepower is correct. If we assume no frictional losses in the power train, we will get out all of the power we put in. This can be demonstrated by using the horsepower equation to solve for the power available at shaft A.

$$P_{Hp} = \frac{To \times N}{5252} = \frac{44 \text{ lb-ft} \times 6{,}000 \text{ rpm}}{5252} = 50.26\ldots \text{ or } 50 \text{ hp}$$

6.10. Pneumatic and Hydraulic Systems

Pneumatic and hydraulic systems are based on the behavior of a flowing liquid. The primary difference is that hydraulic oil is considered noncompressible and air is not. Hydraulic systems are used on more agricultural equipment than pneumatic. Therefore, the remaining discussion will concentrate on hydraulic systems. All pneumatic and hydraulic systems have five required components, Figure 6.10. The heart of both systems is a pump. Many different types of pumps are used, but they have a common job—energize the system. The system also requires a reservoir to store the oil, control valves to control the amount and direction of flow, some type of actuator to perform the work and hoses and pipes to connect all of the components. Other components are included as necessary to achieve the desired performance of the system.

Hydraulic systems operate on two principles, pressure and flow. Pressure is used to cause force or torque, and flow is used to cause movement. The pressure in a system is governed by Pascal's Law, which states that the pressure on a confined

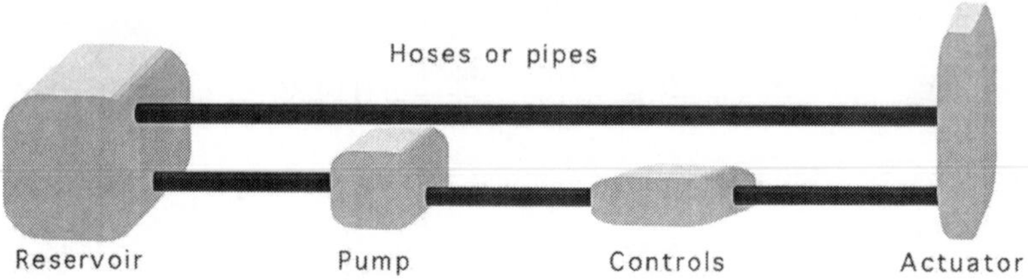

FIGURE 6.10. Five required components of a hydraulic system.

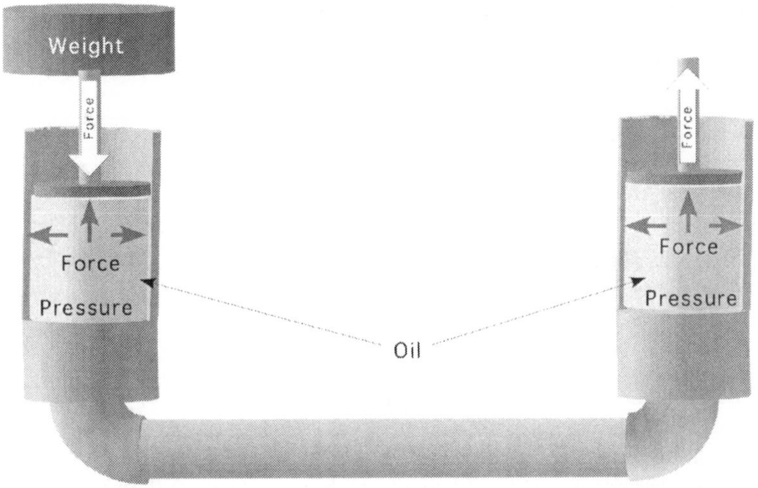

FIGURE 6.11. Force—the first principle of hydraulics.

liquid is transmitted equally in all directions and results in equal force at right angle to the container walls, Figure 6.11.

In Figure 6.12, force is used to pressurize the liquid and pressure is used to transmit and change the intensity of the resultant force. Changing the intensity of the resultant force is accomplished by changing the pressure or the area, Figure 6.12. The amount of force produced is determined by multiplying the pressure times the area.

$$P = F \times A$$

where P = Pressure, psi or lb/in^2; F = Force, lb; A = Area, in^2.

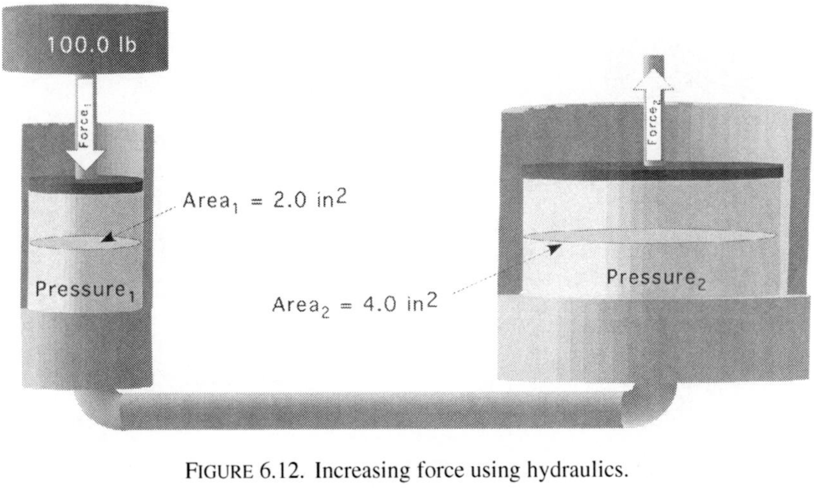

FIGURE 6.12. Increasing force using hydraulics.

Problem: How much force is produced on the second piston in Figure 6.12?

Solution: The first step is to determine the pressure in the system.

$$\begin{aligned} P_1 &= F_1 \times A_1 \\ &= 100.0 \text{ lb} \times 2.0 \text{ in}^2 \\ &= 200 \frac{\text{lb}}{\text{in}^2} \end{aligned}$$

But $P_1 = P_2$ The pressure is the same at all locations. Therefore:

$$\begin{aligned} F_2 &= P_1 \times A_2 \\ &= 200 \frac{\text{lb}}{\text{in}^2} \times 4 \text{ in}^2 \\ &= 400 \text{ lb} \end{aligned}$$

This example shows that when a 100.0 lb weigh applies force to an area of 2.0 square inches a pressure of 200 lb per square inch is produced. When this pressure is contact with an area of 4.0 square inches, a force of 400 lb will be produced. The increase or reduction of the force is proportional to the area.

The second principle of hydraulic systems is flow. The flow of the fluid determines the speed and amount of movement for the actuators in the system. In a two piston system as illustrated in Figure 6.13 the total flow (in^3) from the first piston to the second can be determined by multiplying the area (in^2) of the piston times the length (in) of the stroke (S). When the second piston has a larger area, the distance moved will be less because the volume moved is the same for both and a larger area requires less stroke length to produce the same volume. This relationship can be expressed mathematically:

$$A_1 \times S_1 = A_2 \times S_2$$

Problem: How far will the second piston in Figure 6.13 move?

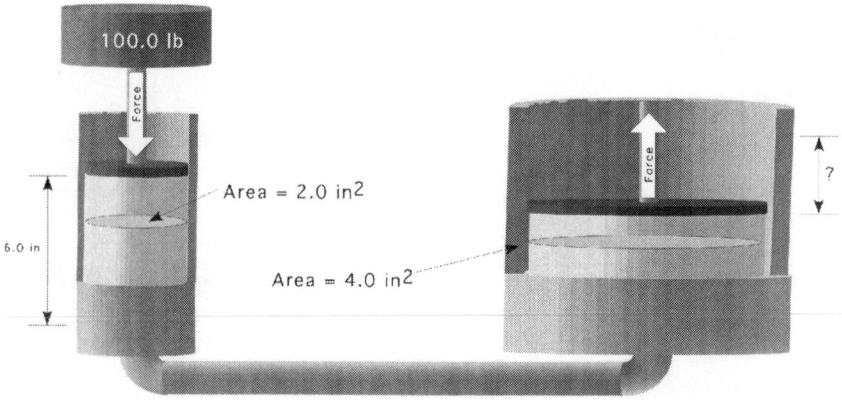

FIGURE 6.13. Determining flow.

Solution:

$$
\begin{aligned}
A_1 \times S_1 &= A_2 \times S_2 \\
S_2 &= \frac{A_1 \times S_1}{A_2} \\
&= \frac{2.0\ \text{in}^2 \times 6.0\ \text{in}}{4.0\ \text{in}^2} \\
&= 3.0\ \text{in}
\end{aligned}
$$

In most mobile applications of hydraulics the pressure is produced by a pump and regulated by a pressure relief valve. In these applications the system pressure determines the force produced by the actuator. When linear actuators, cylinders, are used, the diameter of the cylinder and the pressure of the fluid at the cylinder determine the force produced. The distance the cylinder extends or retracts is determined by the amount of fluid that moves and the speed at which the cylinder extends or retracts is determined by the flow rate of the fluid.

Problem: Determine the maximum force possible, amount of extension (in) and rate of extension (in/min) for a 3.50 in diameter cylinder when the system pressure is set at 2,500 lb per square inch (psi), the flow rate is 3.25 gallons per minute (gpm) and the control valve is activated for 15.0 sec.

Solution: The process for determining the maximum force produced is the same as the one used in the previous problem.

$$
\begin{aligned}
F &= P \times A \\
&= 2{,}500\ \frac{\text{lb}}{\text{in}^2} \times \left(\pi \times (3.50\ \text{in})^2\right) \\
&= 2{,}500\ \frac{\text{lb}}{\text{in}^2} \times 38.484\ldots \\
&= 96211.27\ldots\ \text{or } 96{,}200\ \text{lb}
\end{aligned}
$$

The amount of extension or stroke is a function of the cross section area of the cylinder and the volume of fluid flow. This problem can be solved using units cancellation method.

$$
\begin{aligned}
S_{\text{in}} &= \frac{231\ \text{in}^3}{1\ \text{gal}} \times \frac{3.25\ \text{gal}}{1\ \text{min}} \times \frac{1\ \text{min}}{60\ \text{sec}} \times \frac{15.0\ \text{sec}}{1} \times \frac{1}{38.5\ \text{in}^2} \\
&= \frac{11261.25}{2310} = 4.875 \text{ or } 4.88\ \text{in}
\end{aligned}
$$

The flow rate of the fluid (Q) and the cross section area of the cylinder determines the rate of movement. This problem can also be solved using the units cancellation method.

$$S\left(\frac{\text{in}}{\text{min}}\right) = \frac{231\ \text{in}^3}{1\ \text{gal}} \times \frac{3.25\ \text{gal}}{1\ \text{min}} \times \frac{1}{38.5\ \text{in}^2} = \frac{750.75}{38.5} = 19.5\ \frac{\text{in}}{\text{min}}$$

6.11. Metric Problems

Problem: What diameter of pulley is needed if the desired speed is 500 rpm and the driven pulley is 6.0 cm and turns at 1725 rpm?

Solution:

$$D_1 \times N_1 = D_2 \times N_2$$

$$D_2 = \frac{D_1 \times N_1}{N_2} = \frac{6.0\ \text{cm} \times 1725\ \frac{\text{rev}}{\text{min}}}{500\ \frac{\text{rev}}{\text{min}}} = \frac{10350}{500} = 20.7 \text{ or } 21\ \text{cm}$$

Problem: Determine the torque available at shafts A and B in Figure 6.14.

Solution: The first step is to determine the speeds of shafts A and B. Shaft B is the first one in the series so it will be used first.

$$T_1 \times N_1 = T_2 \times N_2$$

$$N_B = \frac{T_1 \times N_1}{T_2} = \frac{24\ \text{T} \times 2{,}200\ \frac{\text{rev}}{\text{min}}}{12\ \text{T}} = \frac{52800}{12} = 4{,}400\ \text{rpm}$$

The speed of shaft A:

$$N_A = N_1 \times \frac{T_1}{T_2} \times \frac{T_3}{T_4} \times \frac{T_4}{T_5}$$

$$= 2{,}200 \times \frac{24}{12} \times \frac{16}{8} \times \frac{8}{12}$$

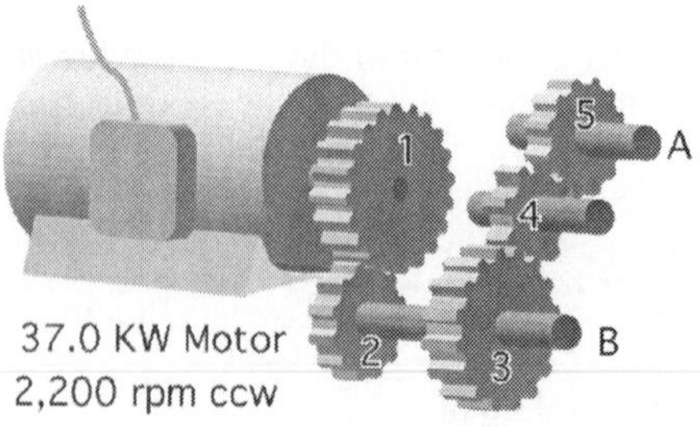

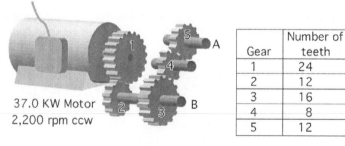

Gear	Number of teeth
1	24
2	12
3	16
4	8
5	12

FIGURE 6.14. Complex power train with metric values.

$$= \frac{6{,}758{,}400}{1{,}152}$$

$$= 5866.66\ldots \text{ or } 5{,}900 \text{ rpm}$$

The maximum torque will be available at only at one shaft. The maximum torque will be calculated for A and B. The maximum torque delivered by the motor is:

$$P_{\text{kW}} - \frac{\text{To} \times N}{9549}$$

Solving for To

$$\text{To} = \frac{9549 \times P_{\text{kW}}}{N}$$

$$= \frac{9549 \times 37.0}{2200}$$

$$\text{To} = 160.6 \text{ Nm}$$

The motor is producing 160.6 Nm or torque.

The torque at shaft A is:

$$\text{To}_1 \times N_1 = \text{To}_2 \times N_2$$

$$\text{To}_2 = \frac{\text{To}_1 \times N_1}{N_2}$$

$$= \frac{160.6 \text{ Nm} \times 2{,}200 \dfrac{\text{rev}}{\text{min}}}{5{,}900 \dfrac{\text{rev}}{\text{min}}}$$

$$= \frac{353{,}320}{5900} = 59.884\ldots \text{ or } 60 \text{ kN}$$

The maximum torque at B is:

$$\text{To}_2 = \frac{\text{To}_1 \times N_1}{N_2}$$

$$= \frac{160.6 \text{ Nm} \times 2{,}200 \dfrac{\text{rev}}{\text{min}}}{4{,}400 \dfrac{\text{rev}}{\text{min}}}$$

$$= \frac{353{,}320}{4400} = 80.3 \text{ Nm}$$

Problem: How much force will be produced by the hydraulic system in Figure 6.15 when the diameter of the first cylinder is 5 cm, the diameter of the second cylinder is 15 cm and the force on the first cylinder is 50 kN?

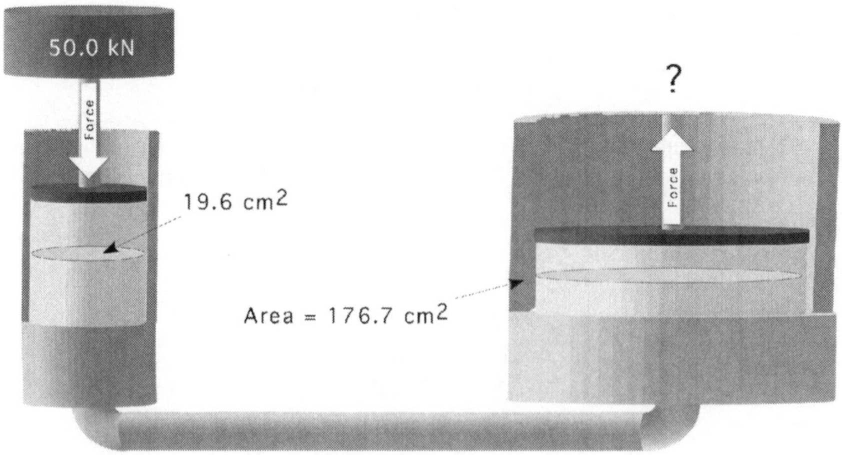

FIGURE 6.15. Force produced using SI units.

Solution: Force divided by area equals pressure and pressure multiplied by area equals force.

$$P(\text{kPa}) = \frac{F\ (\text{N})}{A\ (\text{cm})}$$

$$= \frac{50\ \text{kN}}{\dfrac{\pi \times (5\ \text{cm})^2}{4}} \times \frac{10{,}000\ \text{cm}^2}{1\ \text{m}^2} \times \frac{1{,}000\ \text{N}}{1\ \text{kN}} \times \frac{1\ \text{Pa}}{\text{N/m}^2} \times \frac{1\ \text{kPa}}{1{,}000\ \text{Pa}}$$

$$= \frac{500{,}000{,}000}{19634.95\ldots} = 25464.79\ldots \text{ or } 25{,}000\ \text{kPa}$$

The pressure on the fluid will be 25,000 kPa. The force on the second cylinder is:

$$F(\text{N}) = P(\text{kPa}) \times A(\text{cm}^2)$$

$$= 25{,}000\ \text{kPa} \times \frac{\pi \times (15\ \text{cm})^2}{4} \times \frac{1\ \text{m}^2}{10{,}000\ \text{cm}^2}$$

$$\times \frac{1{,}000\ \text{Pa}}{1\ \text{kPa}} \times \frac{\text{N/m}^2}{1\ \text{Pa}} \times \frac{1\ \text{kN}}{1{,}000\ \text{N}}$$

$$= \frac{417{,}864{,}669}{10{,}000{,}000} = 441.786 \text{ or } 440\ \text{kN}$$

Problem: How far will the second piston, in the previous problem, move when the piston in the first cylinder moves 5 cm?

Solution: Area $\times$ stroke distance = area $\times$ stroke distance

$$A_1 \times S_1 = A_2 \times S_2$$

$$S_2 = \frac{A_1 \times S_1}{A_2}$$

$$= \frac{19.6\ \text{cm}^2 \times 5\ \text{cm}}{176.7\ \text{cm}^2}$$

$$= \frac{98}{176.7} = 0.554\ldots \text{ or } 0.55\ \text{cm}$$

7
Tractors and Power Units

7.1. Objectives

1. Be able to describe the common designs of tractors.
2. Be able to estimate draw bar and PTO power using the 86% rule.
3. Be able to derate a stationary power unit for the intended work environment.
4. Be able to explain the concept of lugging ability.
5. Be able to describe the principles of tractor testing.
6. Understand the ASABE and OECD tractor testing procedures.

7.2. Introduction

Tractors are very versatile machines, but the range of uses is too wide for one machine to be successful in all of the possible jobs. Tractor manufacturers target tractor designs for different categories of use. These category boundaries are not ridged, but if owner/operators push the boundaries too far, the tractor can fail which may lead to an accident which damages the tractor or implements. It can also result in injuries for the operator or bystander. It is important, therefore, for the owner/operator to have a basic understanding of the common categories of tractors. This chapter will also explain how stationary engines and tractors are derated, and how tractors are tested.

7.3. Categories of Tractors

The diversity of modern agriculture requires many tractor designs. Historically, utility (use) has been the basis of tractor classification schemes. Based on utility, there are six categories of tractors: general purpose, row crop, orchard, vineyard, industrial, and garden. The designers of modern tractors have attempted to produce tractors with the broadest possible uses, but these categories are still applicable if a sub-category is added for each type of propulsion system: rear wheel drive (RWD), four wheel drive articulating steering (4WDAS), four wheel drive four wheel steer (4WD), tracks (T), and rear wheel drive with front wheel assist (FWA).

7.3.1. General Purpose

The general purpose tractor is the traditional design with the rear wheels and the front wheels spaced the same distance apart. This type of tractor usually is built closer to the ground than the row crop design. The power range of tractors of this type is very broad; sizes range from about 25 to over 400 horsepower (19 to 300 kW). The use of a general purpose tractor is influenced by its power. The smaller sizes are very popular for smaller agricultural enterprises and for mowing. The mid-range sizes are used extensively for cultivating, spraying, tilling, and mowing, and for mobile or stationary power take off (PTO) power. The larger sizes normally are used for primary tillage and to provide PTO power for larger mobile and stationary machines such as large balers, forage harvesters, emergency generators and silage blowers. Some of the tractors at the top of the power range are designed for tillage only and may not have a PTO or three point hitch.

General purpose tractors are available with all five types of propulsion systems. Historically, this tractor category has been dominated by the rear wheel drive, but in recent years the situation has changed. The propulsion system also is influenced by the power. Smaller tractors use the rear wheel drive or the front wheel assist; in the middle of the power range, all types can be found; and in the largest engine sizes, the most common type is the articulating four wheel drive.

7.3.2. Row Crop

Row crop tractors are designed with greater ground clearance than the general purpose tractors have. This gives them the ability to straddle taller crops with less plant damage. The size range of row crop tractors is narrower than that of the general purpose, as these tractors are usually 50 to 100 horsepower (37 to 75 kW). Historically many were built with the front wheels closer together than the rear wheels (tricycle style). The narrow front wheels eliminate the use of tracks, front wheel assist, and articulating steering; but row crop tractors without narrow front wheels can be found with these configurations.

7.3.3. Orchard

Orchard tractors are not a separate type of design as general purpose and row crop tractors are, but are tractors that have been modified to reduce the possibility of tree limbs catching on them. Modifications usually include changing the location of the exhaust and the air intake and the addition of shields around the tires and other protuberances. Orchard work is not as power-demanding as primary tillage; therefore, these tractors are usually in the medium power sizes.

7.3.4. Vineyard

The vineyard tractor also is found in the smaller power range. It has been designed or modified to reduce its width so that it can pass between narrow rows. It also may use shielding similar to that of the orchard tractor. The narrower size limits the propulsion system to the conventional rear axle drive, front wheel assist, and tracks.

7.3.5. Industrial

These tractors look like general purpose agricultural tractors, but they have important differences. They are equipped with tires designed for use on hard, smooth surfaces and the front axle and frame will be designed to withstand the weight and shocks of front loaders, backhoes, and other industrial equipment.

7.3.6. Garden

This tractor category has the greatest amount of variation in mechanical construction. These tractors usually are less than 25 horsepower. Some are manufactured to look like a small tractor, but others may look more like a riding lawn mower. They generally use the rear wheel drive system, with or without front wheel assist.

7.4. Tractor Power Ratings

Different power ratings are used to evaluate the size of tractors and engines. It is important to remember that manufactures like to advertise their engine power with the highest possible number to make them competitive in the marketplace. The useable power is usually less than the advertised power. The type of value used to express the power must be known before the useable power can be determined, Figure 7.1.

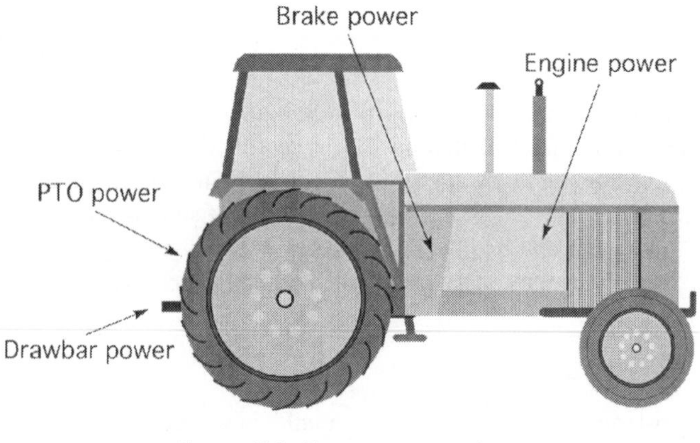

FIGURE 7.1. Tractor power ratings.

7.4.1. Engine Power

Engine power is a calculated power based on the bore, cylinder pressure, and speed of the engine. Engine power is not a useful rating for determining the useable power available from a tractor or stationary engine.

7.4.2. Brake Power

The term brake power comes from the early use of the Prony brake to measure engine power. Brake power is the power available at the flywheel of the engine. It is not a useful rating for tractors, but it is the power rating used to size stationary engines for irrigation pumps and other similar uses.

7.4.3. PTO Power

Power take-off power (ptohp) is a rating of the power available at the power take-off of an agricultural tractor. Power take-offs are used to supply rotary power to many different types of machines such as balers, pumps, and mowers. To determine PTO power either power equation using torque can be used.

$$\text{Power (hp)} = \frac{\text{To} \times N}{5252}$$

where hp = horsepower; To = torque (lb-ft); N = speed (rpm)

$$\text{Power (kW)} = \frac{\text{To} \times N}{9549}$$

where kW = kilowatts; To = torque (Nm); N = speed (rpm).

7.4.4. Drawbar Power

Drawbar power (Dbhp) is power measured at the point implements are attached to the tractor, drawbar or 3 point hitch. During tractor testing a load cell is placed between the tractor and a load. The drawbar pull and the speed of travel are recorded and used to calculate drawbar power. Using measurements of pull (pounds) and speed (miles per hour) eliminates the PTO power equations. Because force is measured in pounds and velocity (V) measures both distance traveled and the time required to travel that distance (miles per hour), the best equation to use is:

$$\text{Db}_{\text{Hp}} = \frac{F \times V}{375}$$

where F = Force (lb); V = Speed (mph); 375 = Units conversion constant.

This equation is derived by modifying the standard power equation.

$$P_{\text{Hp}} = \frac{F \times D}{t \times 33{,}000}$$

Because $V = \frac{D}{t}$ and $1\ \frac{\text{mile}}{\text{hr}} = 88\frac{\text{ft}}{\text{min}}$, then $1\ \text{hp} = \frac{F \times V \times 88}{33,000} = \frac{F \times V}{375}$

Problem: How much drawbar power is a tractor producing if it develops 1,500 lb of force at a speed of 5.5 miles per hour?

Solution:

$$\mathrm{Db_{Hp}} = \frac{F \times V}{375} = \frac{1{,}500\ \mathrm{lb} \times 5.5\ \frac{\mathrm{mile}}{\mathrm{hr}}}{375} = \frac{8250}{375} = 22\ \mathrm{hp}$$

7.5. Converting Tractor Power Ratings

Manufactures can use anyone of the four power ratings to advertise their tractors. When manufactures use engine or brake power to rate a tractor they are not providing any useful information because the useable power will be less than that. When the tractor is rated by engine or brake power, the PTO power will be less because of the losses in the power train. The drawbar power will be less because there are additional losses in the transmission and drive train and a tractor is not able to apply all of the torque of the drive wheels to the soil.

Actual PTO and drawbar power ratings can be determined if the tractor has been tested by the Nebraska Tractor Test (NTT) Station or by the Organisation for Economic Cooperation and Development (OECD). When actual PTO and drawbar ratings are not available, the 86% rule can be used to estimate them from the engine rating.

Studies have shown that if a large number of tractors are compared over many different traction conditions, the PTO power will be approximately 86% of the engine power, and the drawbar power when tested on a concrete track will be approximately 86% of the PTO. The 86% rule permits an estimate of PTO and drawbar power. It is not intended to be a substitute for actual data.

Problem: Estimate the amount of power available at the PTO and the drawbar for a tractor rated at 125.0 engine power (E_{Hp}).

Solution: Using the 86% rule:

$$\begin{aligned}
\mathrm{PTO_{Hp}} &= E_{\mathrm{Hp}} \times 0.86 = 125.0\ \mathrm{hp} \times 0.86 = 107.5 \text{ or } 110\ \mathrm{PTO_{Hp}} \\
\mathrm{Db_{Hp}} &= \mathrm{PTO_{Hp}} \times 0.86 \text{ or } E_{\mathrm{Hp}} \times 0.86 \times 0.86 \text{ or } \mathrm{E_{Hp}} \times 0.7396 \\
\mathrm{Db_{Hp}} &= E_{\mathrm{Hp}} \times 0.7396 = 125\ \mathrm{hp} \times 0.7386 = 92.45 \text{ or } 92.4\ \mathrm{Db_{Hp}}
\end{aligned}$$

The 86% rule can be used in other ways also. For example, if the drawbar power is known, the PTO and engine power can be estimated.

Problem: What is the estimated PTO power if the tractor produces 50.0 power at the drawbar?

Solution: From the discussion of the 86% rule, we know that the PTO power will be larger than the drawbar. Therefore:

$$\mathrm{PTO_{Hp}} = \frac{\mathrm{Db_{Hp}}}{0.86} = \frac{50.0\ \mathrm{Db_{Hp}}}{0.86} = 58.139\ldots \text{ or } 58\ \mathrm{PTO_{Hp}}$$

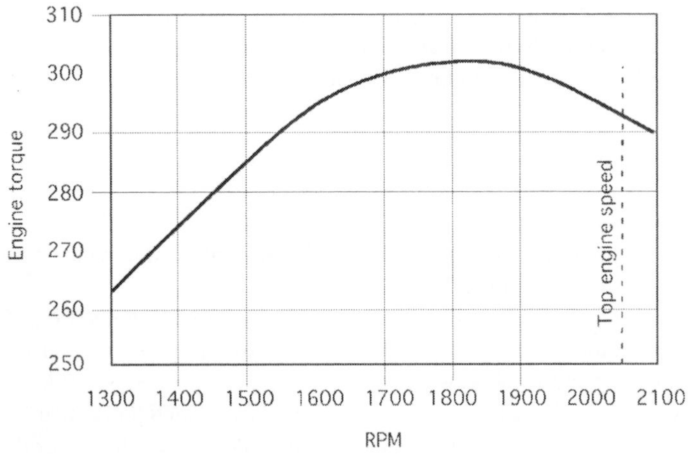

FIGURE 7.2. Engine torque curve with lugging ability.

7.6. Lugging Ability

An important feature of tractors that will be used for primary tillage or as a stationary power supply is their lugging ability. Lugging ability is a measure of the amount of temporary overload the tractor can withstand. The lugging ability for a tractor or stationary engine can be illustrated with a horsepower curve. Figure 7.2 shows a power curve for an engine designed with lugging ability. Notice that as the engine is overloaded and the rpm drops from the top engine speed of 2,050 rpm to 1,850 rpm the power increases.

The amount of increase in torque is a measure of the engines lugging ability. Compare the torque curve in Figure 7.2 with the one in Figure 7.3.

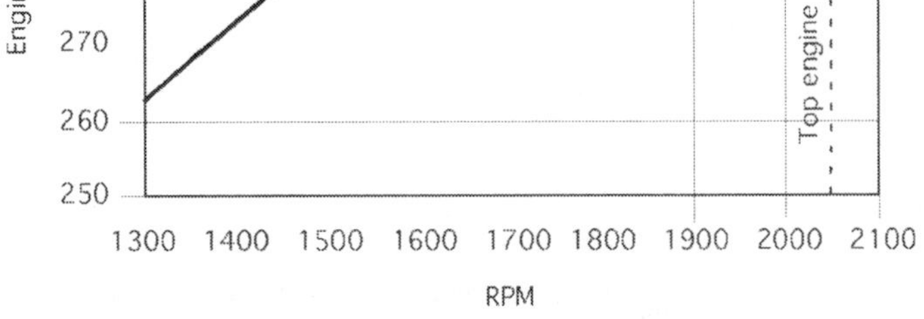

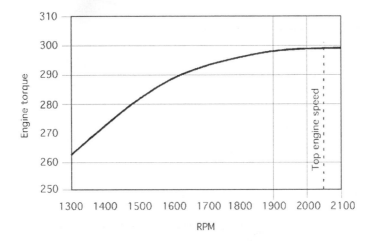

FIGURE 7.3. Engine torque curve with no lugging ability.

This engine does not have any lugging ability because the torque decreases as soon as the rpm decreases. There is no temporary increase in torque to help the tractor pull through a temporary overload.

7.7. Derating Power Units

Manufacturers of engines intended for stationary use, for example, to power a pump or a generator, usually can supply the purchaser of an engine with a performance curve for the engine, Figure 7.4.

The performance curve will show the rated power for an engine for different engine speeds ranging from the minimum speed recommended for a load to the maximum. However, the usable power is less than the rated power if one of several conditions exists, as explained below. The adjustment from rated power to usable power is called derating. Failure to derate an engine could shorten the life of the engine.

The factors used for derating depend on the engine type and the operating environment. Spark ignition and diesel engines are derated to allow for the effect of accessories, temperature, altitude, and type of service.

7.8. Accessories

If the power rating is reported for a basic engine, sometimes called the gross power rating, it was tested with all accessories removed. These accessories include the air cleaner, muffler, generator, governor, and fan and radiator. If the power rating is

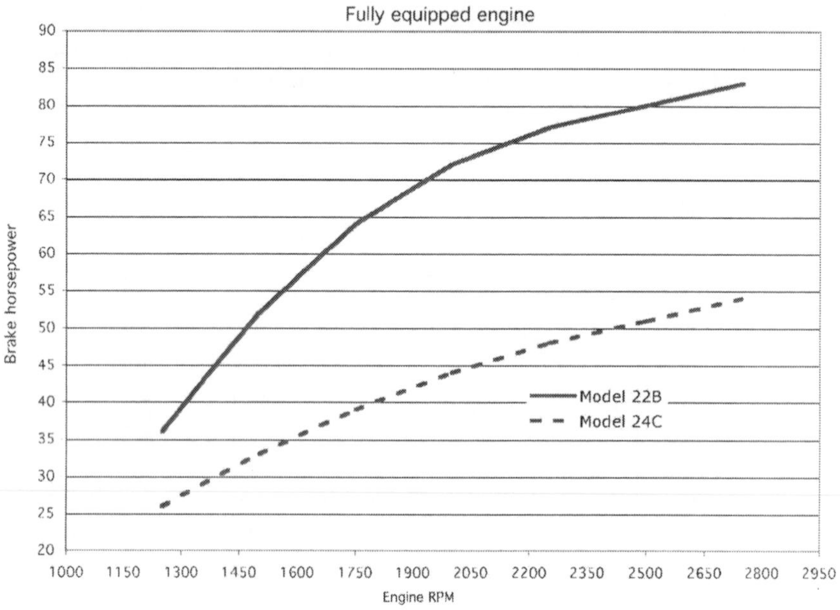

FIGURE 7.4. Engine performance curves.

for a fully equipped engine, sometimes called a net power rating, these accessories were on the engine when it was tested.

If basic engine power is reported, the engine must be derated 10% for all accessories other than the fan. An additional 5% must be deducted if the engine will have a fan and a radiator instead of a heat exchanger.

Problem: How much usable power is available if the basic engine rating is 75.0 hp, and the engine will use a fan and a radiator for cooling?

Solution: Derating 10% for the accessories and 5% for the fan:

$$\text{hp} = 75.0\ \text{hp} \times (1.0 - (0.10 + 0.05)) = 75.0 \times 0.85 = 63.75 \text{ or } 64\ \text{hp}$$

Note: the same answer can be determined by subtracting the lost power.

$$\begin{aligned}\text{hp} &= 75.0\ \text{hp} - \text{Loss} = 75.0\ \text{hp} - (75.0 \times (0.10 + 0.05)) \\ &= 75 - 11.25 = 63.75 \text{ or } 64\ \text{hp}\end{aligned}$$

The effect of the accessories is to reduce the useable power from 75 to 64 horsepower.

7.9. Temperature

The ambient temperature must be considered because as the temperature of the air that the engine breathes increases, the density decreases, and there is less oxygen per cubic foot of air. The decrease in oxygen decreases the efficiency of the engine. The important air density is the density of the air as it enters the engine. For turbocharged and supercharged engines the density of the air has been increased above ambient removing the effect of temperature. Therefore no adjustment for temperature needs to be completed. For naturally aspirated spark ignition engines the power rating must be reduced 1% for each 10°F (5.6°C) ambient temperature rise above 85°F (29°C). For diesel engines the adjustment is 1% for each 5°F (2.7°C above 85°F (29°C).

Problem: How much usable power will a 165 Bhp spark ignition engine produce if it will be operating in 100°F air temperature?

Solution: Using the recommended derating for temperature:

$$\begin{aligned}\text{hp} &= 165.0\ \text{hp} \times \left(1 - \left(0.01 \times \frac{100^\circ\text{F} - 85^\circ\text{F}}{10^\circ\text{F}}\right)\right) = 165.0 \times 0.985 \\ &= 162.525 \text{ or } 160\ \text{hp}\end{aligned}$$

The effect of the 100°F temperature is to reduce the useable power from 165 to 160 horsepower.

7.10. Altitude

As the altitude increases, the barometric pressure decreases. This reduction in pressure reduces the efficiency of naturally aspirated engines because as the elevation of the engine increases, the pressure difference is less between the air pressure inside the cylinder, on the intake stroke, and the atmospheric pressure. Decreasing the pressure difference decreases the amount of air that will flow into the engine on the intake stroke.

Altitude is not a problem for engines with turbochargers or superchargers because they increase the intake system pressure. For naturally aspirated spark ignition and diesel engines the power rating must be reduced by 3% for each 1,000 ft (305 m) of elevation above 500 ft (152 m) above sea level.

Problem: What is the usable power rating for a 225 hp diesel engine operating at an elevation of 4,500 ft?

Solution: Using the derating recommendation for altitude (*Note*: the method used to solve this example is different from the previous one. Either method can be used.)

$$E_{\mathrm{Hp}} = 225\ \mathrm{hp} \times \frac{100\% - \left(3.0\% \times \left(\frac{4{,}500\ \mathrm{ft} - 500\ \mathrm{ft}}{1{,}000}\right)\right)}{100}$$
$$= 225\ \mathrm{hp} \times 0.88$$
$$= 198\ \mathrm{hp}$$

This example was worked using percentages (%).

7.11. Type of Service

The type of engine service is determined by the load–intermittent or continuous. An intermittent load on an engine is a load that varies in torque and speed; for example, tractors in tillage operations are subject to intermittent loads. Continuous loads provide little variation in the torque and speed demands placed on the engine; irrigation pumps are an example of a continuous load. Some manufacturers may indicate the type of service for the rated power. If not, the engine should be derated for the type of service. The available power must be reduced by 10% for intermittent loads and 20% for continuous loads.

Problem: What is the usable power for an engine that has been derated for accessories, temperature, and altitude to 115.3 hp if it will be used for continuous duty?

Solution: Derating for continuous duty:

$$\mathrm{hp} = 115.3\ \mathrm{hp} \times \frac{100\% - 20\%}{100} = 115.3 \times 0.80 = 92.48 \text{ or } 92\ \mathrm{hp}$$

From this discussion it is evident that an important difference may exist between rated power and usable power for stationary engines. Also note that if a tractor is being used for stationary power for an extended period of time, it should be derated. When derating a tractor, remember that the PTO power will be equal to that of a fully equipped engine.

The previous section explained the procedure for derating an engine for individual factors. One question that remains is how is an engine derated when more than one factors is involved. This is accomplished by summing up the reduction for each individual factor and multiplying this number by the appropriate power rating.

Problem: Determine the usable power for a fully equipped, non-turbocharged, diesel engine rated at 115.0 power when it is operated on continuous duty, at 3,350 ft elevation and with an air temperature of 98.5 degrees.

Solution: For this example the engine must be derated for temperature, altitude and duty. The example was worked using the decimal form of percentages.

Temperature:

$$0.01 \times \left(\frac{98.5^\circ\text{F} - 85^\circ\text{F}}{5^\circ\text{F}}\right) = 0.027$$

Altitude:

$$0.03 \times \left(\frac{3{,}350 - 500}{1{,}000}\right) = 0.0855$$

Duty: Continuous duty requires a derating of 20%.

Total adjustment:

$$\text{Total} = \text{temperature} + \text{altitude} + \text{duty} = 0.027 + 0.0855 + 0.20 = 0.3125$$

$$E_{\text{Hp}} = 115.0\ \text{hp} \times (1 - 0.3125) = 79.0625 \text{ or } 79.1\ \text{hp}$$

A 115.0 horsepower, non-turbocharged diesel engine operating under these conditions will have 79.1 useable power.

7.12. Tractor Testing

Early tractors were designed to deliver power in three ways: by belt pulley, drawbar, and power take-off. On newer tractors the belt pulley has been eliminated. One problem the owner/manager has faced since the first tractor was designed is lack of information on the power ratings of tractors. As a rule, tractor manufacturers do not advertise the drawbar or PTO ratings of their tractors; instead many use brake power or engine power. If engine power is used, it could be a theoretical power rating for the engine.

In 1918, the Nebraska legislature passed a law that established the Nebraska Tractor Test Station. This law mandated that a typical model of tractor must be tested before it could be sold within the state of Nebraska, and that the manufacturer must maintain a part supply depot within the state. This testing station has provided the only independent evaluation of tractors in the United States.

As time passed and a large percentage of tractor manufacturing moved outside the United States, a change was needed. Two changes have occurred: the Nebraska

Law was changed to allow tractors to be sold in Nebraska with either a Nebraska test or an Organisation for Economic Cooperation and Development (OECD) test, and the Nebraska test was changed to match the standards established by the Society of Automotive Engineers (SAE) and the American Society of Agricultural and Biological Engineers (ASABE). Currently the Nebraska Tractor Test Station is qualified to conduct either test at the request of the tractor manufacturer. A full review of both testing standards is not possible in this text; instead the following is a discussion of the general principles of testing and the two test methods. Individual reports and additional information can be obtained from the testing stations.[1,2]

7.13. Principles of Testing

Because of the important role tractors play in agricultural production, accurate information is a must for efficient management of the modern farm and ranch. Any information is usable, but the main objective of testing is to provide standardized results so that comparisons can be made between different models and different years. The primary purpose of tractor testing standards is to establish the test conditions and the rules of behavior for the manufacturer and the testing station.

It should be evident that engines are complex mechanisms with many different factors that influence the power produced and fuel efficiency. Because of the widespread use of the data, manufacturers want to be sure they get the best possible results. To provide standardized information, the testing environment must be consistent for every test, day after day, year after year, or changes must be thoroughly investigated so that the appropriate adjustments in the results can be made. In both tests, these factors are called test conditions.

Test conditions include the rules for the selection of the tractor and the control and/or recording of environmental conditions. During the PTO test, air temperature, barometric pressure, fuel type, fuel temperature, fuel measurement, lubricants, and accessory equipment on the tractor must be either controlled or recorded to ensure that they are within acceptable limits. Strict control of these factors is important because many tests measure the maximum power of the tractor. Each one of these factors will influence the power produced.

For the drawbar test all of the environmental factors in the PTO test must be accounted for, plus those that affect traction. The latter include tractor ballast, tires, and testing surface. The ballast, tires, and testing surface are critical because

[1] *OECD Standard Codes for the Official Testing of Agricultural Tractors*, Organisation for Economic Cooperation and Development, 1988. 2, rue Andre-Pascal, 75775 Paris CEDEX 16, France.

[2] Nebraska Test Station, Agricultural Engineering Department, University of Nebraska–Lincoln, Lincoln, Nebraska 68583.

the amount of pull a tractor can produce at the drawbar is greatly influenced by the amount of traction. Both the NTT station and the OECD station have strict standards on how all these factors are controlled and/or recorded.

In addition to the test conditions, the testing codes must establish rules governing such things as breakdowns, repeating tests, testing accessory systems such as hydraulic systems, safety standards, noise standards, and the method of reporting and publishing the results.

7.14. Metric Problems

Problem: Determine the power being produced by a tractor if it developed 6.67 kN of force while traveling at a speed of 9.0 km/hr.

Solution:

$$\begin{aligned} P(\text{kW}) &= \frac{F(\text{kN}) \times V(\text{km/hr})}{3.6} \\ &= \frac{6.67\ \text{kN} \times\ 9.0\ \text{km/hr}}{3.6} \\ &= 16.7\ \text{kW} \end{aligned}$$

Note: Appendix II includes conversions from customary to SI units. The answer to this problem could be converted to horsepower by using the appropriate conversion. Horsepower equals:

$$\text{Hp} = \frac{16.68\ \text{kW}}{0.7456999\ \dfrac{\text{kW}}{\text{hp}}} = 22.368\ldots\ \text{or}\ 22.37\ \text{hp}$$

Problem: Determine the useable power for a 120 kW naturally aspirated natural gas irrigation engine that will be operated with a fan and radiator. It is located at an elevation of 1,000 m and will operate in an air temperature of 38°C and on continuous duty.

Solution: The process is the same as customary units. The only difference is in the values used. For this example direct conversions are used, Table 7.1.

TABLE 7.1. Direct conversions for derating factors.

Accessories	10% other than fan
	5% fan and radiator
Temperature	1% or each 5.6°C above 29°C for gasoline
	1% for each 2.7°C above 29°C for diesel
Altitude	3% for each 305 m above 152 m
Type of service	10% for intermittent loads
	20% continuous loads

The useable power is determined by determining the total amount of power loss.

$$\text{Power loss }(\%) = L_{\text{acc}} + L_{\text{alt}} + L_{\text{temp}} + L_{\text{duty}}$$

$$L_{\text{acc}} = 0.10 + 0.05 = 0.15$$

$$L_{\text{alt}} = 0.03 \times \left(\frac{1{,}000\text{ m} - 152\text{ m}}{305\text{ m}}\right) = 0.03 \times 2.78\ldots = 0.083$$

$$L_{\text{temp}} = 0.01 \times \left(\frac{38 - 29}{4.72}\right) = 0.01 \times 1.906\ldots = 0.019$$

$$L_{\text{duty}} = 0.20$$

$$\text{Total loss} = 0.05 + 0.083 + 0.019 + 0.20 = 0.662$$

$$\text{Useable power} = 120\text{ kW} - (120\text{ kW} \times 0.662) = 120\text{ kW} - 79.44 = 40.55\text{ kW}$$

All of the power losses for this engine are 79.44 kW, leaving 40.55 kW of useable power.

8
Machinery Calibration

8.1. Objectives

1. Understand the need for evaluating the performance of machines.
2. Understand the principles of calibration.
3. Be able to calibrate a broadcast type fertilizer applicator.
4. Be able to calibrate a grain drill.
5. Be able to calibrate a row crop planter.
6. Be able to calibrate an air seeder.
7. Be able to calibrate a field sprayer.
8. Be able to prepare the proper mix of chemicals and water for a sprayer.

8.2. Introduction

One role of agricultural machines is the dispensing of materials such as seeds, fertilizers, and sprays. These machines are designed to dispense the material at a fixed, or variable rate, and in a fixed, or variable pattern. A machine that fails to dispense the material at the desired rate and pattern should not be used. An insufficient amount of material will not produce the desired results, and excessive amounts are a lost resource that may result in crop damage and/or contribute to contamination of the environment. A pattern that is not correct may cause streaking or skips. Calibration charts or tables are usually supplied with or are available from the manufacturers of the machines in the owner's manuals or from the supplier of the material being dispensed. Charts and tables become lost or damaged, leaving the operator with no or incomplete information.

Incorrect application of materials can have several causes. A machine may have been damaged or modified. Variations in the weight, size, moisture content, and cleanliness of the seed or variations in the physical condition (lumpiness, flow ability) of the fertilizer or chemical granules can all contribute to an inaccurate dispensing rate and distribution pattern. The economic penalties and potential environmental damage from incorrect application of materials warrant the time and effort required to ensure that the machine is dispensing the desired amount and that the distribution pattern is acceptable.

Checking a machines application rate and patterns of dispersal is called calibration. Although the exact procedure used to calibrate machines and other measuring devices varies from one situation to another, this chapter will illustrate the calibration of four common agricultural material dispensing machines.

8.3. Principles of Calibration

In Chapter 1, two characteristic of mathematical problem where discussed. Both of these must be applied when calibrating dispensing machines. It would be a rare occurrence for a dispensing machine to distribute the desired amount of material. In this case perfection is not expected. Therefore levels of acceptability must be established. The amount of error that would be considered acceptable varies for each machine and situation. In some cases guidelines or standards have been established, but in others it is up to the owner/operator to determine how much error will be accepted. Standards of acceptability for each machine will be discussed in more detail for each machine.

All of the methods used to calibrate agricultural machines are based on three principles: (1) all dispensing machines meter (control the rate) the flow of material at a predetermined rate selected by the operator, and (2) calibration occurs by collecting material dispensed by the machine in units of volume, weigh or mass, or number of granules or seeds per unit area. For example, to calibrate a row crop planter it is necessary to determine the seeds per acre or seeds per hectare. To calibrate a sprayer, the application rate is determined in units of gallons per acre or liters per hectare. Seeding rate of certain grasses is in bushels per acre or liters per hectare. Lime is applied in tons per acre or metric tons per hectare. The unit of area used during calibration will usually be an acre or hectare or some fraction of these. (3) There are two methods for collecting the material, stationary or mobile. With the stationary method, the wheel that drives the machine metering unit must be elevated so it doesn't contact the surface and then it is turned for a fixed number of revolutions. In the mobile calibration method, the flow of material is interrupted into a container as the machine is pulled through the field. The advantage of the stationary method is that the machine can be calibrated in off productive time. The mobile method is best for machines that have a large number of metering points. The mobile method may be the only option when methods other than a ground driven wheel are used to drive the metering mechanism. It is important to remember that the larger the area used during calibration, the greater the accuracy of the calibration procedure.

8.4. Calibrating Fertilizer Applicators

Fertilizers, which are mainstays of modern agriculture, are applied in liquid, gaseous, or granular form. In this section we will discuss the calibration of granular

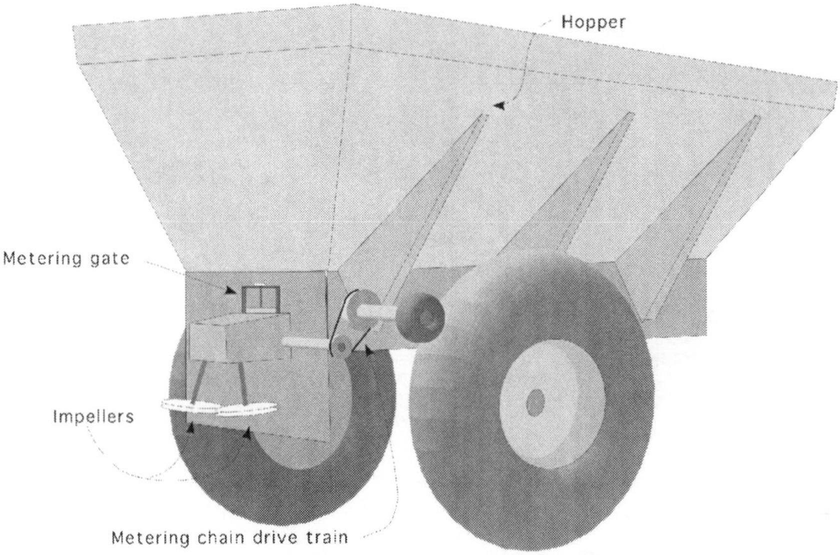

FIGURE 8.1. Broadcast fertilizer spreader.

applicators. The application of liquid fertilizers follows the same principles as spraying chemicals and sprayers are discussed in a separate section. Gaseous fertilizers require specialized equipment and are very hazardous to use. The equipment to dispense gaseous fertilizers is not included in this text.

Most granular fertilizer applicators will be one of two designs, broadcast or gravity flow. As with any agricultural machinery, designs vary, but a common tractor powered spreader consist of a hopper that holds the material, a metering mechanism, and a spreading mechanism, Figure 8.1. The hopper capacity will vary depending on the size of the machine. Metering usually is accomplished through the use of a variable-speed chain belt in the bottom of the hopper and an adjustable opening. The chain belt generally is operated by a drive train powered by one of the wheels of the spreader. Because a ground wheel drives the chain belt, the amount of material dispensed changes as the speed of the spreader changes. The impellers are usually powered by the tractor PTO and must turn at a constant speed for the spreader to have a consistent dispersal pattern.

In this design the application rate of the material is set by changing the speed ratio of the drive train that powers the metering device and/or changing the size of the opening at the metering gate. The process of using a variable opening for metering is called bulk metering. Bulk metering units meter the volume of material applied. The material usually is spread as it drops onto one or more rotating disks, impellers, which sling it out in a wide pattern either behind or in front of the spreader. An alternative design powers the impellers from one of the ground wheels, eliminating the need for a PTO or additional power source.

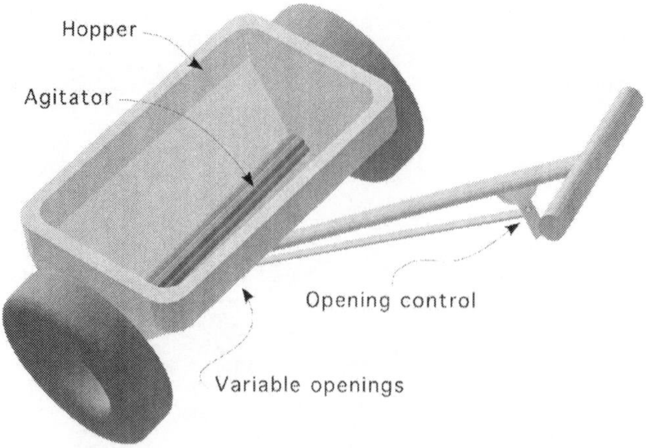

FIGURE 8.2. Gravity fertilizer spreader.

Gravity flow applicators also use a hopper, Figure 8.2, but in this type of applicator the hopper extends across the width of the machine. The material is dispensed through adjustable openings along the bottom of the hopper. The rate of material flow is set by the size of the opening. This design uses an agitator along the bottom to prevent the material from bridging across the hopper and to break up lumps, thus improving the uniformity of flow. The material flows through the openings and onto the soil surface.

At one time gravity flow fertilizer spreaders were popular in agriculture, but limits on practical widths has almost eliminated agricultural uses. Small ones are still used for lawn care and other situations where 18 to 24 inch widths can be effective.

Both types of spreaders may be calibrated using either the stationary or mobile method. For both methods, the application rate, R, (lb/ac) is equal to the pounds of material collected per revolution of the drive wheel, divided by the area covered. The area covered is determined by multiplying the width (w) × distance traveled. The distance traveled is determined by calculating the perimeter of the wheel ($2\,\pi r$) and multiplying it by the number of revolutions of the drive wheel (n_r). In equation form this relationship is:

$$R = \frac{\frac{W}{n_r}}{\frac{A}{n_r}}. \text{ This can also be expressed as: } R = \frac{W}{n_r} \times \frac{n_r}{A}$$

where R = application rate (lb/ac); W = weight of material (lb); n_r = number of revolutions of drive wheel; A = area (acres).

To calibrate the spreader using the stationary method, the drive wheel for the metering mechanism is elevated and the material is collected as the drive wheel is rotated a predetermined number of turns. The weight of the material collected divided by the number of turns of the drive wheel equals the pounds per revolution. Acres per revolution are determined by multiplying the circumference of the drive wheel by the number of turns and the effective width of the spreader, and then converting this area to acres, or:

$$\frac{A(\text{ac})}{n_r(\text{rev})} = \frac{\dfrac{2\pi r}{1\ \text{rev}} \times w}{43{,}560\ \dfrac{\text{ft}}{\text{ac}}}$$

where $\pi = 3.14$; r = Effective radius of drive wheel (ft); n_r = number of revolutions of the wheel; w = Effective width of the spreader (ft); 43,560 = Number of square feet per acre.

Combining the unchanging units to form a constant gives:

$$\frac{A(\text{ac})}{n_r(\text{rev})} = \frac{2\pi}{43{,}560} \times r \times w = 0.0001442 \times r \times w \text{ or } 1.442\ E - 4 \times rw$$

Therefore, when a spreader is calibrated in a stationary position, the pounds per acre can be determined by:

$$R = \frac{W}{A}$$

where R = application rate (lb/ac); W = weight of material collected (lb); A = area (ac).

$$R\left(\frac{\text{lb}}{\text{ac}}\right) = \frac{W\ (\text{lb collected})}{1.4420\ \text{E} - 4 \times r\ (\text{radius}) \times w\ (\text{width}) \times n\ (\text{turns})}$$

Problem: You set your neighbor's spreader to apply 1,200 lb/ac of fertilizer. During a stationary calibration of the applicator 160.0 lb of fertilizer were collected as the drive wheel was turned 15 revolutions. The effective radius of the drive wheel is 21.0 inches and the effective width of the spreader is 30.0 ft. Is the application rate correct?

Solution:

$$R\left(\frac{\text{lb}}{\text{ac}}\right) = \frac{W}{0.0001442 \times r \times w \times n_r}$$

$$= \frac{160.0\ \text{lb}}{1.442\ \text{E} - 4 \times \dfrac{21.0\ \text{in} \times \dfrac{1\ \text{ft}}{12\ \text{in}}}{1\ \text{rev}} \times 30.0\ \text{ft} \times 15\ \text{rev}}$$

$$= \frac{160.0}{0.1135575} = 1408.977\ldots \text{ or } 1,400\ \frac{\text{lb}}{\text{ac}}$$

In this problem the spreader is applying 200 lb/ac more than the desired amount. Is this an acceptable level of accuracy? Should the spreader metering unit be adjusted and rechecked? No consistent standard has been established to determine the acceptable error for broadcast fertilizer spreaders. One way to evaluate the error is to calculate the percentage (%) of error. In this example the amount of error was 17% [(200 lb/1200 lb) × 100]. It would be up to the owner/operator to determine if 17% was an acceptable level of error.

The one major disadvantage of the stationary method of calibrating broadcast applicators is the problem of collecting the fertilizer during the calibration process (160 lb in this example). The quantity of material that must be collected will be greater if the desired application rate (lb/ac) is increased, or if the number of revolutions is increased. The primary advantage of this method is that it can be performed when it is impossible to have the applicator in the field, for instance, because of muddy ground or inclement weather.

The alternative is to complete a mobile calibration. This can be accomplished in two ways. In the first method, collectors (flat pans or pieces of plastic or tarp) are randomly placed in the distribution path of the spreader or an alternative is to place the collection devices in a row perpendicular to the line of travel of the spreader. After the spreader is driven over the collectors, the material is weighed. The application rate (lb/ac) is computed by dividing the amount of material collected, by the area of the collectors, and then converting the answer to the desired units of volume weight or mass divide by the area. An advantage of the perpendicular row of collectors is that the distribution pattern can be evaluated by plotting the amount collected from each collector on a graph.

Problem: What is the application rate of a spreader that deposited 23.5 lb of material on two tarps, each measuring 10.0 ft × 12.0 ft?

Solution:

$$R\left(\frac{\text{lb}}{\text{ac}}\right) = \frac{W\,(\text{lb})}{A\,(\text{ac})} = \frac{23.5\ \text{lb}}{2 \times 10.0\ \text{ft} \times 12.0\ \text{ft} \times \dfrac{1\ \text{ac}}{43{,}560\ \text{ft}^2}} = \frac{23.5}{0.005509\ldots}$$

$$= 4265.25 \text{ or } 4,260\ \frac{\text{lb}}{\text{ac}}$$

[*Note*: This is an example where the number of significant figures in the answer could be ambiguous because of leaving a zero when rounding. To clarify the number of significant figures, the answer should be written as 4.260 E3.]

This method does not require a spreader wheel to be elevated, but you must be able to drive the spreader in the field or over an area as wide as the dispersal pattern. If the spreader is dispensing an excessive amount of material, the test area will be over fertilized.

Problem: Determine the application rate and distribution pattern for a broadcast spreader that was calibrated by placing 20–2 ft^2 collectors perpendicular to the direction of travel. The desired rate is 800 lb/ac. Table 8.1 shows the pounds of fertilizer collected.

Solution: The first step is to determine the application rate by summing the quantity of fertilizer in the 18 collector pans (n_c) containing fertilizer W_t. Then determine

TABLE 8.1. Fertilizer collected during mobile calibration of broadcast fertilizer spreader.

Collector	lb
1	0.000
2	0.008
3	0.014
4	0.016
5	0.020
6	0.023
7	0.030
8	0.038
9	0.050
10	0.062
11	0.062
12	0.060
12	0.057
14	0.052
15	0.048
16	0.041
17	0.035
18	0.026
19	0.017
20	0.000

the total area of the collector pans containing fertilizer (A_t). The application rate is:

$$R\left(\frac{\text{lb}}{\text{ac}}\right) = \frac{W_t}{n_u \times A_c}$$

$$= \frac{0.659\ \text{lb}}{18\ \text{collectors} \times 2\dfrac{\text{ft}^2}{\text{collector}} \times \dfrac{1\ \text{ac}}{43560\ \text{ft}^2}}$$

$$= \frac{0.659}{0.0008264\ldots} = 797.39 \text{ or } 797\ \frac{\text{lb}}{\text{ac}}$$

The spreader is applying 797 lb/ac. [*Note*: Only 18 collectors were used to determine the area because two collectors did not have any measurable material.] In this case the percentage of error was

$$\%(\text{error}) = \frac{800 - 797}{800} \times 100 = 0.38\%$$

This is an acceptable level of error. The calibration is completed by plotting the distribution pattern. A plot of the distribution is shown in Figure 8.3.

In the graph of the distribution the dashed line represents the ideal or acceptable distribution for an overlapping distribution pattern. The solid line is the distribution for this spreader. It is clear that this spreader has a problem. Even thought the

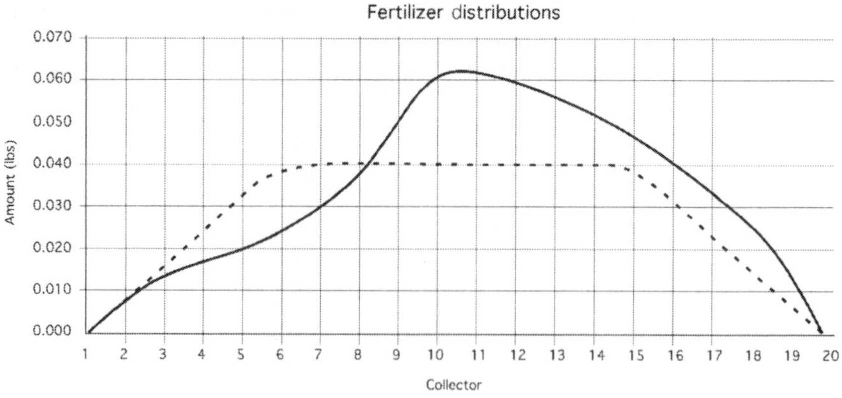

FIGURE 8.3. Distributor graph for spreader.

application rate is acceptable, the distribution of the material is not. The cause of the poor distribution should be determined and corrected before it is used.

Another mobile method can be used if it is certain that the spreader pattern is reasonably accurate. First, weigh the spreader, and then drive it over a measured area. Then reweigh the spreader. The difference between the before and after weights is the amount of material applied. The area is already known, so it is easy to find the application rate using mobile calibration equation.

Both of these methods can be used to calibrate a fertilizer spreader. The choice and the materials used are up to the operator.

8.5. Calibrating Grain Drills

The traditional end wheel grain drill includes a hopper, a metering unit, seed tube and furrow opener for each row, and the metering unit drive train, Figure 8.4. Grain

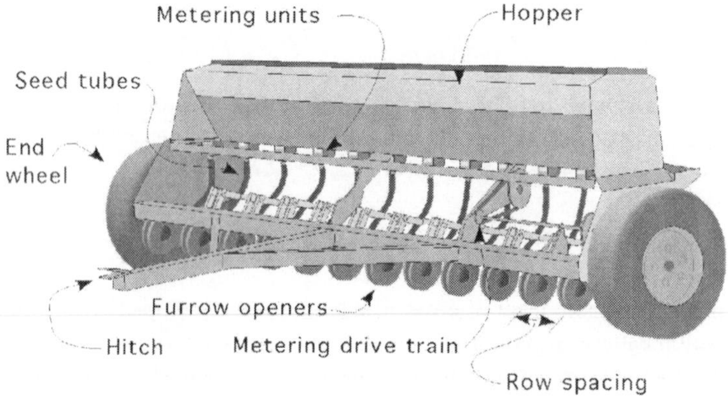

FIGURE 8.4. End wheel grain drill.

drills also use bulk or volume metering. The rows are usually spaced 6–10 inches (15–25 cm) apart. Combining the row spacing and number of metering units is the traditional method for indicating the width of the drill. A drill identified as 13–6 would have 13 metering units spaced 6 inches apart or a width of $18 \times 6/12 = 9$ ft. The calibration of grain drills is more critical than the calibration of fertilizer spreaders because drills dispense seeds. A small error in seeding rate can have a greater impact on the yield than an error in fertilizer application. In addition, it is more important that the seeds are planted uniformly. Grain drills can be calibrated stationary or mobile. The units cancellation method can be used in either situation. For both the stationary and mobile methods a container is attached to each metering unit, the drill is driven a measured distance (mobile), or the drill is jacked up and the drive wheel is turned (stationary) for a selected number of revolutions.

Problem: An 18–6 (18 metering units spaced 6 inches apart) end wheel drill is set to apply 1.0 bu/ac of wheat. The quantity of seeds collected during calibration is shown in Table 8.2. The diameter of the drive wheel is 26.0 inches, and the drive wheel was turned 25 revolutions. Is the drill planting the correct amount of seed (bu/ac)? (*Note*: Because each wheel of an end wheel grain drill powers half of the metering units, only nine units are shown.)

Solution: The first step is to determine the total weight of seed collected.

$$W_t = 0.10 + 0.10 + 0.12 + 0.10 + 0.11 + 0.11 + 0.11 + 0.15 + 0.13 = 1.03\,\text{lb}$$

The next step is to determine the bushels per acre. To complete this step, the problem can be broken down into several steps or the units cancellation can be used in one continuous equation. A previous section explained that calibration is determining a volume of material for an area. Using this concept, the drill can be calibrated by determining the volume (bu) and the area (ac).

It is common to use a volume measure, bushels, to express the desirable seeding rate for a grain drill. This means that when the seeds are measured in pounds, a conversion from weight to volume must be made. The relationship between weight and volume is the specific weight (γ) where:

$$\gamma = \frac{W}{V}.$$

where γ = specific weight (weight/volume); W = weight (lb); V = volume (bu).

TABLE 8.2. Pounds of wheat collected from nine grain drill metering units.

Unit	1	2	3	4	5	6	7	8	9
lb	0.10	0.10	0.12	0.10	0.11	0.11	0.11	0.15	0.13

The standard specific weight of wheat is $\frac{60\text{ lb}}{1\text{ bu}}$. Therefore:

$$\text{Vol(bu)} = \frac{1.03\text{ lb}}{60\,\dfrac{\text{lb}}{\text{bu}}} = 0.0171667\text{ bu}$$

The area is:

$$\begin{aligned} A(\text{ac}) &= \frac{1\text{ ac}}{43{,}560\text{ ft}^2} \times \left(25\text{ rev} \times \frac{\pi \times 26\text{ in}}{\text{rev}} \times \frac{1\text{ ft}}{12\text{ in}}\right) \\ &\quad \times \left(9\text{ unit} \times \frac{6\text{ in}}{\text{unit}} \times \frac{1\text{ ft}}{12\text{ in}}\right) \\ &= \frac{1\text{ ac}}{43{,}560\text{ ft}^2} \times 170\text{ ft} \times 4.5\text{ ft} = 0.017561983\text{ ac} \end{aligned}$$

The seeding rate in bushels per acre is:

$$R\left(\frac{\text{bu}}{\text{ac}}\right) = \frac{0.0171667\text{ bu}}{0.017561983\text{ ac}} = 0.97749\ldots\text{ or } 0.98\,\frac{\text{bu}}{\text{ac}}$$

In this example, breaking the problem down into parts requires less math ability than trying to determine the answer by using one equation. The basic equation is relatively simple:

$$R\left(\frac{\text{bu}}{\text{ac}}\right) = \frac{\text{Vol (bu)}}{n_r\text{ (rev)}} \times \frac{n_r\text{ (rev)}}{A\text{ (ac)}}$$

But when the values are included to arrive at bu/rev and rev/ac the math is more complicated. The first variable is determined by:

$$\frac{\text{Vol (bu)}}{n_r\text{ (rev)}} = \frac{1\text{ (bu)}}{\gamma\text{ (lb)}} \times \frac{W\text{ (lb)}}{n_r\text{ (rev)}}$$

the second variable is determined by:

$$\frac{n_r\text{ rev}}{A\text{ (ac)}} = \frac{1\text{ rev}}{\left(\pi \times \text{dia} \times \dfrac{1\text{ ft}}{12\text{ in}}\right) \times \left(n_u \times \dfrac{\text{width}}{\text{unit}} \times \dfrac{1\text{ ft}}{12\text{ in}}\right) \times \dfrac{1\text{ ac}}{43{,}560\text{ ft}^2}}$$

where n_u = number of metering units. Putting the two parts together produces:

$$\begin{aligned} R\left(\frac{\text{bu}}{\text{ac}}\right) &= \frac{W\text{ (lb collected)}}{\gamma\left(\dfrac{\text{lb}}{\text{bu}}\right) \times n_r\text{ (rev)}} \\ &\quad \times \frac{1\text{ rev}}{\left(\pi \times \text{dia} \times \dfrac{1\text{ ft}}{12\text{ in}}\right) \times \left(n_u \times \dfrac{\text{width}}{\text{unit}} \times \dfrac{1\text{ ft}}{12\text{ in}}\right) \times \dfrac{1\text{ ac}}{43{,}560\text{ ft}^2}} \end{aligned}$$

The solution for the sample problem is:

$$R\left(\frac{\text{bu}}{\text{ac}}\right) = \frac{1.03\ \text{lb}}{60\left(\frac{\text{lb}}{\text{bu}}\right) \times 25\ \text{rev}}$$
$$\times \frac{1\ \text{rev}}{\left(\pi \times 26 \times \frac{1\ \text{ft}}{12\ \text{in}}\right) \times \left(9 \times \frac{6.0\ \text{in}}{\text{unit}} \times \frac{1\ \text{ft}}{12\ \text{in}}\right) \times \frac{1\ \text{ac}}{43{,}560\ \text{ft}^2}}$$
$$= \frac{0.000687\ \text{bu}}{\text{rev}} \times \frac{1\ \text{rev}}{0.000703\ \text{ac}} = 0.9767\ldots \text{ or } 0.98\frac{\text{bu}}{\text{ac}}$$

The desired planting rate is 1 bu/ac resulting in an error of 0.02 bu/ac. Is this acceptable for planting wheat? This is an example when perfection is not expected, therefore the level of acceptability must be established. Some drill manufacturers publish acceptable seeding rates for their products, or this information may be found in an extension bulletin. A standard has been established for grain drills. The seeding rate should be within plus or minus of 5% of the desired rate. When this method is used acceptable limits are set on each side of the desired rate. If the actual rate falls within this limit, it is acceptable. In this example the upper limit (L_u) is:

$$L_u = 1.0 + (1.0 \times 0.05) = 1.05\ \frac{\text{bu}}{\text{ac}}$$

and the lower limit (L_l) is:

$$L_l = 1.0 - (1.0 \times 0.05) = 0.95\ \frac{\text{bu}}{\text{ac}}$$

Using this standard, the accuracy of the grain drill is acceptable ($1.05 > 0.98 > 0.95$).

The calibration of the drill is not completed until the uniformity of distribution is also checked. This is accomplished by using the same rule and setting the limits around the mean amount of seeds collected from the metering units. This will give:

$$\text{Mean} = \frac{W_t}{n_u} = \frac{1.03\ \text{lb}}{9\ \text{units}} = 0.1144\ldots\ \text{lb}$$

The distribution upper limit is:

$$\text{DL}_u = 0.114\ldots + (0.114\ldots \times 0.05) = 0.1197 \text{ or } 0.12\ \frac{\text{bu}}{\text{ac}}$$

The distribution lower limit is:

$$\text{DL}_l = 0.114\ldots - (0.114\ldots \times 0.05) = 0.1083 \text{ or } 0.108\ \frac{\text{bu}}{\text{ac}}$$

A comparison of these limits to the calibration results shown in Table 8.2 indicates that although the grain drill seeding rate is acceptable, the distribution is not. The rate for metering unit #8 and #9 is above the upper limit. Both metering units should be repaired before the drill is used.

8.6. Calibrating Row Crop Planters

Row crop planters are used to plant crops in wider rows than those planted by grain drills. Row crop planters are commonly used to plant large seeds, such as corn, soybeans, and sunflowers, but are used for small seeded vegetable crops such as radishes. Row crop planters will have a speed hopper, a metering unit, a seed tube, a furrow opener and some type of closing or press wheels. Additional hoppers are sometimes used for fertilizer or insecticides, Figure 8.5. The common metering units for row crop planters are plate, disc, drum, and volume or bulk, but other mechanisms are available. Plates are the traditional method for metering seeds, but for most crops they have been replaced with discs. One style of antique plate planter was designed to plant hills of seeds, but most plate and the modern disc metering units singulate seeds. Singulating means each seed is selected from a small volume of seeds and transferred to the furrow. When plates are used the seeds are deposited in holes (cells) on the rim of a ground driven metering plate, and as the metering plate rotates, the seeds leave the plate and fall into a furrow that has been opened in the soil. The seeding rate is changed by using a plate with a different number of cells, or by changing the drive train speed ratio between the drive wheel and the metering unit.

Disc metering relies on air to hold the seeds in the cells. Two different designs are used, based on air pressure either greater than or less than atmospheric. One type uses pressurized air to hold the seed in a vertical seed plate and when the air is shut off, the seeds drop into the opened furrow or are blown into tubes that deliver them to the opened furrow. The other type of air planter uses a vacuum to hold the seeds in the cells in the seed plate.

The drum type of metering unit works on the same principles as the pressurized air disc planter. The primary difference is that the drum has a ring on holes for each row instead of a separate metering unit for each row. Each drum usually singulates seeds for either 6 or 8 rows. The metering rate can be adjusted by changes drums and varying the speed between the drum and the drive wheel.

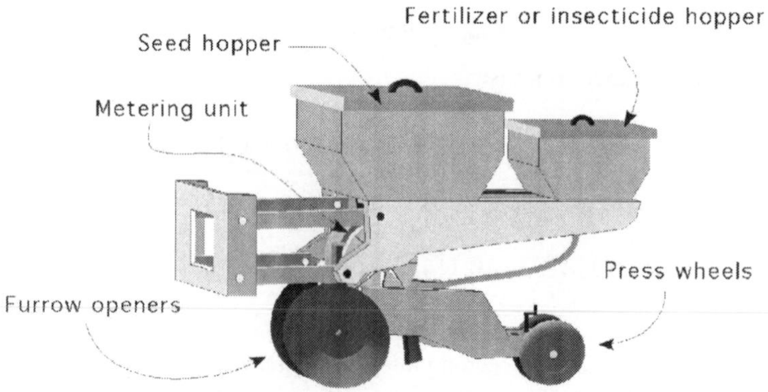

FIGURE 8.5. Single unit of row crop planter.

A few manufacturers produce a volume type metering unit that looks like and operates like a grain drill metering unit. The seeding rate (seeds/acre) for the volume metering unit is determined by the speed at which the metering unit operates in relation to the ground speed of the planter.

8.6.1. Stationary Calibration

Both bulk and singulating row crop planters can be calibrated stationary by following the procedure used for stationary calibration of fertilizer applicators and grain drills. When the stationary method is used, the seeding rate is calculated by dividing the seeds planted per revolution of the drive wheel by the simulated acres covered per revolution of the drive wheel, when the stationary method is used or:

$$R = \frac{\frac{\text{sd}}{n_r}}{\frac{A}{n_r}}$$

This equation can also be written as:

$$R = \frac{\text{sd}}{n_r} \times \frac{n_r}{A}$$

where R = seeding rate (seeds/ac); sd = number of seeds; n_r = number of revolutions of drive wheel; A = area.

Problem: What is the planting rate for corn when 0.22 lb of seeds were weighted after the 18.0 inch drive wheel was turned 25 revolutions? The planter is set for 36 inch rows and the seed size is 1,500 seeds per pound.

Solution: When calibrating row crop planters the area used is determined by the row spacing and the distance traveled during the calibration, or the simulated distance traveled.

$$A = w \times d$$

where A = area (ft^2); w = row spacing (ft); d = distance (ft).

$$R\left(\frac{\text{lb}}{\text{ac}}\right) = \frac{W\ (\text{lb})}{n_r\ (\text{rev})} \times \frac{n_r\ (\text{rev})}{A\ (\text{ac})}$$

$$= \frac{0.22\ \text{lb}}{25\ \text{rev}} \times \frac{1\ \text{rev}}{18.0\ \text{in} \times \frac{1\ \text{ft}}{12\ \text{in}} \times \pi \times \frac{1\ \text{ac}}{43560\ \text{ft}^2} \times 36\ \text{in} \times \frac{1\ \text{ft}}{12\ \text{in}}}$$

$$= \frac{0.22}{25} \times \frac{1}{3.245\ldots\ \text{E}-4} = \frac{0.22}{8.1136\ldots\ \text{E}-3} = 27.114\ldots \frac{\text{lb}}{\text{ac}}$$

The planting rate in seeds per acre is:

$$R\left(\frac{\text{sd}}{\text{ac}}\right) = \frac{27.114\ldots\ \text{lb}}{\text{ac}} \times \frac{1{,}500\ \text{sd}}{1\ \text{lb}} = 40672.36 \text{ or } 40{,}700\ \frac{\text{sd}}{\text{ac}}$$

8.6.2. Mobile Calibration

All row crop planters are easy to calibrate using the mobile method because for each row spacing and planting rate there is a unique spacing for the seeds in the row. To complete a mobile calibration of a row crop planter seeds are added to the hopper, the planter is lowered into the ground and driven for a short distance to insure the system has stabilized. Next carefully uncover several seeds in a row and measure the distance between the seeds. The average distance is used to determine the seeding rate.

Problem: The calibration of a row crop planter, set for 20 inch rows, determined soybean spacing of 2.24, 2.23, and 2.25 inches. What is the planting rate (sd/ac)?

Solution: First determine the mean distance between seeds in the row.

$$\text{Mean distance} = \frac{2.24 + 2.27 + 2.25}{3} = 2.24 \text{ in}$$

$$R\left(\frac{\text{sd}}{\text{ac}}\right) = \frac{1 \text{ sd}}{2.24 \text{ in}} \times \frac{12 \text{ in}}{1 \text{ ft}} \times \frac{43560 \text{ ft}^2}{1 \text{ ac}} \times \frac{1 \text{ row}}{20 \text{ in}} \times \frac{12 \text{ in}}{1 \text{ ft}}$$

$$= \frac{6272640}{44.8} = 140014.28\ldots \text{ or } 140{,}000 \text{ or } 1.40 \text{ E5 } \frac{\text{sd}}{\text{ac}}$$

Sometimes it is useful to know the in-the-row seed spacing for the planting rate. This can also be determined using by using units cancellation.

Problem: What is the spacing in the row (in/seed) for 40,000 plants per acre when the planter plants rows 28.0 inches apart?

Solution: Using units cancellation:

$$\frac{\text{in}}{\text{seed}} = \frac{1 \text{ ac}}{40{,}000 \text{ seed}} \times \frac{43{,}560 \text{ ft}^2}{1 \text{ ac}} \times \frac{144 \text{ in}^2}{1 \text{ ft}^2} \times \frac{1 \text{ row}}{28 \text{ in}}$$

$$= \frac{6272640}{1120000} = 5.60057\ldots \text{ or } 5.60 \frac{\text{row-in}}{\text{seed}}$$

Is this the correct answer? The units in the answer do not match the desired units. It is correct because it is common practice to drop off the unit of row in the answer and just use in/sd.

When doing a few calculations it would be appropriate and preferred to complete the problem as in this example by using the units cancellation method. This would not be true if the in-the-row spacing needed to be calculated many times. In this situation it is easier to use an equation. In Chapter 1 the source of many equations was identified as units cancellation calculations where the consistent numbers were reduced to a constant. Combining the consistent units in the previous problem (43,560 ft^2 $\times$ 144 in^2) produces the constant 6.273 E6. The seed spacing in the row can be found by the equation:

$$SS = \frac{6.273\ E6}{POP \times RS}$$

where SS = Seed spacing in the row (in); POP = Population, planting rate (seeds/ac); RS = Row spacing (in).

If the seeding rate is not correct, further evaluation of the planter must be made. Because the metering unit is ground driven, the source of the error may be in the drive train ratio between the drive wheel and the metering unit.

8.7. Air Seeders

Air seeders, Figure 8.6, have become the machine of choice for planting wheat and other crops in large fields. Seeders with effective widths of 70 to 100 ft are being used. The basic design is a seed cart pulled behind a field cultivator or other secondary tillage machine that has soil engaging tools mounted on shanks. The seed cart may have one or more hoppers. When multiple hoppers are used one hopper will be for seed and the other(s) can be used for fertilizer or other materials. Each hopper will have either a single ground driven metering unit that splits into multiple air streams, or multiple air streams with a metering unit for each. At the metering unit the seeds are divided into several tubes and flowing air delivers the seeds to a splitter manifold. The splitter manifolds divide the seeds into a stream for each shank. As air seeder technology has developed, more sophisticated soil opening devices and metering units have been developed, but the basic design is the same.

Most manufacturers state that air seeders can be calibrated using stationary or mobile methods, but the seeding rate for some air seeders calibrated stationary may be less than the actual rate in the field. For this reason the mobile calibration is recommended.

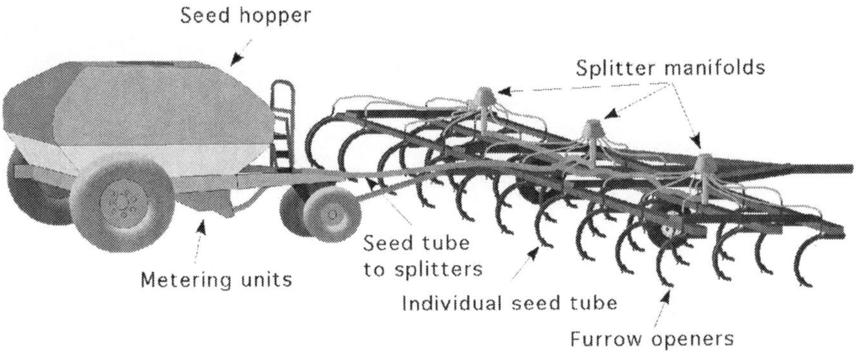

FIGURE 8.6. Air seeder.

8.7.1. Mobile Calibration

Mobile calibration methods are the same for air seeders as for grain drills. One difference is the number of rows that must be collected. Air seeders are so large that it is not practical to collect all of the rows. Recommendations for the number of rows that should be collected vary, but 25% is a general recommendation. The appropriate number of seed hoses are disconnected from the individual seed tubes and connected to a bag or other collection device. The seeder is driven a fixed distance, 100 to 200 ft and the material is weighed from each bag. With this information and knowing the effective width of each shank the seeding rate or application rate can be calculated.

Problem: Determine the seeding rate (lb/ac) for a 50-ft air seeder with a 10.0 inch spacing. The following quantities of seeds were collected from every 4th shank on a single row during 200 ft of travel, Table 8.3.

Solution: The first step is to determine the total amount of seeds collected.

$$\begin{aligned}\text{Total Pounds} &= 0.34 + 0.33 + 0.33 + 0.36 + 0.34 + 0.35 + 0.35 + 0.33 + 0.34 \\ &\quad + 0.34 + 0.35 + 0.35 + 0.33 + 0.34 + 0.34 \\ &= 5.12 \text{ lb}\end{aligned}$$

The next step is to determine the area: [*Note*: For calibration purposes the effective

TABLE 8.3. Quantity of seeds collected during calibration of air seeders.

Collector	lb
1	0.34
2	0.33
3	0.33
4	0.36
5	0.34
6	0.35
7	0.35
8	0.33
9	0.33
10	0.34
11	0.34
12	0.35
13	0.35
14	0.33
15	0.34
16	0.34

area is the number of shanks times the spacing between shanks.]

$$\text{Area (ac)} = \frac{10\text{ in}}{\text{row}} \times \frac{1\text{ ft}}{12\text{ in}} \times 16\text{ row} \times 200\text{ ft} \times \frac{1\text{ ac}}{43560\text{ ft}^2}$$

$$= \frac{32000}{522720} = 0.06121\ldots\text{ or } 0.061\text{ ac}$$

The air seeder applied 5.45 lb of seeds in 0.61 acre. The last step is to determine the pounds of seed per acre.

$$R = \frac{\text{Pounds}}{\text{Acre}} = \frac{5.12\text{ lb}}{0.061\text{ ac}} = 83.934\ldots\text{ or } 83.9\ \frac{\text{lb}}{\text{ac}}$$

The air seeder is applying seeds at the rate of 83.9 lb/ac. It is up to the owner/operator to determine if this is an acceptable level of performance.

8.8. Calibrating Sprayers

Accurate calibration of spray equipment is very important because with only slight changes the application rate may cause chemical damage to the crop or the environment, be wasteful of materials, or be ineffective. There is one important difference between the design of fertilizer spreaders, grain drills, row crop planters and sprayers. The method used to control the flow rate of the material for a sprayer is not normally ground driven. The application rate (gal/ac) is a function of the flow rate of the nozzles (gal/min) and the speed of the sprayer (mi/hr). The flow rate from the nozzles does not change as the ground speed changes. The exception to this statement occurs when an application rate controller is added to the sprayer. Most application rate controllers vary application rate to compensate for changes in applicator speed. This is generally done by changing the operating pressure of the sprayer. When a rate controller is not used, the speed of the sprayer must be considered in calculating the application rate.

Figure 8.7 illustrates the common parts for the typical overlapping boom sprayer. Overlapping boom sprayers consist of tank, filter, pump, means to control the pressure, multiple sections of nozzles, and boom selection valve. The mixture in the tank flows through the filter to remove any particles that might plug the orifice in the nozzles, and then to the pump. One of several types of pumps is used, depending on the flow (gal/min) and the pressure (psi) needed by the system. From the pump the mixture goes to the pressure regulating valve. In some sprayer designs, a portion of the flow from the pump is returned to the tank for agitation to prevent the spray materials from separating. Any excess flow produced by the pump also is returned to the tank by the pressure regulating valve. From this valve the fluid is pumped to the boom selector valve. At some point in this line a connection is made for a pressure gage. The boom selector valve directs the flow of the mixture to the different sections of the boom. In some types of valves provision is made for a handgun. A handgun is very useful for spraying skips or along fencerows and other obstructions.

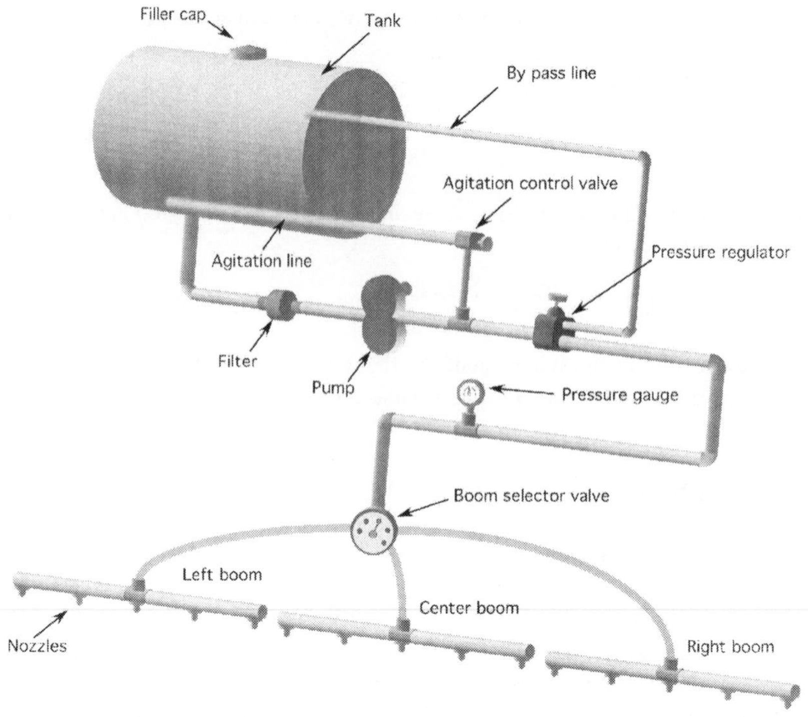

FIGURE 8.7. Parts of a typical boom sprayer.

One popular selector valve design has seven positions:

1. All outlets off
2. All booms on
3. Left boom on
4. Center boom on
5. Right boom on
6. Handgun on
7. Booms and handgun on.

In the typical design, the pump can be engine, power take off (PTO), or hydraulic motor driven. As long as the pump is operating and the selector valve is in one of the on positions, fluid will flow. It is important to understand the effect ground speed has on the spray application rate (gal/ac). When a sprayer operating in the field slows down, the application rate increases. This occurs because the nozzle flow rate (gal/min) is constant. When the travel speed is reduced the same amount of material is applied to a smaller area.

$$\frac{\text{gal}}{\text{ac}} = \frac{\text{gal}}{\text{min}} \times \frac{\text{min}}{\text{ac}}$$

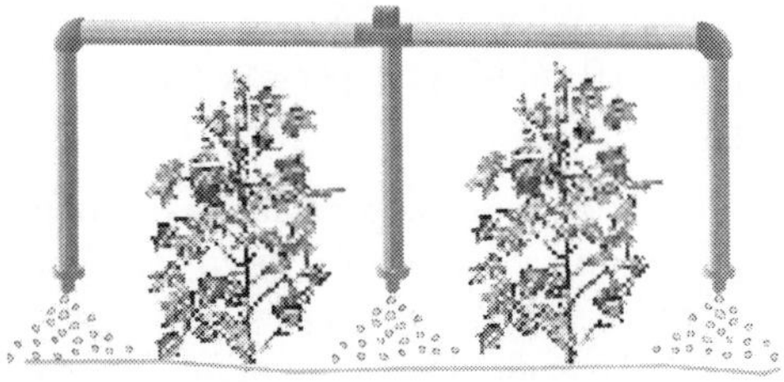

FIGURE 8.8. Nozzles arranged for banding between the rows.

When the travel speed is reduced, the acres per minute (ac/min) are reduced. Conversely, when the sprayer travel speed increases, the application rate decreases. Therefore, precise control of the sprayer speed is very important.

The design shown in Figure 8.7 often is modified to meet the demands of different types of plants or application methods. Two additional examples are shown in Figure 8.8 and Figure 8.9.

The application rate (gal/ac) of a field sprayer is controlled by three factors:

1. The speed of the sprayer (mi/hr).
2. The rate of discharge from the nozzle (gal/min).
3. The width covered by one nozzle (in).

Arranging these variables into one equation produces the standard sprayer equation:

$$R = \frac{5940 \times Q}{V \times w}$$

where R = application rate (gal/ac); 5940 = units conversion constant; Q = flow rate per nozzle (gal/min); V = travel speed (mi/hr); w = nozzle spacing (in).

Note: Use either the flow rate (gal/min) from one nozzle and the spacing between two adjacent nozzles, or the flow from all nozzles (gal/min $\times$ number of nozzles) and the total width of the sprayer ($w \times$ number of nozzle). In either case, the application rate (gal/ac) will be the same. Do not interchange these values.

The calibration of a sprayer is a multiple-step process. In addition, the process can be started at different points, depending on which one of the three variables (speed, nozzle flow rate, or application rate) is selected first.

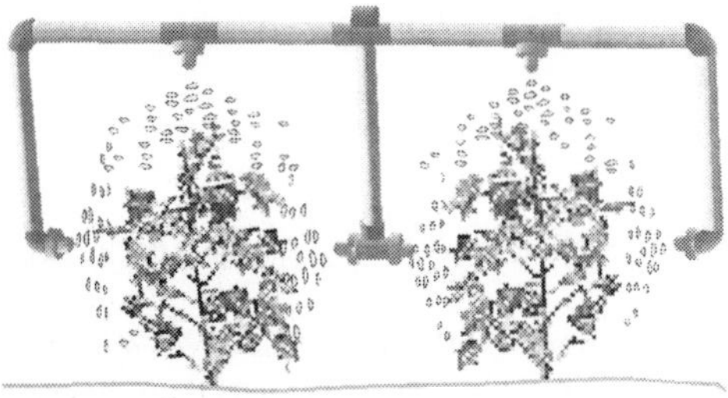

FIGURE 8.9. Nozzles arranged for row crop sprayer.

Problem: What size of nozzles (gal/min) is required for a boom type sprayer to apply 20.0 gal of spray per acre? The sprayer has 24 nozzles spaced 18.0 inches apart.

Solution: Because the application rate of field sprayers is speed-dependent, begin by selecting a reasonable speed that can be maintained in the field, and then determine the size of nozzles needed. For this problem we will use the typical speed (Appendix IV) of 6.5 mi/hr. The required flow rate for the nozzles can be determined by rearranging the standard sprayer equation to solve for the flow rate (gal/min).

$$R\left(\frac{\text{gal}}{\text{ac}}\right) = \frac{5940 \times Q\left(\frac{\text{gal}}{\text{min}}\right)}{V\left(\frac{\text{mi}}{\text{hr}}\right) \times w\,(\text{in})}$$

$$Q\left(\frac{\text{gal}}{\text{min}}\right) = \frac{R \times V \times w}{5940} = \frac{\frac{20.0\ \text{gal}}{1\ \text{ac}} \times \frac{6.5\ \text{mi}}{\text{hr}} \times 18.0\ \text{in}}{5940}$$
$$= \frac{2340}{5940} = 0.3939\ldots \text{ or } 0.39\ \frac{\text{gal}}{\text{min}}$$

For this application rate and nozzle spacing, nozzles with a capacity of 0.39 gal/min should be installed on the sprayer. Before it is used, the sprayer should be calibrated to ensure that the application rate is correct because small variations in the construction of the nozzles or in the pressure at the nozzles can cause an unacceptable error in the application rate.

Assume the operator installed the 0.39 gal/min nozzles on the sprayer and proceeded with the calibration.

Problem: A container placed under all 24 nozzles of the sprayer collected 14.40 gal of spray in 2.0 min of operation. The desired application rate was 20.0 gal/ac. Is the sprayer accurate?

Solution: In this example only the total volume is known. One alternative is to determine the average flow rate per nozzle and then use the sprayer equation, but this process will not be as accurate as using the units cancellation method and the total flow rate:

$$R\left(\frac{\text{gal}}{\text{ac}}\right) = \frac{14.40\ \text{gal}}{2.0\ \text{min}} \times \frac{60.0\ \text{min}}{1\ \text{hr}} \times \frac{1\ \text{hr}}{6.5\ \text{mi}} \times \frac{1\ \text{mi}}{5{,}280\ \text{ft}} \times \frac{43{,}560\ \text{ft}^2}{1\ \text{ac}} \times \frac{12.0\ \text{in}}{1\ \text{ft}}$$
$$\times \frac{1\ \text{nozzle}}{18.0\ \text{in}} \times \frac{1}{24\ \text{nozzles}}$$
$$= \frac{451630080}{29652480} = 15.2307\ldots \text{ or } 15\ \frac{\text{gal}}{\text{ac}}$$

The desired application rate was 20 gal/ac, but the calibration indicates that the application rate is 5 gal/ac less than this (20 gal/ac −15 gal/ac). It is important to check the label of the chemical to determine if this is an acceptable application rate. If the error is unacceptable, how do we reduce it? The first step is to check the filters and nozzles of the sprayer to make sure that one or more were not slightly restricted. If all of the nozzles are in proper working order, the sprayer must be adjusted to apply the correct rate.

Adjustments can be made in the speed and/or the system pressure. To adjust the speed the sprayer equation can be rearranged to calculate a new speed of travel. Adjusting the pressure is not as effective because only a small amount of change in the application rate can be made by adjusting the pressure. The pressure must be doubled to increase the flow rate by 41%, and modern boom type sprayers operate within a narrow pressure range. For this problem we will adjust the speed of travel.

Units cancellation could be used, but this time we will rearrange the sprayer equation to solve for the speed of travel. Because the sprayer equation requires the flow rate per nozzle, we will determine the average flow rate per nozzle. This average provides an acceptable level of accuracy.

$$\frac{Q}{\text{nozzle}} = \frac{\text{gal}}{\text{No. nozzle}} = \frac{14.4\ \text{gal}}{24\ \text{nozzles}} = 0.60\ \frac{\text{gal}}{\text{nozzle}}$$

and because the spray was collected for 2 min:

$$Q\left(\frac{\dfrac{\text{gal}}{\text{min}}}{\text{nozzle}}\right) = \frac{0.60\ \dfrac{\text{gal}}{r}}{2.0\ \text{min}} = 0.30\ \text{gal/min/nozzle}$$

For convenience, the unit "nozzle" usually is not used; gpm = 0.30. Then:

$$R\left(\frac{\text{gal}}{\text{min}}\right) = \frac{5{,}940 \times Q\left(\dfrac{\text{gal}}{\text{min}}\right)}{V\left(\dfrac{\text{mi}}{\text{hr}}\right) \times w\ (\text{in})}$$

$$V\left(\frac{\text{mi}}{\text{hr}}\right) = \frac{5{,}940 \times Q\left(\dfrac{\text{gal}}{\text{min}}\right)}{R\left(\dfrac{\text{gal}}{\text{ac}}\right) \times w\ (\text{in})}$$

$$= \frac{5{,}940 \times 0.30 \dfrac{\text{gal}}{1\ \text{min}}}{20.0\ \dfrac{\text{gal}}{\text{ac}} \times 18\ \text{in}} = \frac{1782}{360} = 4.95\ \text{or}\ 5.0\ \text{mph}$$

If the speed of the sprayer is changed from 6.5 to 5.0 mi/hr, the sprayer will apply the correct rate.

Although the preferred way of calibrating a sprayer is to determine the nozzle size first, this method requires the purchase of a new set of nozzles if the correct size is not available. In some situations the nozzle size is selected first (the best available), and then the required speed of travel is determined. When this method is used, the calculated speed of travel may be unrealistic.

Problem: We need to apply 15.0 gal of spray per acre. Only one set of nozzles is available, and they have a capacity of 0.25 gal/min. The sprayer has 35 nozzles spaced 24.0 inches apart. What speed will be required to apply the correct rate?

Solution: The sprayer equation could be rearranged to solve for speed. Instead we will use the units cancellation method.

$$V\left(\frac{\text{mi}}{\text{hr}}\right) = \frac{1\text{ mi}}{5{,}280\text{ ft}} \times \frac{43{,}560\text{ ft}^2}{\text{ac}} \times \frac{1\text{ ac}}{15.0\text{ gal}} \times \frac{0.25\ \dfrac{\text{gal}}{1\text{ min}}}{\text{nozzle}} \times \frac{60\text{ min}}{1\text{ hr}} \times \frac{12\text{ in}}{1\text{ ft}} \times \frac{1}{24.0\text{ in}}$$

$$= \frac{7{,}840{,}800}{1{,}900{,}800} = 4.125 \text{ or } 4.1\ \frac{\text{mi}}{\text{hr}}$$

In this problem, if the sprayer is operated at 4.1 mi/hr, the correct rate will be applied.

Other types of sprayers can be calibrated by using these methods if the appropriate adjustments are made for differences in how the area and the application rate are determined. For example, to calibrate the row crop sprayer in Figure 8.9, the width becomes the distance between the rows, and each nozzle should apply one third of the required flow (gal/min) per row.

Sprayers used for banding also can be calibrated. In checking a sprayer used for banding, the nozzle spacing (w) becomes the width of the band, Figure 8.8.

8.9. Preparing Spray Mixes

A very important part of chemical application is the proper preparation and mixing of the chemical and the carrier (usually water). The application rates of most chemicals are given in terms of pounds of active ingredient to be applied per acre. The concentration for a liquid chemical is usually printed on the label. A typical material might contain 4 lb of active ingredient per gallon. Wettable powders (WPs), on the other hand, are specified as a certain percent strength, such as a 50% or an 80% WP, which means that 50% or 80% of the weight of the material in the container is the active ingredient. The rest is inert material (carrier).

Problem: You need to apply 2.0 lb of active ingredient of an 80% WP in a 30.0 gal/ac dilution. The sprayer has a 100.0-gal tank, how many pounds of WP are required to mix 100.0-gal of spray?

Solution: To obtain 2.0 lb of active ingredient, 2.5 lb of powder would be required (80% of 2.5 = 2.0). If the solution is to be applied at 30 gal/ac, mix 2.5 lb of 80% WP in each 30 gal of water. The tank holds 100 gal or 3.33 units of 30 gal. Because 2.5 lb of WP should be added to each 30 gal of water, you should add 3.33 × 2.5

or 8.33 lb of WP to each 100 gal. Or by ratio:

$$\frac{2.5 \text{ lb WP}}{30.0 \text{ gal } H_2O} = \frac{\text{W lb WP}}{100.0 \text{ gal } H_2O}$$

$$\text{W lb WP} = \frac{2.5 \times 100.0}{30.0} = 8.33\ldots \text{ or } 8.33 \text{ lb } 80\% \text{ WP}$$

To mix the spray, add 8.33 lb of WP to a partially filled tank, mix it thoroughly, and add water to make 100 gal of mixture. During spraying this mixture must be continuously agitated to prevent the WP from settling.

Preparing a mix using liquid chemicals can be accomplished with the same procedures.

Problem: A liquid concentrate contains 2.0 lb of active ingredient per 5.0 gal. The desired rate is 1.0 lb of active ingredient per acre at a rate of 20.0 gal of chemical and water (solution) per acre. If the sprayer tank holds 180.0 gal, how much water and how much concentrate should be used for each tank full?

Solution: If the liquid contains 2.0 lb of active ingredient per 5.0 gal, then 2.5 gal would contain 1.0 lb of active ingredient. Thus, for each acre 2.5 gal is mixed with 17.5 gal of water (20–2.5) to get 20 gal of spray. By ratio:

$$\frac{2.5 \text{ gal concentrate}}{20.0 \text{ gal solutiion}} = \frac{\text{V gal concentrate}}{180.0 \text{ gal solution}}$$

$$\text{V gal water} = \frac{2.5 \times 180.0}{20.0} = 22.5 \text{ gal concentrate}$$

V = volume in gallons and 180 gal − 22.5 gal = 157.5 gal. Thus, for every tank 22.5 gal of chemical is mixed with 157.5 gal of water.

8.10. Metric Problems

Problem: Determine the application rate (kg/ha) for a fertilizer spreader that applied 52.5 kg of material during 15 revolutions of the drive wheel, which has a diameter of 0.53 m. The effective width of the spreader is 0.76 m.

Solution:

$$R\left(\frac{\text{kg}}{\text{ha}}\right) = \frac{W\text{ (kg)}}{n_r\text{ (rev)}} \times \frac{n_r\text{ (rev)}}{A\text{ (ha)}}$$

$$= \frac{52.5 \text{ kg}}{15 \text{ rev}} \times \frac{1 \text{ rev}}{3.14 \times 0.53 \text{ m}} \times \frac{10{,}000 \text{ m}^2}{\text{ha}} \times \frac{1}{0.76 \text{ m}}$$

$$= \frac{525{,}000}{18.98\ldots} = 27672.53\ldots \text{ or } 27{,}000 \frac{\text{kg}}{\text{ha}} \text{ or } 27 \frac{t}{\text{ha}}$$

Problem: Determine the acceptability of the seeding rate for a grain drill when the desired rate is 9.0 kg/ha. The drill has 25 metering units with a 0.20 m spacing. The drive wheel has an effective diameter of 80 cm. During the stationary calibration the

TABLE 8.4. Data for grain drill calibration in SI units.

Unit	1	2	3	4	5	6	7	8	9	10	11	12
Grams	22.6	22.4	22.0	22.9	22.7	22.6	22.0	22.8	22.8	22.7	22.6	22.8

drive wheel was rotated 50 times. The following amounts of seeds were collected from 12 of the metering units, Table 8.4.

Solution: The seeding rate is determined by calculating the quantity of seed for the area and then converting the results to the desired units.

The quantity collected is:

$$\begin{aligned}\text{Total} &= 22.6 + 22.4 + 22.0 + 22.9 + 22.7 + 22.6 + 22.0 + 22.8 + 22.8 \\ &\quad + 22.7 + 22.6 + 22.8 \\ &= 270.9 \text{ g or } 0.2709 \text{ kg}\end{aligned}$$

The area is:

$$\begin{aligned}A\ (\text{ha}) &= \text{No. Units} \times w(\text{m}) \times d(\text{cm}) \times \pi \times n_r \\ &= 12 \text{ units} \times 0.20\ \frac{\text{m}}{\text{unit}} \times \frac{80 \text{ cm} \times \pi}{\text{rev}} \times 50 \text{ rev} \times \frac{1 \text{ m}}{100 \text{ cm}} \times \frac{1 \text{ ha}}{10{,}000 \text{ m}^2} \\ &= \frac{30159.28\ldots}{1.0 \text{ E6}} = 0.03015\ldots \text{ha}\end{aligned}$$

The seeding rate is:

$$R\ (\text{kg/ha}) = \frac{0.2709 \text{ kg}}{0.03015\ldots \text{ha}} = 8.982\ldots \text{ or } 8.9\ \frac{\text{kg}}{\text{ha}}$$

The acceptability of the drill performance is determined by establishing the upper and lower limits using the 10% rule.

$$\begin{aligned}\text{Upper} &= 9.0 \times 1.05 = 9.4 \\ \text{Lower} &= 9.0 \times 0.95 = 8.6\end{aligned}$$

Based on the 10% rule the performance of the grain drill is acceptable because $9.4 > 8.9 > 8.6$.

Even though the seeding rate is acceptable, the distribution pattern of the drill should also be checked.

$$\begin{aligned}\text{Mean} &= \frac{270.9 \text{ g}}{12} = 22.575 \text{ g} \\ \text{Upper limit} &= 22.575 \times 1.05 = 23.703\ldots \text{ or } 23.7 \text{ g} \\ \text{Lower limit} &= 22.575 \times 0.95 = 21.44\ldots \text{ or } 21.4 \text{ g}\end{aligned}$$

The distribution pattern is acceptable because all of the quantities for each metering unit are within the upper and lower limit.

Problem: Determine the plants per hectare for a row crop planter when the average seed spacing is 11.5 cm in the row and the row spacing is 70 cm.

Solution:

$$R\left(\frac{\text{sd}}{\text{ha}}\right) = \frac{10{,}000\ \text{m}^2}{1\ \text{ha}} \times \frac{100\ \text{cm}}{\text{m}} \times \frac{1\ \text{sd}}{11.5\ \text{cm}} \times \frac{100\ \text{cm}}{\text{m}} \times \frac{1}{70\ \text{cm}}$$

$$= \frac{1.0\ \text{E8}}{805} = 124223.6025 \text{ or } 124{,}000 \text{ or } 1.24\ \text{E5}\ \frac{\text{sds}}{\text{ha}}$$

Problem: Determine the seeding rate for a 10.8-m air seeder with 18 cm row spacing when 13 kg of seeds were collected during 50 m of travel.

Solution: Since we have collected seed from all rows we do not need to consider the row spacing.

$$R\left(\frac{\text{kg}}{\text{ha}}\right) = \frac{W\ (\text{kg})}{A\ (\text{ha})} = \frac{W\ (\text{kg})}{l\ (\text{m}) \times w\ (\text{m})} \times \frac{10{,}000\ \text{m}^2}{\text{ha}}$$

$$= \frac{13\ \text{kg}}{50\ \text{m}} \times \frac{1}{10.8\ \text{m}} \times \frac{10{,}000\ \text{m}^2}{1\ \text{ha}}$$

$$= \frac{1.3\ \text{E5}}{540} = 240.74\ldots \text{ or } 241\ \frac{\text{kg}}{\text{ha}}$$

Problem: Determine the size of nozzles (L/min) required to apply spray at the rate of 185 L/ha when the nozzle spacing is 0.4545 m and the sprayer will be traveling at a velocity of 4.0 km/hr.

Solution:

$$Q\left(\frac{\text{L}}{\text{min}}\right) = 185\ \frac{\text{L}}{\text{ha}} \times \frac{1\ \text{ha}}{10{,}000\ \text{m}^2} \times 4\ \frac{\text{km}}{\text{hr}} \times \frac{1\ \text{hr}}{60\ \text{min}} \times \frac{1{,}000\ \text{m}}{\text{km}} \times 0.45\ \text{m}$$

$$= \frac{3.33\ \text{E7}}{6.0\ \text{E7}} = 0.555 \text{ or } 0.56\ \frac{\text{L}}{\text{min}}$$

9
Equipment Efficiency and Capacity

9.1. Objectives

1. Understand the concept of efficiency and be able to apply it to agricultural operations.
2. Understand the concept of capacity and be able to apply it to agricultural machines.
3. Be able to calculate effective field capacity.
4. Be able to calculate the throughput capacity of agricultural machines.

9.2. Introduction

To be efficient means being able to produce the desired results with a minimum of effort, resources, or waste. This is a concept that permeates our lives. A part of every occupation is the desire to improve the product or services by producing more for less or a better product for the same input. Engineers are constantly trying to improve the efficiency of operations by reducing the energy requirements and/or wastes from agricultural and manufacturing processes. When referring to machinery, efficiency is an evaluation of how well a machine does the tasks that it is designed to perform.

Capacity is a measurement of the amount of performance that has occurred. The evaluation of a machine's capacity is an important evaluation because under utilization of a machine increases the production costs and over utilization can lead to increase repair, maintenance costs, and shorten machine life.

9.3. Efficiency

In this chapter, we will use the concept of efficiency to evaluate how well a machine performs its designed task in terms of quantity and/or quality of performance. Owners and managers of farm enterprises are deeply concerned with efficient operation of equipment and other resources because inefficient operation leads to greater operating expenses and reduced profits. Efficiency is usually expressed as a percentage. A percentage is calculated by comparing to quantities and multiplying by 100. Because efficiency is a ratio of two quantities having the same units, the units cancel.

Efficiency can be expressed mathematically in several forms. In the most general terms, efficiency can be expressed as:

$$\text{Efficiency } (E) = \frac{\text{output}}{\text{input}} \times 100$$

Efficiency is the ratio of what we get out of something relative to what we put in. If the output is 9 units (pounds, hours, etc.) and the input 10 units, the efficiency is:

$$\%E = \frac{\text{output}}{\text{input}} \times 100 = \frac{9}{10} \times 100 = 90\%$$

Or, if the output is 5 units and the input 10 units, the efficiency is:

$$\%E = \frac{\text{output}}{\text{input}} \times 100 = \frac{5}{10} \times 100 = 50\%$$

Efficiency can also be determined by comparing the actual performance to the theoretical performance. This equation is:

$$\%E = \frac{\text{actual}}{\text{theorectical}} \times 100$$

It is important to remember that an efficiency calculation provides a mathematical answer to a problem. It is only a tool or information that can be used to make a decision. For example, if you determine that the fuel efficiency of an automobile is 20 miles per gallon that is not sufficient information to determine if the automobile is performing satisfactory. This number must be compared to the historical performance, manufacturer's guidelines or other data to make a decision on its acceptability.

9.3.1. Mechanical Efficiency

Mechanical efficiency has to do with how well machines convert energy from one form to another. For example, an internal combustion engine converts the chemical energy in fuel into power. Internal combustion engines are not 100% efficient because all the energy in fuel is not converted power (the majority of the heat produced escapes through the radiator and out the exhaust). A typical gasoline engine is about 22% efficient; a diesel engine is 30 to 33%. An electric motor converts electrical energy into power with an efficiency of 95 to 98%. The efficiency of mechanical power trains and other mechanical devices can be determined if accurate numbers for the power in and power out can be determined.

9.3.2. Performance Efficiency

Performance efficiency refers to the quality of work done by a machine. The importance of a performance evaluation is not the same for all machines. For example, the quality of the job for primary tillage, like plowing, is not as critical to the profitability of the farming enterprise as the quality of the job for a combine. In addition, it would be very difficult to mathematically evaluate the quality of plowing.

For a harvesting machine, performance efficiency is a measure of the actual performance of the machine compared to the desired performance. For example, if the machine was a combine, we could measure the bushels of grain harvested compared to the total bushels of grain in the field. Combines also could be evaluated according to the amount of damaged grain. Other harvesting machines could be evaluated on the basis of the amount of bruising of fruit or on the number of cracked shells. An example is the small grain combine.

A combine can lose grain in three different ways: the gathering unit can shatter grain from the head or drop heads, the threshing unit can fail to remove grain from the head as it passes through the machine, and the separating and cleaning units can fail to separate the grain from the material other than grain (MOG). The losses are usually expressed as a percentage of the yield of the field.

Evaluating combine losses is a multiple-step problem. What we want to know is the amount of grain that the combine fails to put into the grain bin. The first step in evaluating the performance of a combine is to determine if a problem actually exists. This is accomplished by determining the total loss behind the machine.

To determine total loss, a known area is marked out, and grain is counted on the ground behind the combine. Losses at this point include grain on the ground before the combine started (preharvest loss) and grain loss by the combine (machine loss). For most cereal grains, losses are determined by counting the number of grains in a known area, for example, 10 or 100 square feet. This will provide data with units of seeds per unit area. The table in Appendix III can be used to convert seeds per square feet (sds/ft^2) to bushels per acre (bu/ac) or seeds/m^2 to kilograms/ha.

Problem: What is the performance efficiency (percent) in lost grain for the combine shown in Figure 9.1?

Solution: Figure 9.1 represents the results of measuring the losses from a combine. The first step is to determine the total losses. In the example illustrated in Figure 9.1, 162 seeds of wheat were counted in a 10-square foot area behind the machine.

By referring to Appendix III, we can convert the total losses from the measured quantity of seeds per 10 square feet to bushels per acre. From Appendix III, 20 seeds per square foot equals one bushel per acre. Applying the units cancellation method:

$$\frac{\text{bu}}{\text{ac}} = \frac{162\ \text{seeds}}{10\ \text{ft}^2} \times \frac{1\ \frac{\text{bu}}{\text{ac}}}{19\ \frac{\text{seeds}}{\text{ft}^2}} = \frac{162}{190} = 0.8526 \text{ or } 0.85\ \frac{\text{bu}}{\text{ac}}$$

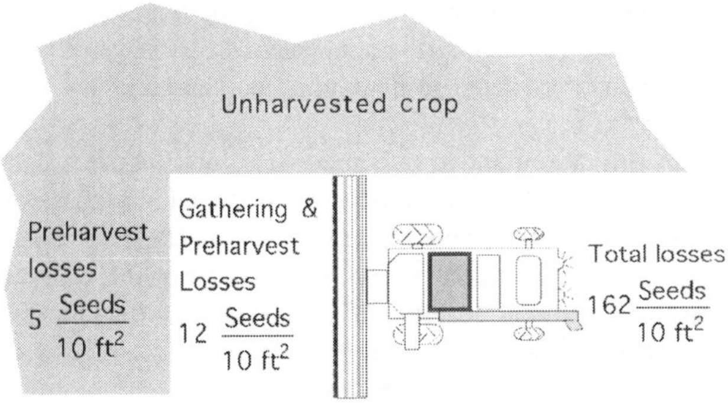

FIGURE 9.1. Determining performance efficiency of a combine.

Note: The previous equation may be easier to understand by remembering that when dividing fractions, inverting and multiplying is the same operation. The previous equation can be written as:

$$\frac{\text{bu}}{\text{ac}} = \frac{162 \text{ seeds}}{10 \text{ ft}^2} \times 1 \frac{\text{bu}}{\text{ac}} \times \frac{1 \text{ ft}^2}{19 \text{ seeds}}$$

Is a loss of 0.85 bu/ac an acceptable level of loss for this machine? As noted earlier, the mathematical calculations do not usually answer the question; they just provide information to make a more informed decision. The two common methods used to make informed decisions are to compare the results with the historical performance of the machine and to compare the results to an accepted standard. Research has shown that under favorable conditions an expert combine operator should be able to keep losses to less than 1%, but that losses up to 3% are typical. Therefore to determine the acceptability of the combine performance, the total losses must be compared to the standard. To determine the loss as a percentage two values with the same units must be used. This is usually done with combines by comparing the loss and the crop yield (bu/ac).

For this example we will assume that the wheat is yielding 30 bu/ac. The total loss percentage ($L_t\%$) is:

$$L_t\% = \frac{L_t}{Y_t} = \frac{0.85 \dfrac{\text{bu}}{\text{ac}}}{30 \dfrac{\text{bu}}{\text{ac}}} \times 100 = 2.833\ldots \text{ or } 2.8\%$$

where Y_t = the total yield in sd/ac, bu/ac, sd/m^2, or kg/ha

$$Y_L = \text{yield loss}, \frac{\text{bu}}{\text{ac}} \text{ or } \frac{\text{kg}}{\text{ha}}$$

$$Y_t = \text{total yield}, \frac{\text{bu}}{\text{ac}} \text{ or } \frac{\text{kg}}{\text{ha}}$$

Based on the less than 3% standard, this machine's performance would be considered acceptable. If the 1% standard was the desired performance level, additional calculations must be completed to determine the source of the excessive grain loss.

At this point it is important to remember that total loss is a combination of machine loss and preharvest loss. An unacceptable total loss does not necessarily mean there is a problem with the combine. The source of the unacceptable total loss may be preharvest losses. The combine has no influence on these losses. Preharvest loss is measured by counting the grain on the ground in the standing crop that has not been harvested. In Figure 9.1, we note that preharvest losses are 5 seeds per 10 square feet. Therefore, the machine loss in bushels per 10 square feet is:

$$\begin{aligned}\text{Machine loss } (L_{\text{m}}) &= \text{total loss } (L_{\text{t}}) - \text{preharvest loss } (L_{\text{p}})\\ &= \frac{162 \text{ sds}}{10 \text{ ft}^2} - \frac{5 \text{ sds}}{10 \text{ ft}^2}\\ &= 157 \frac{\text{sds}}{10 \text{ ft}^2}\end{aligned}$$

Converting this value to bushels per acre:

$$L_{\text{m}} = 157 \frac{\text{seeds}}{10 \text{ ft}^2} \times \frac{1 \dfrac{\text{bu}}{\text{ac}}}{19 \dfrac{\text{seeds}}{1 \text{ ft}^2}} = \frac{157}{190} = 0.826 \ldots \frac{\text{bu}}{\text{ac}}$$

Converting this to a percent:

$$L_{\text{m}} = \frac{0.83 \dfrac{\text{bu}}{\text{ac}}}{30 \dfrac{\text{bu}}{\text{ac}}} \times 100 = 2.754 \ldots \text{ or } 2.8\%$$

Rounding to two significant figures, the total loss and machine loss are both 2.8%. The preharvest loss is too small of a percentage to be measured.

It can now be concluded that, if the 1% standard is used, the losses are excessive and the machine is the source of the excessive losses. When this occurs, it is necessary to determine which component of the machine is the source of the losses.

The first step is to check the gathering unit. The performance of the gathering unit (header) is checked by counting the seeds on the ground between the header and the uncut crop. 9.1 shows that the losses in this area are 12 seeds per 10.0 square feet. (Remember that this also includes the preharvest losses.) The gathering unit (L_{g}) losses are determined by subtracting the preharvest loss from seeds counted in the area between the header and the uncut crop.

$$L_{\text{g}} = \frac{12 \text{ seeds}}{10.0 \text{ ft}^2} - \frac{5 \text{ seeds}}{10.0 \text{ ft}^2} = \frac{7 \text{ seeds}}{10 \text{ ft}^2} = 0.7 \frac{\text{seeds}}{\text{ft}^2}$$

Converting this value to bushels per acre:

$$\frac{\text{bu}}{\text{ac}} = 0.7\,\frac{\text{seeds}}{\text{ft}^2} \times \frac{1\,\dfrac{\text{bu}}{\text{ac}}}{19\,\dfrac{\text{seeds}}{\text{ft}^2}} = \frac{0.7}{19} = 0.0368 \text{ or } 0.04\,\frac{\text{bu}}{\text{ac}}$$

or as a percent:

$$\frac{0.04\,\dfrac{\text{bu}}{\text{ac}}}{30\,\dfrac{\text{bu}}{\text{ac}}} \times 100 = 0.13\%$$

This shows that the gathering unit losses are 0.13%, a very small part of the total machine losses.

When the preharvest loss and the gathering unit losses are both low, the threshing and separating units must be checked.

Threshing losses are represented by heads or cobs on the ground behind the machine with grain still attached. Cleaning and separating losses are represented by grain on the ground behind the machine that is not preharvest or gathering unit loss. Cleaning and separating losses can be determined by subtraction:

$$L_{\text{c\&s}} = L_{\text{t}} - L_{\text{p}} - L_{\text{g}} - L_{\text{th}}$$

where $L_{\text{c\&s}}$ = Cleaning and separating losses; L_{t} = Total losses; L_{p} = Preharvest losses; L_{g} = Gathering unit losses; L_{th} = Threshing losses.

$$\begin{aligned} L_{\text{c\&s}} &= L_{\text{t}} - L_{\text{p}} - L_{\text{g}} - L_{\text{th}} \\ &= 162\,\frac{\text{sds}}{10\text{ ft}^2} - 5\,\frac{\text{sds}}{10\text{ ft}^2} - 7\,\frac{\text{sds}}{10\text{ ft}^2} - 0\,\frac{\text{sds}}{10\text{ ft}^2} \\ &= 150\,\frac{\text{sds}}{10\text{ ft}^2} \end{aligned}$$

Note that the threshing losses are listed as zero. This is because the information is not available to determine this loss. The threshing and separating loss percentage is:

$$\frac{\text{bu}}{\text{ac}} = \frac{150\text{ sds}}{10\text{ ft}^2} \times 1\,\frac{\text{bu}}{\text{ac}} \times \frac{1\text{ ft}^2}{19\text{ sds}} = \frac{150}{190} = 0.789\ldots \text{or } 0.79\,\frac{\text{bu}}{\text{ac}}$$

$$\% = \frac{0.79\,\dfrac{\text{bu}}{\text{ac}}}{30\,\dfrac{\text{bu}}{\text{ac}}} \times 100 = 2.631\ldots \text{ or } 2.6\%$$

Clearly, if the desire is to improve the performance of the combine the cleaning and separating units should be adjusted or repaired.

9.3.3. *Field Efficiency*

Field efficiency is usually used to evaluate the performance of tillage or harvesting machines. It is a comparison of the actual amount of "work" (volume of activity, not Force × Distance) done by a machine compared to what it would do with no lost time or capacity. The maximum rate that a machine can perform is determined by the width of the machine and the speed of travel. When a machine operates with a constant width and travels at a constant speed, it will perform at 100% field efficiency. A machine is capable of operating at 100% field efficiency for short periods of time, but as soon as the speed changes (slow down for turns, etc.), or the width changes (overlap width of the machine to prevent skips), the efficiency drops below 100%. The primary cause of loss efficiency is lost time (unproductive time) and a working width of the machine less than the maximum. Typical field efficiencies for common machines can be found in Appendix IV. This concept is illustrated in more detail in the next section, on capacity.

9.4. Capacity

The term capacity is used to evaluate the productivity of a machine. In agriculture, two types of capacity are commonly used, *field capacity* and *throughput capacity*. Field capacity is used to evaluate the productivity of machines used to work the soil, such as plows, cultivators, and drills, sprayers, and harvesting machines. Throughput capacity is used to describe machines that handle or process a product, such as grain augers, balers, forage harvesters, and combines.

An additional concept relating to both types of capacity is the difference between theoretical and actual productivity. If a tillage machine operates at 100% efficiency, it is operating at 100% capacity. This is called the *theoretical field capacity*. Theoretical field capacity is determined using the width of the machine and the speed of travel. It can be calculated using units cancellation, but an equation is commonly used.

$$\text{Theoretical field capacity}(C_T) = \frac{S \times W}{8.25}$$

where C_T = Theoretical field capacity ($\frac{ac}{hr}$); S = Speed of travel ($\frac{mi}{hr}$); W = Width of the machine (ft); 8.25 = Units conversion constant (43,560 $\frac{ft^2}{ac}$) ÷ (5,280 $\frac{ft^2}{mi}$).

This equation can be used as long as the unit used for speed is miles per hour and the unit used for the width of the machine is feet.

Problem: Determine the theoretical capacity for a machine that travels at 5.0 mph and has an operating width of 20.0 ft.

Solution:

$$C_T = \frac{S \times W}{8.25} = \frac{5.0\,\frac{mi}{hr} \times 20.0\text{ ft}}{8.25} = 12.12\ldots \text{ or } 12\,\frac{ac}{hr}$$

When this machine travels at a constant speed and uses a constant width, it has a theoretical capacity of 12 ac/hr.

Effective field capacity is the amount of productivity that actually occurred not what is theoretical possible. Lost capacity is an important concern for the machine operator and/or manager because it represents lost revenues or resources. Usually lost capacity is caused by lost time, time not operating, and operating the machine with less than the maximum working width. Common causes of lost time include:

1. Mechanical breakdowns.
2. Taking time to adjust the machine.
3. Stopping to fill seed hoppers, spray tanks, etc.
4. Slowing down to turn at the end of the row or crossing waterways, etc.
5. Operator rest stops.

The equation for effective capacity is the same as theoretical capacity with a field efficiency added. A range and typical field efficiency values for common machines can be found in Appendix IV. The common equation for effective field capacity is:

$$C_{\mathrm{E}} = \frac{S \times W \times E_{\mathrm{f}}}{8.25}$$

where C_{E} = Effective field capacity (ac/hr); S = Average speed of travel (mph); W = Effective width of the machine (ft); E_{f} = Field efficiency (decimal form).

Problem: Assume that the operator in the previous problem averages 0.75 hr of lost time per 10.0-hr day. What is the effective field capacity?

Solution: The first step is to determine the time efficiency:

$$E_{\mathrm{f}} = \frac{\text{output}}{\text{input}} \times 100 = \frac{10.0\text{ hr} - 0.75\text{ hr}}{10.0\text{ hr}} = 92.5\%$$

The second step is to determine the effective capacity:

$$C_{\mathrm{E}} = \frac{S \times W \times E_{\mathrm{f}}}{8.25} = \frac{5.0\text{ mph} \times 20.0\text{ ft} \times 0.925}{8.25} = \frac{95.5}{8.25} = 11.21\ldots\text{or } 11\ \frac{\text{ac}}{\text{hr}}$$

Now the effects of lost productivity are apparent. The theoretical capacity is 12 ac/hr, but because of lost time, the effective capacity is 11 ac/hr.

The concept of effective capacity also can be used to determine the amount of time it would take a machine to cover a field.

Problem: How many hours will it take to cultivate 200.0 acres with a field cultivator that is 24.0 ft wide?

Solution: This is an example of a problem with a hidden intermediate step. Before the hours can be determined, the effective capacity of the machine must be calculated. *Note*: The effective capacity equation requires three values. Two of these, speed and field efficiency, are not given in the problem. If the actual speed and the field efficiency are unknown, the typical values found in Appendix IV can be used. In Appendix IV, the typical field efficiency for a field cultivator is 85%, and the typical speed is 5.5 mph. With these values, the effective capacity can be calculated:

$$C_E = \frac{S \times W \times E_f}{8.25} = \frac{5.5\,\frac{\text{mi}}{\text{hr}} \times 24.0\text{ ft} \times 0.85}{8.25} = \frac{112.2}{8.25} = 13.6 \text{ or } 14\,\frac{\text{ac}}{\text{hr}}$$

Once the effective capacity is known, the time required to work the field can be calculated. Using units cancellation:

$$\text{hour} = \frac{1\text{ hr}}{14\text{ ac}} \times 200\text{ ac} = 14.28\ldots \text{ or } 14\text{ hr}$$

If the average field speed is 5.5 mph and the operator can maintain an 85% field efficiency, it will take 14 hr to cultivate the 200 acre field.

9.5. Throughput

The concepts of theoretical and effective capacity also are applicable to throughput capacity. Throughput is based on time, but because throughput usually refers to the flow of material through a machine, the units may be different from those used for capacity. For example, the performance of a hay baler could be evaluated using units of bales per hour or tons per hour.

Problem: What is the throughput of a baler if it baled 150.0 tons in one week while operating an average of 6.0 hr per day?

Solution: With the information given, different units could be used for the output. These include tons/week, tons/day, or bales/day. The "correct" units are the ones that match the input units. For this example, assume that the manufacturer advertises that the baler has a capacity of 6 tons/hr. This means we need to determine the output (effective throughput) in units of ton/hr.

$$\frac{\text{ton}}{\text{hr}} = \frac{150.0\text{ ton}}{\text{week}} \times \frac{1\text{ week}}{5\text{ day}} \times \frac{1\text{ day}}{6.0\text{ hr}} = \frac{150}{30} = 5.0\,\frac{\text{ton}}{\text{hr}}$$

Knowing the actual throughput for the baler, the throughput efficiency can be determined. The throughput efficiency is:

$$E = \frac{\text{output}}{\text{input}} \times 100 = \frac{5.0\,\frac{\text{ton}}{\text{hr}}}{6.0\,\frac{\text{ton}}{\text{hr}}} \times 100 = 83.3\ldots \text{ or } 83\%$$

Assuming the advertised throughput of the baler is a reasonable value, this operation is only 83% efficient in baling hay.

The throughput of a baler also can be evaluated in units of bales per hour (bales/hr). To use these units additional information is required, including the weight of the hay (lb/bale) and two unit conversion values.

Problem: Determine the effective throughput in units of bales per hour when each bale weighs 1,200.0 lb.

Solution : No equation has been presented to solve this problem. This is an example where units cancellation should be used.

$$\frac{\text{bales}}{\text{hr}} = \frac{150.0\ \text{ton}}{\text{week}} \times \frac{1\ \text{week}}{30\ \text{hr}} \times \frac{2{,}000\ \text{lb}}{1\ \text{ton}} \times \frac{1\ \text{bale}}{1{,}200.0\ \text{lb}}$$

$$= \frac{3.0\ \text{E5}}{3.6\ \text{E4}} = 8.333\ldots \text{ or } 8.33\ \frac{\text{bales}}{\text{hr}}$$

Obviously, throughput can be expressed in many different ways depending on the values being compared. Remember that efficiency is a ratio—both values must have the same units. In addition, these examples illustrate the usefulness of units cancellation for solving problems of this type.

9.6. Metric Problems

Problem: Determine the theoretical capacity for a 20.0-m wide cultivator traveling at 9.6 km/hr.

Solution:

$$C_T\left(\frac{\text{ha}}{\text{hr}}\right) = S\left(\frac{\text{km}}{\text{hr}}\right) \times w(\text{m}) \times \frac{1{,}000\ \text{m}}{\text{km}} \times \frac{1\ \text{ha}}{10{,}000\ \text{m}^2}$$

$$= 9.6\ \frac{\text{km}}{\text{hr}} \times 20.0\ \text{m} \times \frac{1{,}000\ \text{m}}{\text{km}} \times \frac{1\ \text{ha}}{10{,}000\ \text{m}^2}$$

$$= \frac{192{,}000}{10{,}000} = 19.2\ \frac{\text{ha}}{\text{hr}}$$

Problem: What is the effective capacity for the cultivator in the previous problems if the operator wasted 0.75 hr per 10.0 hr working day?

Solution: The first step is to solve for the time efficiency.

$$\text{Efficiency } (\%) = \frac{\text{output}}{\text{input}} \times 100$$

$$= \frac{10.0\ \text{hr} - 0.75\ \text{hr}}{10.0\ \text{hr}} \times 100 = \frac{9.25}{10.0} \times 100 = 92.5\%$$

The effective capacity is:

$$C_E\left(\frac{\text{ha}}{\text{hr}}\right) = S\left(\frac{\text{km}}{\text{hr}}\right) \times w(\text{m}) \times \frac{1{,}000\ \text{m}}{\text{km}}\ \frac{1\ \text{ha}}{10{,}000\ \text{m}^2} \times E_F$$

$$= 9.6\ \frac{\text{km}}{\text{hr}} \times 20.0\ \text{m} \times \frac{1{,}000\ \text{m}}{\text{km}} \times \frac{1\ \text{ha}}{10{,}000\ \text{m}^2} \times 0.925$$

$$= \frac{177{,}600}{10{,}000} = 17.8\ \frac{\text{ha}}{\text{hr}}$$

Problem: How many hours will it take to cultivate 20.0 ha with an 8.0-m cultivator traveling at 10 km/hr? Assume a field efficiency of 85%.

Solution: The first step is to determine the effective field capacity.

$$C_E\left(\frac{\text{ha}}{\text{hr}}\right) = S\left(\frac{\text{km}}{\text{hr}}\right) \times W(\text{m}) \times \frac{1{,}000\ \text{m}}{\text{km}} \times \frac{1\ \text{ha}}{10{,}000\ \text{m}^2} \times E_F$$

$$= 10\ \frac{\text{km}}{\text{hr}} \times 8.0\ \text{m} \times \frac{1{,}000\ \text{m}}{\text{km}} \times \frac{1\ \text{ha}}{10{,}000\ \text{m}^2} \times 0.85$$

$$= \frac{68{,}000}{10{,}000} = 6.8\ \frac{\text{ha}}{\text{hr}}$$

The second step is to calculate the amount of time it will take to cultivate 20.0 ha with a field capacity of 6.8 ha/hr.

$$\text{Time (hr)} = \frac{1\ \text{hr}}{6.8\ \text{ha}} \times 20.0\ \text{ha} = 0.3095\ldots\ \text{or } 2.9\ \text{hr}$$

10
Economics of Agricultural Machinery

10.1. Objectives

1. Be able to list the criteria for selecting tractors and machines.
2. Be able to determine optimum machine size.
3. Be able to calculate ownership and operating costs of agricultural equipment.
4. Understand ways to reduce the costs of owning and operating tractors and machinery.
5. Understand the concept of break-even use (BEU) for a machine.
6. Be able to calculate BEU for a machine.
7. Understand the importance of regular maintenance of agricultural machinery.

10.2. Introduction

Machinery is one of the largest investments for agricultural enterprises. The selection process may include manufacturer, design, size, and options. A bad decision on any one of these factors will have a serious effect on the profitability of the enterprise. After studying this chapter, you should have a better understanding of the criteria to use when selecting tractors and machines, and be able to match tractor and machine sizes.

10.3. Selection Criteria

Some of the characteristics or capabilities of a tractor or machine that make it more attractive than another are abstract, but they can have a tremendous bearing on the quality of the machine and/or the quality of its performance. The following sections will discuss some of the criteria that should be used when selecting a tractor or machine and illustrate their importance.

10.3.1. Company Name

The company name should be considered in machinery selection. Manufacturers spend years and multitudes of resources establishing a reputation. A company's reputation is based on the quality and durability of its products, service to its customers, or a combination of both. It is important to know if a manufacturer will stand behind its product and warranties.

The importance of selecting tractors and machinery from reputable companies cannot be overstated. In some situations, the best economic decision would be to choose one tractor or machine over another, even though it did not have the better durability record, just because the reputation of the manufacturer was better.

An important factor to consider on brand name is the availability of parts. A low cost tractor or machine that has an extended down time each time it needs a part is not a good investment.

10.3.2. Cost

Cost is always an important consideration when purchasing any item. It is important to remember that the least cost tractor or machine may not be the least expensive when both ownership and operating expenses are considered. The least cost machine may have the lowest ownership costs, but the operating costs may offset this advantage. *Note*: these costs are explained in a later section.

An important aspect of the total cost of a machine is the down time. Whenever a machine is not available because of the need for repairs, the cost to the enterprise is not just the cost of the repairs; it also includes lost capacity or lost time. When the crop is ready for harvest and the combine is not available because it needs repairs, there is a cost associated with leaving the crop in the field after it is ready for harvest, because some of the crop will be lost.

10.3.3. Repairs

All machines have breakdowns, but the quality of the construction will dramatically influence the repair costs for a tractor or machine. There is also a correlation between the complexity of a tractor or machine and the cost of repairs. It would be expected that a combine would have more repair costs that a cultivator. The cost of repairs is also influenced by the use and abuse of the tractor or machine. Any equipment that is not maintained or used in any manner not recommended by the manufacturer will have greater repair costs. As discussed in the previous section, downtime for repairs is also important because of the lost capacity. Evaluate the equipment not only on its quality and durability but also on the amount of time and money that will be required to have it repaired.

10.3.4. Design

The design of the machine is an important consideration. Because of the diversity of agricultural machinery, several designs might produce an acceptable level of performance, yet small variations in design may make one more suitable than the others for a given situation. For most tillage equipment, the cost of the machine is primarily determined by the quantity of the steel it contains. A tillage machine designed for use in fields that are large and relatively flat will have wide rigid sections. But a machine that is designed for terraced or undulating fields will require a frame with more flexibility. More flexibility will require a more complicated frame that will cost more money. It is sometimes difficult to know what features of a design improve the value of a machine for a particular situation without an understanding of the advantages and disadvantages of the different designs.

10.3.5. Capacity

The trend in agriculture has been to larger farms. This trend has increased the demand for machines with larger and larger capacity. A previous chapter discussed that the options for increasing capacity are: using a wider width, operating at a faster speed (both requiring more power) and increasing efficiency of operation.

It also is important to be able to complete the operation in a timely fashion. Unfortunately, information on the timeliness of agricultural operations is limited. Some concepts of timeliness and the effect of machine size are discussed in the next section.

10.4. The Optimum Machine Size

Optimizing the machine size is one technique for reducing production costs. When the machine is too small, the operating costs may be higher, and its reduced capacity will require a greater number of hours to complete an operation. It is also possible to exceed the design forces when the machine size is small for the available power. When the machine is too large, power is the limiting factor and the machine may not produce the desired results, or the life of the tractor may be shortened, because modern tractors are designed to operate at lighter drafts and faster speeds.

The optimum machine size can be selected from two different points of view: it can be based on the amount of time available to complete an operation, or it can be based on the amount of power available from the tractor. The process used depends on the situation.

When the limiting factor is time, the recommendation is to determine the available time (timeliness) for completing the operation and then determine the size of machine required. Once the machine size is determined, the required tractor size can be determined. A high priority on timeliness may result in a calculation that would require a larger tractor to power the machine. A full discussion of timeliness is beyond the scope of this text, but we will illustrate this concept for determining the size of machine and tractors.

When the limiting factor is power, the recommended procedure is to determine the maximum size of machine that can be powered by the tractor. When power is the limiting factor, the calculations may result in a time requirement for completion of the activity that is not realistic. The first example problem will match tractor and machine size base on timeliness.

Problem: A general recommendation is that the primary tillage for planting corn should not take longer than 1 week. If you anticipate planting 500 acres of corn and only work 12.0 hr per day and 6 days per week, what size of moldboard plow do you need if you can plow at typical speed and efficiency?

Solution: The first step is to determine the capacity of the plow, C_E (ac/hr) required to plow the ground in the available time. This can be accomplished by using units cancellation:

$$C_E\left(\frac{\text{ac}}{\text{hr}}\right) = \frac{500\text{ ac}}{1\text{ wk}} \times \frac{1\text{ day}}{12.0\text{ hr}} \times \frac{1\text{ wk}}{6\text{ day}} = 6.944\ldots \text{ or } 6.9\ \frac{\text{ac}}{\text{hr}}$$

To plow 500 acres in 1 week, the plow must have a width and be operated at a speed that will produce a capacity of 6.9 ac/hr. The second step is to determine the width of the plow. This is accomplished by remembering that the capacity of field equipment can be determined by using the effective capacity equation, Chapter 9.

$$C_E = \frac{V \times w \times E_f}{8.25}$$

Rearranging this equation to solve for width (w) and looking up the typical speed and efficiency in Appendix IV results in the width of the plow:

$$w = \frac{C_E \times 8.25}{V \times E_f} = \frac{6.90\ \frac{\text{ac}}{\text{hr}} \times 8.25}{4.5\ \frac{\text{mi}}{\text{hr}} \times 0.80} = \frac{56.925}{3.6} = 15.8125 \text{ or } 16\text{ ft}$$

To be able to plow the field in 1 week, assuming that the plow will be operated at typical speed and efficiency, you will require a plow with an effective width of 16 ft. Plows are not usually sized by feet of width; instead they are sized by the effective width of each plow bottom and the number of bottoms. Sixteen feet is the same as a 12–16 plow (12 bottoms with 16 inches per bottom).

The next step is to determine the size of the tractor needed to pull the plow. This is accomplished by determining the drawbar horsepower required. The first step is to replace the term force in the drawbar power equation with the term draft (D_F) or draft force.

$$P_{\text{DbHP}} = \frac{D_F \times V}{375}$$

The term draft is used to describe the amount of force required to pull a machine. Appendix V gives the American Society of Agricultural and Biological Engineers (ASABE) standard values for the draft of agricultural machines. Notice that more than one type of unit is used for draft. That is, the draft of plows is listed as pounds of force per square inch (cross sectional area) plus a speed factor, spring tooth harrows are listed as pounds per foot of width, and some rotary power machines are given in horsepower. It is important to pay close attention to the units being used so that the equation can be modified. The draft is calculated by the following equation:

$$D_F = F_i \times \left[C_1 + (C_2 \times V) + \left(C_3 \times V^2\right)\right] \times w \times T_d$$

where D_F = Draft in lb or N; F_i = Soil texture factor. For fine soils $i = 1$, for mediums soils $i = 2$, and for coarse textured soils $i = 3$; C_1, C_2, C_3 = machine specific parameters from Appendix V; W = machine width, ft or m; T_d = Tillage depth, in or cm.

Problem: What size tractor (Dbhp) will be required to pull the 12–16 (16.0 ft) plow that was calculated in the previous problem?

Solution: The first step is to find the draft of the plow in Appendix V. Notice that the draft for a plow varies with soil type. We will assume that the plow will be used in a clay soil, which is a fine textured soil.

From Appendix V, for customary units we find the following parameters:

$$F_i = 1.0$$
$$C_1 = 113$$
$$C_2 = 0$$
$$C_3 = 2.3.$$

The tillage depth, $T_d = 6$ in, and the plow width, $w = 16$ ft.

Selecting a depth of 6.00 inches and solving for draft:

$$\begin{aligned} D_F &= F_i \times [C_1 + (C_2 \times V) + (C_3 \times V^2)] \times w \times T_d \\ &= 1.0[113 + (0 \times 4.5) + (2.3 \times 4.5^2)]12 \text{ bottoms} \times \frac{16 \text{ in}}{\text{bottom}} \times \frac{1 \text{ ft}}{12 \text{ in}} \times 6.0 \text{ in} \\ &= \frac{183{,}830.4}{12} \\ &= 15{,}319.2 \text{ or } 15{,}000 \text{ lb} \end{aligned}$$

Solve for drawbar power in hp

$$\begin{aligned} P_{DbHP} &= \frac{D_F \times V}{375} \\ &= \frac{15{,}000 \times 4.5}{375} \\ &= 180.0 \text{ or } 180 \text{ hp} \end{aligned}$$

With typical speed and efficiency, and operating 6 inches deep in clay soil, the 12–16 plow will require a tractor with 180 drawbar horsepower.

One result of using this method to determine the optimum size of machines is now apparent. The calculated tractor size may be larger than the size of tractor that you have. What are your alternatives if you do not want to purchase a larger tractor? Plowing is usually the highest draft tillage operation on a farm so a tractor could be leased for the time required to do the plowing. The plowing could be hired out. Another option, which might allow you to complete the plowing with the tractor available, is to adjust one or more of the factors in the drawbar horsepower equation. A common practice is to extend the amount of time available to complete the tillage operation. It also is possible to reduce the draft of the plow by reducing the operating depth or to reduce the speed. In some situations one of these two choices is the best option, but for this problem the available time will be adjusted.

Problem: Assume that the largest tractor you have available to pull the plow in the previous example is 100 drawbar horsepower. If this tractor is used, how many 12 hr days will it take to plow the 500 acres?

Solution: The first step is to determine the size of plow the 100 Dbhp tractor can pull. This is accomplished by rearranging the drawbar horsepower equation to solve for width [because Area (A) = Width(w) × Depth (d)]:

$$W\,(\text{in}) = \frac{\text{Dbhp} \times 375}{D \times d \times S} = \frac{100.0\ \text{Dbhp} \times 375}{\dfrac{14.0\ \text{lb}}{\text{in}^2} \times \dfrac{6\ \text{in}}{1} \times \dfrac{4.50\ \text{mi}}{1\ \text{hr}}}$$

$$= \frac{37{,}500}{378} = 99.206\ldots \text{ or } 99.2\ \text{in}$$

Under the conditions set up in this problem, a 100.0 Dbhp tractor is capable of pulling a plow that is 99.2 inches wide. A 99.2-inch wide plow would be 6.2 or 6, 16 inch bottoms. Therefore, a 6–16 plow (8.00 ft) plow is used.

The next step is to determine the capacity of a plow 8.00 ft wide.

$$C_E = \frac{S \times W \times E_f}{8.25} = \frac{4.50\ \dfrac{\text{mi}}{\text{hr}} \times 8.00\ \text{ft} \times 0.80}{8.25} = \frac{28.8}{8.25} = 3.4909\ldots \text{ or } 3.5\ \frac{\text{ac}}{\text{hr}}$$

Determining the number of days using units cancellation:

$$\text{days} = \frac{1\ \text{day}}{12\ \text{hr}} \times \frac{1\ \text{hr}}{3.5\ \text{ac}} \times 500\ \text{ac} = 11.9047\ldots \text{or } 11.9\ \text{days}$$

This means using the 100 Dbhp instead of the 193 Dbhp, will take 6 extra days (12 days – 6 days = 6 days) to plow the field.

Another situation that often arises in agriculture is determining the size of machine that a tractor can power.

Problem: The opportunity to buy a 185 PTO horsepower tractor is too good to pass up. What size (width) of offset disk harrow (W_{dh}) and field cultivator (W_{fc}) will you need for your new tractor? The disc harrow is operated 6 inch deep. Assume a clay soil and the lowest typical value for speed and efficiency. *Note*: remember to convert from PTO power to drawbar power using the 86% rule.

Solution: To solve this problem PTO power must be changed to drawbar power.

$$P_{\text{DbHP}} = \text{PTO}_{\text{HP}} \times 0.86 = 185\ \text{hp} \times 0.86 = 159\ \text{hp}$$

Draft can be calculated by rearranging the drawbar power equation.

$$P_{\text{Dbhp}} = \frac{V \times D_{\text{F}}}{375}$$

$$D_{\text{F}} = \frac{P_{\text{DbHP}} \times 375}{V}$$

$$= \frac{159 \text{ hp} \times 375}{3.5 \text{ mph}}$$

$$= 17{,}035.71\ldots \text{ or } 17{,}000 \text{ lb}$$

The following parameters for the ASABE draft equation were obtained from Appendix V:

$$F1 = 1.0 \text{ (fine textured, clay, soil)}$$

$$C_1 = 62$$

$$C_2 = 5.4$$

$$C_3 = 0.$$

Rearrange the ASABE equation for draft and solve for the machine width (W).

$$W(\text{ft}) = \frac{D_{\text{F}}}{F_{\text{i}} \times \left[C_1 + (C_2 \times V) + \left(C_3 \times V^2\right)\right] \times T_{\text{d}}}$$

$$= \frac{17{,}000}{1.0 \times \left(62 + (5.4 \times 3.5) + \left(0 \times 3.5^2\right)\right) \times 6}$$

$$= \frac{17{,}000}{485.4}$$

$$= 35.022\ldots \text{ or } 35 \text{ ft}$$

The 185 PTO horsepower tractor will be able to pull an offset disk harrow up to 35 ft wide in a clay soil when operating at the low end of the typical speed and efficiency ranges.

The draft of a secondary tillage field cultivator is calculated as pounds of force per tool (shovel, sweep etc.) plus a speed factor. Therefore, the number of shanks determines the size of the cultivator. For a cultivator, the drawbar horsepower equation is converted to:

$$W(\text{shanks}) = \frac{D_{\text{F}}}{F_{\text{i}} \times \left[C_1 + (C_2 \times V) + \left(C_3 \times V^2\right)\right] \times T_{\text{d}}}$$

The first step in determining the size of cultivator than can be pulled by the 159 hp tractor is determining the number of shanks for the cultivator. Assume the tillage depth T_{d} is 3 inches. From Appendix V, the machine parameters for a secondary tillage field cultivator are:

$$F_1 = 1.0$$

$$C_1 = 19$$

$$C_2 = 1.8$$

$$C_3 = 0.$$

Solving the ASABE draft equation for w (in this case number of tools or shovels):

$$
\begin{aligned}
W &= \frac{D_F}{F_i \times \left[C_1 + (C_2 \times V) + \left(C_3 \times V^2\right)\right] \times T_d} \\
&= \frac{17{,}000}{1.0 \times \left(19 + (1.8 \times 6) + \left(0 \times 6^2\right)\right) \times 3} \\
&= \frac{17{,}000}{89.4} \\
&= 190.156 \text{ or } 191 \text{ tools or shanks}
\end{aligned}
$$

The 185 PTO horsepower tractor will be able to pull a field cultivator with 191 shanks. The width of the cultivator depends on the number of rows of shanks and the spacing between each shank. For example, one possible arrangement would be a cultivator with three rows of staggered shanks, the second and third row staggered equal distance between the front row shanks, spaced 21 inches apart. This arrangement would have 191 shanks (64, 63, and 64), and be 112 ft wide (64 shanks or tools multiplied by 21 inches and divided by 12 in/ft). Note: the tape measure width from shank to shank would be (63 × 21 inches) divided by 12 in/ft or 110.25 ft. The effective width includes one half of the spacing on each end of the row. 110.25 + 10.5 + 10.5 = 112 ft.

Appendix V presents the draft of equipment in the form found in the ASABE Standards. Other sources may provide draft figures for machines in pounds per foot. If these values are used, the drawbar horsepower equation does not need to be modified each time, but the calculations will not be as accurate as this example problem.

10.5. Costs of Machinery

This section discusses those elements that contribute to the cost of owning and operating agricultural machinery. An understanding of the types of costs and their impact on the profitability of the enterprise improves our ability to make meaningful decisions regarding the management of agricultural equipment.

It is best to base decisions on the actual costs of the machine. Good managers will record the annual costs for machines, but unfortunately the total costs are not known until a machine has reached the end of its serviceable life, that is, until the machine wears out or is sold or traded in. Therefore, cost determinations must be based on estimations. To make these estimations realistic and reliable, they must be based on past machine performance and cost records. The following sections will discuss the typical costs of owning and operating machinery as well as methods that can be used to estimate the costs of agricultural equipment.

Generally, machinery costs are classified into two groups, ownership costs (fixed costs) and operating costs (variable costs). Fixed and variable costs are:

Fixed costs	Variable costs
1. Depreciation	6. Repair and maintenance
2. Interest on investment	7. Fuel
3. Taxes	8. Oil
4. Shelter	9. Labor
5. Insurance	10. Consumables

10.5.1. Fixed Costs

Fixed costs are independent of machine use, and occur whether or not the machine is used. They are referred to as the cost of ownership. Each fixed cost is estimated on a calendar year or an annual basis. The common fixed costs are:

1. Depreciation: Depreciation is the loss in value of a machine with the passage of time, whether or not it is used. Depreciation can be regarded as the amount of money that should be saved each year as a machine is used so that, at the end of its useful life, this money along with the remaining value of the machine (salvage value) could be used to replace it. Several choices exist in the way by which depreciation can be figured for cost and/or tax purposes. All depreciation methods require an estimation of the machine service life. The simplest method is straight line depreciation:

$$\frac{\$}{\text{yr}} = \frac{P - SV}{\text{yr}}$$

where \$/yr = Annual depreciation; P = Purchase price; SV = Salvage value; yr = Years of service.

Problem: Determine the annual depreciation for a machine with a purchase price of \$15,400, a salvage value of \$800.00 and an expected life of 8 years.

Solution:

$$\begin{aligned}\frac{\$}{\text{yr}} &= \frac{P - SV}{\text{yr}}\\ &= \frac{15{,}400.00 - 800.00}{8}\\ &= 1{,}825.00\ \frac{\$}{\text{yr}}\end{aligned}$$

The annual straight line depreciation is \$1,825.00 per year.

2. Interest on investment: An interest cost should be used because the money tied up in purchasing a machine could have been used for another purpose or invested. When money is owed on a machine the annual interest charges added to the depreciation can be used to estimate yearly costs of ownership. A method that includes the yearly costs of ownership and the time value of money is called

capital recovery factor, CRF. CFR can be calculated by:

$$R = \left\{ (P - S) \times \left[\frac{\left(\frac{i}{q}\right) \times \left(1 + \frac{i}{q}\right)^{nq}}{\left\{\left(1 + \frac{i}{q}\right)^{nq}\right\} - 1} \right] \right\} + \frac{Si}{q}$$

where R = one of a series of equal payments due at the end of each compounding period, q times per year; P = principal amount; i = interest rate as compounded q times per year; n = life of the investment in years; S = salvage value, \$.

Problem: Determine the capital recovery factor for a machine that cost \$125,000.00 at purchase, has a salvage value of \$3,000.00. A 10-year loan is used with an interest rate of 10.5% per year.

Solution:

$$R = \left\{ (P - S) \times \left[\frac{\left(\frac{i}{q}\right) \times \left(1 + \frac{i}{q}\right)^{nq}}{\left\{\left(1 + \frac{i}{q}\right)^{nq}\right\} - 1} \right] \right\} + \frac{Si}{q}$$

$$= \left\{ (\$125{,}000.00 - \$3{,}000.00) \times \left[\frac{\left(\frac{0.105}{1}\right) \times \left(1 + \frac{0.105}{1}\right)^{10\times 1}}{\left\{\left(1 + \frac{0.105}{1}\right)^{10\times 1}\right\} - 1} \right] \right\}$$

$$+ \frac{\$3{,}000.00 \times 0.105}{1}$$

$$= \left\{ \$122{,}000 \times \left(\frac{0.105 \times 2.714\ldots}{2.714\ldots - 1} \right) \right\} + \$\,315.00$$

$$= \{\$122{,}000 \times 0.166\ldots\} + \$315.00$$

$$= 20598.3912 \text{ or } 20{,}600.00\ \frac{\$}{\text{yr}}$$

The CRF for this machine is \$20,600.00 per year.

3. Taxes: Any property and sales taxes paid on the equipment must be included as a fixed cost. An estimate of the annual taxes for agricultural machinery is 1% of remaining value.

Problem: What is the annual tax during the 2 year of ownership for a \$55,000.00 machine when the expected life is 12 years? The salvage value is \$500.00.

Solution: The first step is to determine the value of the machine at the beginning of the second year. The value at the beginning of the second year is determined by the amount of depreciation that occurred during the first year. Assuming straight line method:

$$\frac{\$}{\text{yr}} = \frac{\$\,55{,}000.00 - \$500.00}{12} = 4541.666\ldots \text{ or } 4{,}540.00\ \frac{\$}{\text{yr}}$$

The value of the machine at the beginning of the second year is:

$$\$ = \$55{,}000.00 - \$4{,}540.00 = \$50{,}460.00$$

The annual taxes are:

$$\frac{\$}{\text{yr}} = \$50{,}460.00 \times \frac{0.01}{\text{yr}} = 504.60\ \frac{\$}{\text{yr}}$$

4. Shelter: A cost of shelter should be assigned to machinery. If a shelter is used, the cost can be determined and assigned over the life of the machine. If the machine is not sheltered, a cost still should still be assigned for shelter because the resale value of the machine will be less than that of a sheltered machine. An estimate for the annual cost of shelter is 0.75% of the value of the machine at the beginning of the year.

Problem: What is the annual cost of shelter during the fourth year for a $4,500.00 machine when the expected life is 5 years and the salvage value is $1,500.00?

Solution: The first step is to determine the value of the machine at the beginning of the fourth year. Assuming straight line depreciation the annual depreciation is:

$$\frac{\$}{\text{yr}} = \frac{\$4{,}500.00 - \$1{,}500.00}{5} = 600.00\ \frac{\$}{\text{yr}}$$

The value of the machine at the beginning of the fourth year is:

$$\$ = \$4{,}500.00 - \left(\frac{600.00\ \$}{\text{yr}} \times 3\ \text{yr}\right)$$

$$= \$4{,}500.00 - \$1{,}800.00 = \$2{,}700.00$$

The cost of the shelter during the fourth year is:

$$\frac{\$}{\text{yr}} = \$2{,}700.00 \times \frac{0.0075}{\text{yr}} = 20.25\ \frac{\$}{\text{yr}}$$

A shelter cost of $20.25 should be used during the fourth year for this machine.

5. Insurance: Any liability or replacement insurance carried on the machine should be included. An estimate for the annual cost of insurance is 0.25% of the remaining value at the beginning of the year.

Problem: Determine the lifetime insurance cost for a $125,500.00 machine when the expected life is 15 years and the salvage value is $10,000.00.

Solution: The annual insurance cost for a machine is determined by multiplying the insurance factor time the value of the machine at the beginning of the year. The lifetime insurance cost for a machine is determined by summing the annual insurance costs. Because of depreciation the value of the machine declines each year and therefore the annual insurance cost declines each year. This problem

TABLE 10.1. Spreadsheet for determining lifetime insurance costs.

Purchase price	\$125,500.00	
Insurance factor	0.0025	
Annual depreciation	\$7,700.00	
Year	Value	Insurance cost
1	\$125,500.00	\$313.75
2	\$117,800.00	\$294.50
3	\$110,100.00	\$275.25
4	\$102,400.00	\$256.00
5	\$94,700.00	\$236.75
6	\$87,000.00	\$217.50
7	\$79,300.00	\$198.25
8	\$71,600.00	\$179.00
9	\$63,900.00	\$159.75
10	\$56,200.00	\$140.50
11	\$48,500.00	\$121.25
12	\$40,800.00	\$102.00
13	\$33,100.00	\$82.75
14	\$25,400.00	\$63.50
15	\$17,700.00	\$44.25
	Sum	\$2,685.00

The total cost of insurance over the lifetime of the machine is \$2,685.00.

can be solved by using a complex equation or using the same method as the previous examples, but it is also an example of a problem that can be solved using a spreadsheet, Table 10.1. Assuming straight line depreciation, the life time cost of insurance is \$2, 685.00.

The annual cost of owning a piece of equipment, then, is the sum of the fixed costs listed above. Notice that at this point no mention has been made of machine use.

With accurate records, the annual cost of a machine can be determined by figuring the cost in dollars per year for each of the five fixed costs. Another approach is to combine the five items into an annual fixed cost percentage (FC%). Cost analysis of data indicates that the FC% times the purchase price of the machine is an acceptable estimate for the annual ownership costs of a machine.

$$\text{Annual owinership cost (AOC)} = \text{FC\%} \times P$$

An ownership cost percentage can be estimated by:

$$C_o = 100 \times \left(\left(\frac{1 - S_v}{L} \right) + \left(\frac{1 + S_v}{2} \times l \right) + K_2 \right)$$

where C_o = ownership cost percentage; S_v = salvage value factor (percent of salvage value compared to the purchase price) of machine at end of machine life (year L), decimal; L = machine life, yr; l = annual interest rate; K_2 = ownership cost for taxes, housing, and insurance, decimal.

Problem: What is the ownership cost percentage for a machine that as an expected life of 10 years, cost \$23,000, a cost for taxes, housing and insurance of 2% and has a salvage value of \$3,400.00?

Solution: The ownership cost percentage is determined using the previous equation. This equation requires the salvage factor as a percent. To determine the percent divide the purchase price by the salvage value and multiply times 100. The salvage factor is:

$$\text{Salavage factor }\% = \frac{\$3{,}400.00}{\$23{,}000.00} \times 100 = 14.78\ldots \text{ or } 15\%$$

The fixed cost percent is:

$$\begin{aligned}
C_o &= 100 \times \left(\left(\frac{1 - S_v}{L}\right) + \left(\frac{1 + S_v}{2} \times I\right) + K_2\right) \\
&= 100 \times \left(\left(\frac{1 - 0.15}{10}\right) + \left(\frac{1 + 0.15}{2} \times 0.10\right) + 0.02\right) \\
&= 100 \times (0.085 + 0.0575 + 0.02) \\
&= 100 \times 0.1625 = 16.25 \text{ or } 16\%
\end{aligned}$$

Problem: What is the annual ownership cost for a \$150,000.00 combine when the fixed cost percentage is 18%?

Solution:

$$\text{AOC}\left(\frac{\$}{\text{yr}}\right) = \text{FC}\% \times P = 0.18 \times \$150{,}000.00 = 27{,}000.00\ \frac{\$}{\text{yr}}$$

10.5.2. Variable Costs

Variable costs are associated with the operation of a machine and occur only when the machine is used. The term operating costs frequently is used to describe variable costs. Variable costs are usually figured on an hourly basis, but they can be figured using acres, bales or any other appropriate unit. The common variable costs are:

1. Repairs and maintenance (RM%): The costs of repair and maintenance are related to the type of machine, the purchase price and the hours of use. The repair and maintenance costs can be estimated using the following equation.

$$C_{\text{rm}} = \text{RF1} \times P \times \left(\frac{h}{1000}\right)^{\text{RF2}}$$

where C_{rm} = accumulated repair and maintenance cost (\$); RF1 & RF2 = repair and maintenance factors (Appendix IV); P = machine price in current dollars, multiply original price by $(1 + i)^n$ where i is the average inflation rate and n is the age of the machine to adjust for inflation; h = accumulated use of machine (h).

Problem: Estimate the accumulated repair and maintenance costs for a self-propelled combine that was purchased 4 years ago for a price of \$90,000.00. The machine has 1,200 hr of use and the average inflation has been 6.5%.

Solution: The first step is to adjust the price of the machine for inflation.

$$\begin{aligned}\$ &= \$90{,}000.00 \times (1+i)^n \\ &= \$90{,}000.00 \times (1+0.065)^4 \\ &= \$90{,}000.00 \times 1.286\ldots \\ &= 115781.9\ldots \text{ or } \$120,000\end{aligned}$$

Next the repair and maintenance costs can be estimated:

$$\begin{aligned}C_{rm} &= \mathrm{RF1} \times P \times \left(\frac{\mathrm{h}}{1{,}000}\right)^{\mathrm{RF2}} \\ &= 0.04 \times \$120{,}000.00 \times \left(\frac{1{,}200}{1,000}\right)^{2.1} \\ &= \$4{,}800.00 \times 1.466\ldots \\ &= 7039.1\ldots \text{or } \$7{,}000.00\end{aligned}$$

2. Fuel (F): The fuel consumption for an engine is influenced by the size of the engine and the percent of load. Average annual consumption can be estimated using the procedures printed in ASABE Standard D497.4 FEB03.

$$Q_{\mathrm{avg}} = 0.0305 \times P_{\mathrm{pto}}$$

where Q_{avg} = average gasoline consumption, L/hr; P_{pto} = maximum PTO power, kW
or

$$Q_{\mathrm{avg}} = 0.06 \times P_{\mathrm{pto}}$$

where Q_{avg} = average gasoline consumption, gal/hr; P_{pto} = maximum PTO power, hp.

These equations must be multiplied by 0.73 for diesel and by 1.20 for liquefied petroleum gas (LPG) tractors.

Problems: Estimate the annual fuel consumption for a 125 PTO Hp diesel tractor that will be used for 850 hr per year.

Solution: The first step is to use the equation to determine the hourly use.

$$\frac{\mathrm{gal}}{\mathrm{hr}} = 0.06 \times 125 \times 0.73 = 5.475 \text{ or } 5.5\,\frac{\mathrm{gal}}{\mathrm{hr}}$$

The second step is the multiply the hourly use by the hours per year:

$$\frac{\mathrm{gal}}{\mathrm{yr}} = 5.5\,\frac{\mathrm{gal}}{\mathrm{hr}} \times 850\,\frac{\mathrm{hr}}{\mathrm{yr}} = 4{,}675 \text{ or } 4{,}700\,\frac{\mathrm{gal}}{\mathrm{yr}}$$

Estimating the fuel consumption for a specific operation requires determining the total tractor power for that operation. ASABE standard D497.4 provides a method for estimating fuel consumption for a specific operation.

3. Oil (O): The oil cost in dollars per hour will be the product of the oil consumption (gal/hr or L/hr) during operation and oil changes, and the oil price ($/qt or $/L). The total oil used (gal/hr) on a per hour basis can be estimated assuming manufacturer's recommended change interval by:

Gasoline	$(0.00011 \times P) + 0.00657$
Diesel	$(0.00021 \times P) + 0.00573$
LPG	$(0.00008 \times P) + 0.00755$

where P = rated power, hp.

The total oil used (L/hr) on a per hour bases can be estimated by:

Gasoline	$(0.000566 \times P) + 0.02487$
Diesel	$(0.00059 \times P) + 0.02169$
LPG	$(0.00041 \times P) + 0.02$

where P = rated power, hp.

Problem: Determine the amount of oil that will be used by a 120 PTO horsepower diesel tractor.

Solution:

$$\frac{\text{gal}}{\text{hr}} = (0.00021 \times 120) + 0.00573 = 0.03093 \text{ or } 0.031\ \frac{\text{gal}}{\text{hr}}$$

4. Labor (L): The hourly wage for labor to operate the equipment in dollars per hour.

5. Consumables (C): Some machines, such as balers, have twine, netting or other materials that are consumed as a part of the machine operation.

Thus, the total hourly operating cost (THOC), in dollars per hour, of a machine is the sum of items 1, 2, 3, 4 and 5 above.

Problem: Determine the total costs for a $35,000.00 grain drill that is used 120 hr per year. It has an expected life of 10 years and a salvage value of $1,000.00. It is pulled by a 98 horsepower diesel tractor. Assume an interest rate of 8%.

Solution: The total annual cost is the sum of the annual fixed costs and the annual variable costs. The fixed costs can be determined by calculating each individual cost, or by using the ownership cost equation to determine the ownership cost percentage. For this example the ownership cost equation will be used.

Fixed costs: the ownership cost equation requires salvage value expressed as a percentage.

Salvage value factor:

$$\text{Salvage factor }\% = \frac{\$1{,}000}{\$35{,}000} \times 100 = 2.857\ldots \text{ or } 2.8\%$$

Total fixed cost percent:

$$\begin{aligned}
C_o &= 100 \times \left(\left(\frac{1 - S_v}{L}\right) + \left(\frac{1 + S_v}{2} \times I\right) + K_2\right) \\
&= 100 \times \left(\left(\frac{1 - 0.028}{10}\right) + \left(\frac{1 + 0.028}{2} \times 0.08\right) + 0.02\right) \\
&= 100 \times (0.0972 + 0.04112 + 0.02) \\
&= 100 \times 0.15832 = 15.832 \text{ or } 16\%
\end{aligned}$$

Total fixed costs:

$$\frac{\$}{\text{yr}} = \$35{,}000.00 \times \frac{0.16}{\text{yr}} = 5{,}600.00\ \frac{\$}{\text{yr}}$$

Variable costs: Annual variable costs=repair and maintenance+Fuel+Oil+Labor

Repair and maintenance:

$$\begin{aligned}
C_{\text{rm}} &= \text{RF1} \times P \times \left(\frac{h}{1000}\right)^{\text{RF2}} \\
&= 0.32 \times \$35{,}000.00 \times \left(\frac{120}{1000}\right)^{2.1} \\
&= 0.32 \times \$35{,}000.00 \times 0.0116\ldots \\
&= 130.466\ldots \text{ or } 130.00\ \frac{\$}{\text{yr}}
\end{aligned}$$

Fuel: The fuel cost for the grain drill is determined by the power and fuel efficiency of the tractor. Estimating tractor fuel costs for specific operation is not included in this text. See ASABE D496 for this information. Average fuel consumption based on the fuel type and power of the tractor can be estimated using the information from ASABE EP496.2 FEB03.

The fuel use (gal/hr) is:

$$\begin{aligned}
Q_{\text{avg}} &= 0.06 \times P_{\text{pto}} \times 0.73 \\
&= 0.06 \times 98 \times 0.73 \\
&= 4.2924 \text{ or } 4.3\ \frac{\text{gal}}{\text{hr}}
\end{aligned}$$

The fuel cost is: (assuming fuel is 2.25 \$/gal)

$$\begin{aligned}
\frac{\$}{\text{yr}} &= \frac{4.3 \text{ gal}}{\text{hr}} \times \frac{2.25\ \$}{\text{gal}} \times \frac{120 \text{ hr}}{\text{yr}} \\
&= 1161 \text{ or } 1,200.00\ \frac{\$}{\text{yr}}
\end{aligned}$$

Oil costs:

$$\begin{aligned}
\frac{\text{gal}}{\text{hr}} &= (0.00021 \times P) + 0.00573 \\
&= (0.00021 \times 98) + 0.00573 \\
&= 0.02631\ \frac{\text{gal}}{\text{hr}}
\end{aligned}$$

Assuming an oil cost of 2.00 \$/qt:

$$\frac{\$}{yr} = \frac{2.00\ \$}{qt} \times \frac{4\ qt}{gal} \times \frac{0.02631\ gal}{hr} \times \frac{120\ hr}{yr}$$

$$= 25.25\ldots \text{ or } 25.00\ \frac{\$}{yr}$$

Labor: Labor costs are determined by multiplying the annual use by the tractor operator cost in \$/hr.

$$\frac{\$}{yr} = \frac{120\ hr}{yr} \times \frac{\$10.00}{hr} = 1{,}200.00\ \frac{\$}{yr}$$

The total cost of owning and using the grain drill 120 hr per year is:

$$\text{Total annual cost} = \text{Annual fixed costs} + \text{Annual variable costs}$$

$$= 5{,}600.00\ \frac{\$}{yr} + \left(130.00\ \frac{\$}{yr} + 1{,}200.00\ \frac{\$}{yr} + 25.00\ \frac{\$}{yr} + 1{,}200.00\ \frac{\$}{yr}\right)$$

$$= 5{,}600.00\ \frac{\$}{yr} + 2,555.00\ \frac{\$}{yr}$$

$$= 8{,}155 \text{ or } 8,200\ \frac{\$}{yr}$$

Machinery costs in dollars per year are useful for the owner/operator of a machine, but for custom work the operator needs to know the costs in terms of dollars/acre, dollars/hour, dollars/bale etc. The cost of a machine on a unit volume basis is influenced by the annual use because as the annual use increases, the fixed costs are spread out over more units. This makes the fixed costs per unit less. To illustrate this point Figure 10.1 shows the costs per acre for the grain drill in the previous problem for an annual use ranging from 100 to 1200 hr per year.

Note that the dollars per hour costs decrease until about 700 hr and then they start back up. This is because the equation used by ASABE to estimate repair and maintenance costs is designed to increase repair costs as the machine uses increases.

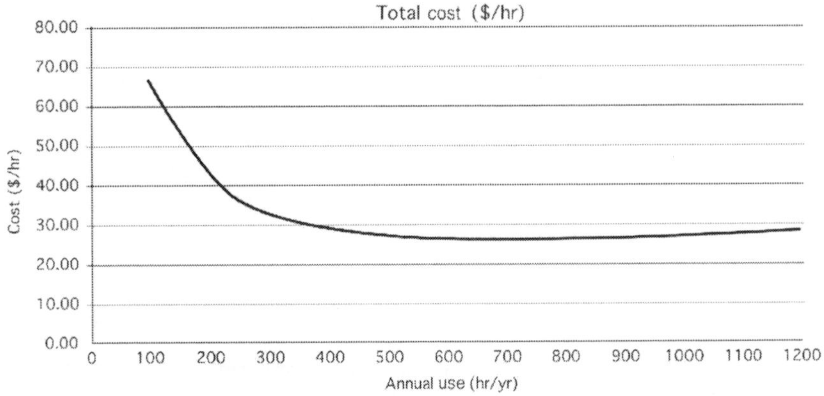

FIGURE 10.1. Total costs per hour of use.

10.6. Ways to Reduce Costs

A feature of agriculture is that the buyer usually fixes the price the producer receives for their products; therefore the best way to increase profits is to reduce production costs. No single factor will keep ownership and operating costs of tractors and machinery to a minimum, but good management of the following factors will reduce the production costs of machinery.

10.6.1. Width Utilization

The effective width is always less than the width of the machine, except for row crops, because most operators overlap each round slightly to prevent skips. Any excessive overlap reduces the effective capacity of the machine and increases the cost per acre.

10.6.2. Time Utilization

Any time that a machine is not performing its designed task, when it could be, means lost productivity. Several practices can be used to reduce lost time. First, adjusting and lubricating of the machine should be kept to a minimum consistent with the need for service. Even better, you should do as much of the maintenance as possible when conditions prevent the machine from being in the field.

Second, breakdowns can be minimized by avoiding overloads and by a complete preventive maintenance program. Any machine is more likely to break down or to need additional maintenance if it is overloaded or misused.

Third, a proper field layout will keep turning at the end of the row or round to a minimum. There is an optimum pattern for every field shape and size of machine. Machinery management texts provide more information on this topic.

Another idea is to reduce the time required to refill seed, fertilizer, and chemicals by using the largest practical size of hopper or tank. In addition, the equipment used to load the tank or the hopper should have the capacity to fill it in a reasonable time.

10.6.3. Matching Tractors and Machines

Tractors furnish power most economically when operated at or near the rated load. If the load is too small, it may be increased by widening the machine or by using two machines in tandem. Increasing the speed also increases the effective tractor load. If all of the tractor power is not being utilized and it is not feasible to increase the speed, fuel costs can be reduced by shifting up into a higher gear and reducing the throttle. (Check the owner's manual before attempting this.)

10.6.4. Reducing Original Investment

The initial cost of a machine is a major factor in determining the fixed cost of owning and operating farm machinery. The initial cost can be reduced by buying used equipment or by building equipment and machinery in the farm shop. Both of these strategies substitute human labor for some of the initial cost.

10.6.5. Increasing Annual Use

As we discovered in the section on calculating costs, the more a machine is used, the less the cost per unit will be. Usage can be increased by joint ownership, by increasing the size of the enterprise, by doing custom work, or by increasing the working season by diversifying production.

10.6.6. Increasing Service Life

Ensuring that the machine lasts as long as possible is one way of reducing the need for the initial cost of a new machine. The service life can be extended through proper maintenance, careful adjustments, avoidance of overloads, and using skilled operators. The useful life also is extended through proper storage during the off-season.

10.7. Break-Even Use

In the previous section, we introduced the notion that the total cost per acre or hour decreases as annual use increases. This is the basis of BEU. For some operations it may be more profitable to consider hiring contract workers instead of owning and operating one's own machine. The decision to hire may hinge on the break-even usage, that is, the amount of use for which the costs of owning a machine are the same as the costs of hiring a custom operator.

Suppose for a given machine that the product of FC% and P equals \$1,000. Also, suppose that the operating costs are \$ 4.00 per acre. The TAC equation would be:

$$\text{TAC} = \$1{,}000.00 + (\$4.00 \times A)$$

where A = acres of use. Next calculate the values of TAC for acre (A) values of 10, 50, 100, 200, and 500 acres. The result is Table 10.2.

Then if we wish to determine the cost of using the machine on a per-acre basis, we divide the TAC by the acres for each situation, and get the values shown in Table 10.3.

A comparison of the various levels of use shows that the cost per acre decreases as use increases. The break-even analysis in Figure 10.2 shows that for a custom

TABLE 10.2. Break-even use.

Acres	Total annual cost
10	\$1040
50	1200
100	1400
200	1800
500	3000

TABLE 10.3. Cost per acre.

Acres	TAC	Cost per acre
10	$1040	$104.00
50	1200	24.00
100	1400	14.00
200	1800	9.00
500	3000	6.00

rate of $21.00 per acre the cost for an annual use of about 60 acres per year it is more economical to hire work. At annual use of over 60 acres per year it is more economical to own the machine and do the work yourself.

This same method can be used to determine the beak-even point for costs with units different than dollars per acre. For the process to result in good information the custom rate and the total costs must be in the same units.

The BEU can be determined by the following equation:

$$\mathrm{BEU} = \frac{\mathrm{AOC}}{\mathrm{CR} - \mathrm{OPC}}$$

where BEU = Break-even use (unit of use); AOC = Annual ownership costs ($/yr); CR = Custom rate ($/unit of use); OPC = Operating costs ($/unit of use).

Although this analysis indicates a hard-and-fast decision for a particular amount of use, the ownership—custom hiring decision should be tempered by other factors. Table 10.4 lists some advantages and disadvantages of custom hiring of farm equipment. It is most important to use consistent units with each of the terms. If

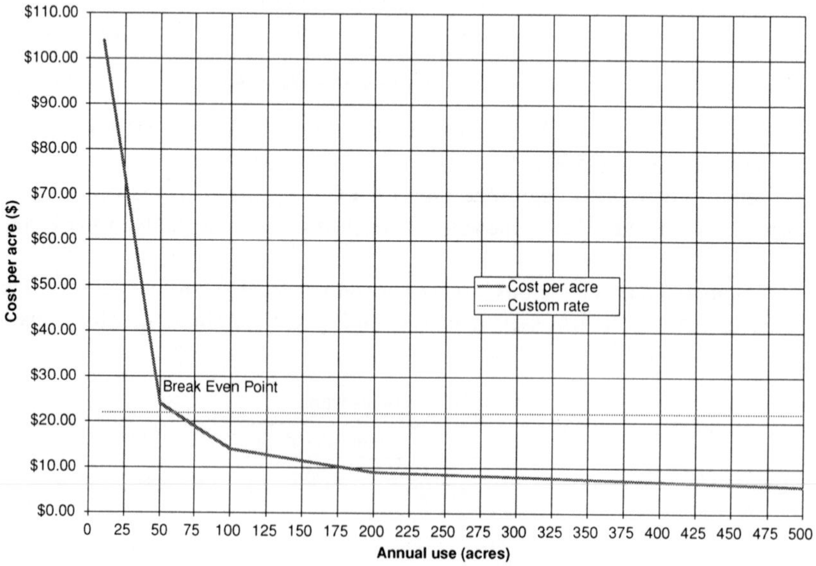

FIGURE 10.2. Break-even analysis for $22.00 per acre custom rate.

TABLE 10.4. Advantages and disadvantages of custom hiring farm equipment.

Advantages	Disadvantages
1. No ownership costs.	1. May not be possible to schedule when needed.
2. Cost of equipment can be invested in other enterprises.	2. Less control of quality of work.
3. Hired equipment usually supplies labor.	3. Increased potential for losses because of delays.
4. Less equipment is needed by owner, particularly specialized.	4. Increased risk of spreading weed seeds and diseases.
5. Owner can take advantage of newest machinery and techniques.	5. Costs for large jobs may be higher than owning machines.
6. Producer with small jobs can gain benefits of large machines.	6. Small jobs have a greater chance of being postponed.
7. Custom operator is responsible for repairs, maintenance, and materials.	7. Producer may not be able to utilize own labor freed up by custom hiring.

the custom rate is in dollars per acre, the operating cost must be in dollars per acre. In this case the units of the answer will be in acres per year.

Problem: What is the BEU in acres for a \$20,000 machine if the FC% is 22%, the operating costs are estimated to be \$9.00 per hour, and the custom rate is \$9.60 per acre?

Solution:

$$\text{BEU (ac)} = \frac{\text{AOC}}{\text{CR} - \text{OPC}} = \frac{\$20{,}000 \times 0.22}{9.60\,\dfrac{\$}{\text{ac}} - 9.00\,\dfrac{\$}{\text{hr}}}$$

The calculation is not completed because the units are not correct. The custom rate and the operating costs must have the same units to be subtracted. When the desired break-even point is in acres, both the custom rate and the operating costs must have the units of dollars per acre. The dollars per hour units must be converted to dollars per acre.

$$\frac{\$}{\text{ac}} = \frac{\$}{\text{hr}} \times \frac{\text{hr}}{\text{ac}}$$

Hours per acre is more commonly expressed as acres per hour, the capacity of a machine. Assume the effective capacity of the machine is 2.50 acres per hour.

$$\text{BEU (ac)} = \frac{\text{AOC}}{\text{CR} - \text{OPC}} = \frac{\$20{,}000 \times 0.22}{9.60\,\dfrac{\$}{\text{ac}} - 9.00\,\dfrac{\$}{\text{hr}}}$$

$$= \frac{\$20{,}000 \times 0.22}{9.60\,\dfrac{\$}{\text{ac}} - \left(9.00\,\dfrac{\$}{\text{hr}} \times \dfrac{1\ \text{hr}}{2.50\ \text{ac}}\right)} = 733.33\ldots \text{ or } 730\ \text{ac}$$

10.8. Maintenance Schedules

Maintenance is the care given to a machine to ensure that it operates correctly and that it receives the required lubrication and adjustments. All machines require maintenance. Failure to provide adequate maintenance can shorten the life of the machine and/or increase its operating costs. Manufacturers specify the amount of resources that a machine will need to maintain its operation. The owner/operator's responsibility is to ensure that the maintenance is accomplished on schedule, based on the manufacturer's recommendations in the owner's manual. This schedule must be followed for the machine to reach its designed potential in performance and longevity.

10.9. Metric Problems

The process used for estimating costs is the same for both the SI and US Customary systems. When estimating the costs for the metric system substitute the appropriate costs. Appendix V provides the parameters needed to calculate draft using the ASABE equation. An example follows for the SI system.

Problem: A chisel plow equipped with 5 cm wide shovels is used for primary tillage in a wheat field. Depth of tillage is 15 cm. The chisel plow width is 13.7 m. The chisel shank spacing is 0.3 m. Plowing speed is 10 km/hr. The soil is a silt loam (medium texture). What is the drawbar power required to pull the chisel plow?

From Appendix V

$$F_2 = 0.7$$
$$C1 = 91$$
$$C2 = 5.4$$
$$C3 = 0.$$

Compute the number of tools (shanks)

$$w_{\text{Tool}} = \frac{w_{\text{mach}}}{w_{\text{tool spacing}}} = \frac{13.7\text{ m}}{0.3\text{ m}} = 45.666 \text{ or } 45 \text{ tools}$$

Calculate the draft in Newtons.

$$\begin{aligned} D_{\text{F}} &= F_{\text{i}} \times \left[C_1 + (C_2 \times V) + \left(C_3 \times V^2\right)\right] \times w \times T_{\text{d}} \\ &= 0.7 \times \left[91 + (5.4 \times 10) + \left(0 \times 10^2\right)\right] \times 45 \times 15 \\ &= 68512.5 \text{ or } 680{,}00\text{ N} \end{aligned}$$

Calculate the drawbar power. Recall that draft is a force.

$$P_{\text{DbkW}} = \frac{F \times V}{60{,}000}$$

$$= \frac{680{,}00 \text{ N} \times 10 \dfrac{\text{km}}{\text{h}}}{60{,}000}$$

$$= 113.333 \text{ or } 113 \text{ kW}$$

The cultivator will require a tractor with at least 113 kW.

11 Sound and Noise

11.1. Objectives

1. Understand the nature of sound and the basis of sound measurement, the decibel (dB).
2. Be able to compare different environmental sounds.
3. Understand how humans are affected by noise.
4. Become familiar with noise exposure standards and protection from excessive noise.

11.2. Introduction

Humans live in a world of sound. Many sounds are quite pleasant and persons who are not hearing impaired enjoy hearing voices, music, and many sounds of nature. People often listen for sounds that could warn of danger or the malfunctioning of equipment. Individuals have differing abilities to detect sound of varying intensities and frequencies, and differ as well in their personal tolerance of and appreciation of sound.

Sound that is harsh and unpleasant (or unhealthy) commonly is referred to as noise. In recent years, the subject of noise has been given considerable attention, especially as it affects human health and behavior. The federal government, through OSHA (Occupational Safety and Health Administration), has established noise exposure limits to prevent worker hearing loss and/or psychological stress due to excessive exposure to noise. OSHA does not have enough inspectors to insure all work environments meet standards. It is up to individuals to understand the nature of sound and noise so they know when unhealthy environments may exist due to undesirable or unwanted sound.

11.3. What Sound Is

The sound that people hear is due to vibrations in air or substances that are transmitted to the hearing organs. What are perceived as sounds are sonic pressure waves that travel through the air (or different substances) and interact with the eardrum by entering the ear canal or by passing through the body. Thus the eardrum responds to "sound pressure." There is a very, very large change in sound pressure (perhaps as much as 10,000,000 times) as sounds vary from the "threshold of hearing" (the intensity of sound just barely detected by an average human ear) to the sound pressure created by a large jet engine operating nearby. People can "feel" sound when they touch a vibrating body and the vibrations pass through the body to the eardrum.

The sense of hearing also responds to sound frequency (or pitch), that is, the number of vibrations or cycles per second—Hertz (Hz).The range of sound frequencies that can be heard varies from about 20 to 20,000 Hz, depending on the individual. As the intensity of sound is increased to an appropriately high level for a given frequency, the hearing sensation becomes painful, and the "threshold of pain" is reached.

Human hearing also can distinguish between sounds that differ in quality, those combinations of frequencies and intensities that produce squealing, grating, grinding, or rasping sounds. When the frequencies and intensities are combined in suitable proportions, pleasant musical or vocal sounds result. Thus, the sounds that are heard can be quite complex, but their effects on humans are well established.

In this chapter, our concern is with that part of sound called noise and how it may affect workers and their work. Generally, excessive noise can lead to hearing impairment, fatigue, annoyance, and interference with performance. Noise also can serve as a warning of equipment malfunction or a signal of needed maintenance. We frequently rely on sounds (or no sound) to tell us that equipment is performing satisfactorily.

11.4. How Sound Is Measured

The intensity of sound is measured in units of decibels (dB). The unit decibel is a ratio of any two components of sound, power, sound pressure, intensity, etc. The human ear has the capability of distinguishing a wide range of sound intensity and frequencies. The range of sound pressure that an individual might be able to distinguish from the threshold of hearing to causing damage to the ears is over a million. Because the decibel scale is a ratio, to measure a single source of sound a base line called a filter is used. Three standard filters are used. They are identified as A, B, and C. Filter A is less sensitive to very high and very low frequencies. The C filter is commonly used for high frequencies. The B filter falls in between the A and C. It is seldom used. When using a sound meter it is important to identify which filter is used. This is commonly done by using the units of dBA or dBC after the reading.

To simplify the measurement of sound, decibel readings are based on an exponential scale of sound pressure levels. Remember that the decibel is not an absolute measure of the sound pressure, but rather is a ratio of a measured sound pressure to a reference sound pressure. Also, because this sound scale is exponential, a sound of 10 dB has an intensity 10 times greater than a sound of 0 dB, and a sound of 20 dB is 100 times more intense than 0 dB. Thus, a 10 dB difference in sound pressure level changes the intensity by 10 times, a 20 dB difference changes the intensity by 100 times, a 30 dB difference changes the intensity by 1,000 times, and so on.

The basic instrument used to measure continuous sound is the sound-level meter. It consists of a microphone to pick up the sound, an amplifier, one or more frequency weighting networks, and a meter to display the sound level. Sound-level meters usually are self-contained (they operate on internal batteries) and are small enough to be hand-held although tripod mounting generally is preferred.

Proper sound-level meter use involves careful placement of the microphone to avoid sound reflections that influence the readings. In measuring the noise exposure of an individual (his or her work environment), the microphone is placed as close as possible to the subject's ear (or where the ear would be in the work environment). Prior to use, the instrument must be calibrated against a known stable sound source.

11.5. Comparing Different Sounds

Table 11.1 shows some common sounds and the approximate sound pressure level associated with them.

TABLE 11.1. Decibel rating of common sounds.

Sound pressure level (dB)		Sound description
188		Apollo lift-off, close
150		Jet engine, 10 ft away
140	Pain threshold	
130		Warning siren
125		Chain saw
120	Discomfort threshold	Loud thunder
115	Max under federal law	
110		Very loud music
105		Loud motorcycle or lawn mower
100	Very loud	Pneumatic air-hammer
90		Cockpit of light plane, heavy truck
85		Average street traffic
80		Lathe, milling machine, loud singing
75		Vacuum cleaner, dishwasher
70		Average radio, noisy restaurant
65	Annoying	
60		Normal conversation, air conditioner
50		Light traffic, average office
40		Library, quiet office
30		Quiet room in home, audible whisper
20		Electric clock, faint whisper
10	Barely detectable	Rustle of leaves
0	Hearing threshold	

This chart may help you identify and classify sounds that you normally encounter; and if they are disturbing or otherwise unhealthy, you can avoid them or properly protect yourself from any of their adverse effects. Note that the values shown in the table are average values. Also, notice that (almost) identical sounds may come from different machines and situations. Further, individuals may differ in their tolerance of and classification of different sounds and noises. In specific situations and when determining if noise levels are acceptable under current standards, the equations in section 11.7 must be used.

11.6. The Effect of Noise

As suggested previously, noise can have both psychological and physiological effects on people. Psychologically adverse noise mainly affects a worker's performance and state of well-being. Direct exposure to excessive noise may cause fatigue, distraction, annoyance, interference with communication, reduction in the memory function, and disturbance of rest and relaxation. Some or all of these effects may be involved in decreased performance in the workplace.

The main physiological effect of adverse noise is noise-induced hearing loss, which is irreversible damage to one or more parts of the hearing organs. However, high sound levels also can induce responses in other parts of the body, such as reduced blood circulation, change on the skin's resistance to electric current and a corresponding activation of the nervous system, increased muscle tension, changes in the breathing pattern, and disturbance of sleep. These non-hearing-related noise responses are considered reversible and soon disappear when the noise source is removed.

Not all adverse noise is encountered in the workplace. We are surrounded by such sound generators such as outdoor and indoor equipment and appliances (lawn mowers, chain saws, sink disposals, blenders, clothes washers, and the like) and recreation and hobby equipment (gas-powered model planes and cars, unmuffled racing car and boat engines, firearms and explosive devices, high power stereo amplifiers and speaker systems, trail motorcycles, snowmobiles, and pleasure aircraft, among other things). Many of these items generate unusually intense sounds, but when we are preoccupied with the utility and the joy of using them, we are seldom aware of the noise they create or their potential for damaging our hearing.

11.7. Determining Noise Exposure

We must become more aware or our "sound" environment, determine acceptable exposure limits, and take any steps necessary to avoid overexposure and potentially adverse psychological and physiological effects. OSHA standards require businesses compute the employee noise exposure. Noise exposure can be measured with an audiodosimeter or with a decibel meter. The following section will explain the recommended method using a decibel meter. This is accomplished using two rules.

1. When the sound level, L,[1] is constant over the entire work shift, the noise dose, D, in percent, is given by:

$$D = 100 \times \frac{C}{T}$$

where D = Dose level (%); C = the total length of the work day (hr); T = the reference duration level corresponding to the measured sound level (L). (L can be determined by table or by equation).

2. When the work shift noise exposure is composed of two or more periods of noise at different levels, the total noise dose over the workday is given by:

$$D = 100 \times \frac{C_1}{T_1} + \frac{C_2}{T_2} + \cdots + \frac{C_n}{T_n}$$

where $C(n)$ = the total time of exposure at a specific noise level, and $T(n)$ = the reference duration for that level as given by Table 11.2 or equation.

The first equation determines the dose (%) for the employee. A dose of 100% is the maximum allowed.

The reference duration level (T) is computed using the following equation:

$$T = \frac{8}{2^{(L-90/5)}}$$

where T = reference duration; L = dBA exposure.

Problem: Determine the sound dose for an individual who was subject to a sound of 55 dBA for 1.25 hr and a level of 105 dBA for 5.0 min.

Solution: The dose equation requires a different reference duration (T) for each sound level and duration. The first step is to determine the value of T for each sound level and duration.

$$T_1 = \frac{8}{2^{(L-90/5)}} = \frac{8}{2^{(55-90/5)}} = \frac{8}{2^{-7}} = \frac{8}{0.0078125} = 1{,}024 \text{ or } 1{,}000$$

and

$$T_2 = \frac{8}{2^{(L-90/5)}} = \frac{8}{2^{(105-90/5)}} = \frac{8}{2^3} = \frac{8}{8} = 1$$

the next step is to determine the value for each exposure using the dose equation:

$$D = 100 \times \left(\frac{C_1}{T_1} + \frac{C_2}{T_2}\right) = 100 \times \left(\frac{1.25}{1{,}000} + \frac{5.0 \text{ min} \times \frac{1 \text{ hr}}{60 \text{ min}}}{1}\right)$$

$$= 100 \times (0.00125 + 0.0833\ldots) = 8.458\ldots \text{ or } 8.4\%$$

The worker dose was 8%, well within the range of the maximum allowed of 100%.

---

[1] OSHA 1910.95 App A, Noise exposure computation.

11.8. Exposure Guidelines

The Occupational Safety and Health Act makes provisions for employers to make accurate sound measurements in their places of business, to determine whether the workplace is safe for workers; the sound level should not exceed 90 dB during an 8 hr work period. For sound levels above 90 dB(A), reductions in exposure time are required, with a limit of 15 min of exposure per day to a sound level of 115 dB. A further aspect of the Act is that no impulsive or impact noise should exceed 140 dB. Also, noise abatement in excessively noisy work areas is required. If a workplace is deemed excessively noisy, workers must be provided with safety equipment and be required to use it.

Besides the OSHA regulations and the regulations set forth under the Federal Noise Control Act (administered by EPA), several other federal agencies oversee noise control in related industries such as aviation, residential and commercial construction, all aspects of ground transportation, and mining. References to these regulations can be found in different volumes of the CFR (Code of Federal Regulations) that involve noise control.

11.9. Controlling Noise

If objectionable, excessive or unhealthy noise situations are identified in any environment, a course of action should be taken to eliminate or reduce the noise (through design changes or the use of sound-absorbing materials), or if that is not possible, to provide human protective equipment such as ear plugs, ear muffs, and specially designed helmets. Further, using Table 11.2 as a guide, potential exposure to excessive noise can be identified and avoided.

TABLE 11.2. Workplace maximum permissible noise exposure level versus daily exposure time.*

Duration (hr/day)	Sound level (dBA)
8	90
6	92
4	95
3	97
2	100
1.5	102
1	105
0.5	110
0.25	115
0.125	120
0.063	125
0.031	130

*CFR 1910.95(b)(2).

11.10. Topics for Discussion

1. In your daily activities, identify and list the places where you believe that sounds (noise) are excessive.
2. For places that you deem to have excessive noise, make an estimate of the time you would want to spend in each place.
3. For places that you deem to have excessive noise, describe any changes that you might make to reduce the noise level.
4. There will be some places that you consider to have excessive noise where you need to be, and where you cannot reduce the noise level in any significant way. Describe how you would protect yourself in such an environment.

11.11. Metric Problems

This information is from OSHA, a United States Agency, and a metric equivalent is not available. However, since measurements are in the form of ratios, the measurements are the same in SI units and countries adopting metric units will likely have similar standards.

12
Measuring Distance

12.1. Objectives

1. Understand the advantages and disadvantages of the six common methods of measuring distance.
2. Be able to use the six common methods of measuring distance.
3. Be able to calibrate distance measuring instruments.
4. Be able to calculate a correction factor for a systematic error.

12.2. Introduction

Measuring distance and angles with simple instruments are two of the most common surveying procedures used in agriculture. Even though the instruments may be simple, a sufficient level of accuracy for many measuring jobs can be achieved with practice and careful work. In this chapter, you will become acquainted with the basic methods and techniques of measuring distances. The layout and measurement of angles is covered in Chapter 13.

12.3. Measuring Distances

Several methods and devices can be used for measuring distance. The principal ones are:

1. pacing
2. odometer
3. taping or "chaining"
4. stadia
5. optical range finder
6. electronic distance measuring (EDM)

One of the decisions that must be made when preparing to measure a distance is selecting the method or device to use. The following sections will discuss each of these devices and help the user make that decision.

First, it will be helpful to review the common units of distance (displacement). The common English units are as follows:

1 foot (ft) = 12 inches (in)
1 yard (yd) = 3 feet (ft)
1 rod = 16.5 feet (ft) = 5.5 yards (yd)
1 mile (mi) = 5280 feet (ft) = 1760 yards (yd) = 320 rods

In the SI system, the units are:

1 meter (m) = 1,000 millimeters (mm)
1 kilometer (km) = 1,000 meters (m)

The units cancellation method is very useful for converting from one unit of measure to another.

12.3.1. Pacing

Pacing is the simplest and easiest method for measuring distance. Pacing is the process of walking the distance and counting the number of steps "paces" to cover the distance. The distance is determined by multiplying the number of steps taken between two points by one's pace factor. A person's pace factor is determined by pacing (walking) a measured distance, usually 300 to 500 ft, several times and determining the average length of pace (step). With practice it is possible to pace a distance with an error of less than 2 ft per 100 ft. To achieve that level of precision a person must learn to adopt a pacing step that is different from their normal walking step because many factors can cause variations in the length of a person's pace. Three of these are the roughness of the surface, the slope of the ground, and the type of vegetation. Care must be taken to ensure that a consistent pace factor is used.

One advantage of pacing for measuring distance is that it doesn't require any specialized equipment. The biggest disadvantage is that it requires being able to walk the route. It is not very useful for measuring distance in rough terrain, across swamps or any other terrain where an individual could not walk the distance or walk in a straight line.

Problem: An individual paces a 200.0 ft distance three times and counts 62, 60, and 64 paces. What is the person's pace factor (PF)?

Solution: To determine a pace factor, divide the measured distance by the average number of paces. The average number of steps ($\bar{x}$) is:

$$\bar{x} = \frac{62 + 60 + 64}{3} = 62$$

and the pace factor is:

$$\text{PF} = \frac{200 \text{ ft}}{62 \text{ paces}} = 3.225\ldots \text{ or } 3.2 \frac{\text{ft}}{\text{pace}}$$

Once the pace factor is known, the length of an unknown distance can be determined

FIGURE 12.1. Odometer wheel.

by counting the number of paces required to cover the distance and multiplying the number of paces by the pace factor.

Problem: An individual with a pace factor of 3.2 ft/pace, measures a distance at 375 paces, what is the distance in feet?

Solution: The distance, in feet, equals the number of paces times the pace factor. Or:

$$\text{ft} = 3.2\ \frac{\text{ft}}{\text{pace}} \times 375\ \text{paces} = 1,200\ \text{ft}$$

12.3.2. Odometer

An odometer is a mechanical device that records revolutions. An odometer wheel has an odometer and a drive train that records the distance traveled, Figure 12.1.

The common odometer wheel counts the revolutions of a wheel and shows the distance traveled on a multiple dial readout, Figure 12.2. The dial can be designed to produce readings in whole feet, decimal feet, feet and inches and meters. The principle of an odometer wheel is circumference of a wheel times the number of

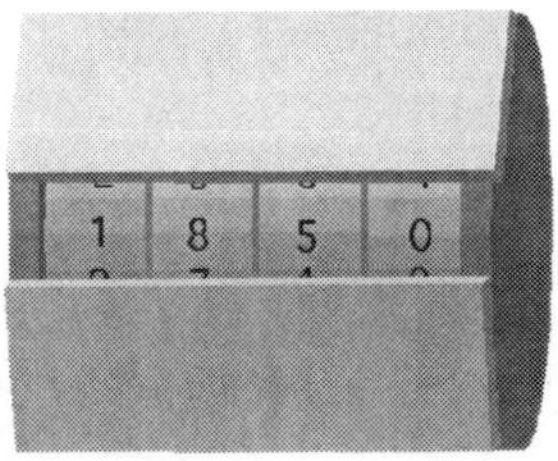

FIGURE 12.2. Odometer wheel reading of 1,850 ft.

revolutions equals distance, therefore any wheel can be used as long as the diameter is known and the number of revolutions are counted.

When used correctly, a wheel is more accurate than pacing. An error of 1 ft per 100 ft can be expected. To be accurate, the wheel must roll along the ground without slipping. Experience has shown that the accuracy of odometer wheels is also influenced by the surface and the topography. An odometer wheel designed to be used on hard surfaces will not be as accurate on grass or in tall weeds. In these situations, the accuracy is improved if the wheel is recalibrated for the surface on which it will be used. See the section 12.4 & 12.5 for an example of the calibration procedure.

When an odometer is used, the readout should be reset to zero before starting; then the distance traveled can be read directly. If another type of wheel is being used, the distance traveled is equal to pi (π) times the diameter multiplied by the number of revolutions:

$$D\ (\text{ft}) = \pi \times D \times N$$

where D = Measured distance; $\pi = 3.14$; D = Wheel diameter (ft); N = Number of wheel revolutions.

Problem: A bicycle wheel completed 155.2 revolutions as it was rolled along the boundary of a field. The wheel diameter is 2.16 ft. What is the length of the field (ft)?

Solution:

$$D(\text{ft}) = \pi \times D \times N = 3.14 \times 2.16\ \frac{\text{ft}}{\text{rev}} \times 155.2\ \text{rev}$$

$$= 1052.628\ldots \text{ or } 1{,}050\ \text{ft}$$

One advantage of odometer wheels over pacing is the increased accuracy, but they still must be rolled along the surface in a straight line to measure the distance.

12.3.3. Taping

The most accurate traditional method of measuring distances uses a steel tape. When proper procedures are followed the error will be less than 1.0 ft in 3,000 ft. The standard equipment for a taping party, usually at least three people, consists of a tape, two range poles, a set of 11 chaining pins, two plumb bobs, a hand level, and a field notebook. Individual items of taping equipment are described in the following paragraphs.

12.3.3.1. Surveyor's Chain

The traditional device for measuring distance accurately is the surveyor's steel chain. Today accurate surveyor's chains constructed of non metalic materials are available. The modern standard steel chain is 100 ft long and 3/8 inch wide, and weighs 2 to 3 lb per 100 ft. The older tapes are graduated with marks set in

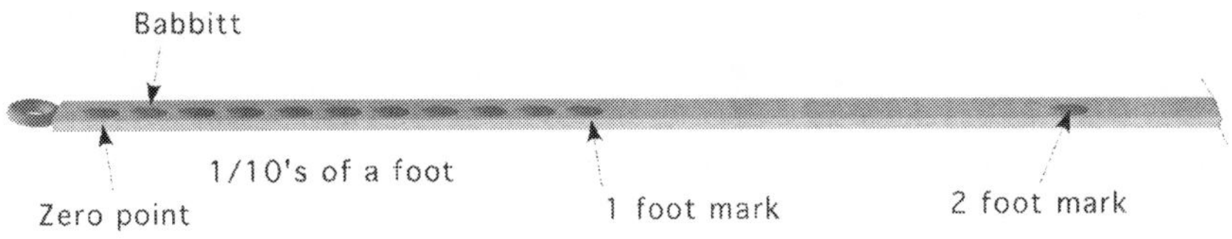

FIGURE 12.3. Segment of Babbitt style tape.

Babbitt-metal bosses with each foot marked, from 0 to 100 ft. Notice that the tape between the first and second foot mark does not have any graduations. This is the common design of surveyor's chains.

Newer surveyor's chains have the marks etched on the face of the tape. Highway chains are fully graduated, but the traditional surveyor's chain only subdivides one foot of the tape. The subdivisions are usually in tenths or hundredths of a foot. The graduated foot may be the first foot, as in Figure 12.3, or in an extended foot, Figure 12.4. A chain with the first foot graduated is called a cut chain and a chain with the extend foot is called an add chain.

Determining the style of chain that is being used is important otherwise it is easy to have an error of one foot when reading an add or cut tape incorrectly, Figure 12.5.

Metal surveyor's chains are manufactured from an alloy of steel. It is very important to wipe them down with a clean rag after each use and lightly oil them periodically to prevent them from rusting. They are also very easy to break by bending them sharply or pulling a loop tight.

12.3.3.2. Tapes

Tapes are popular for making rough measurements or when the accuracy of a chain is not needed. Tapes are constructed of metal and different nonmetallic materials. Tapes are usually fully graduated and range in length from 25 to 200 ft.

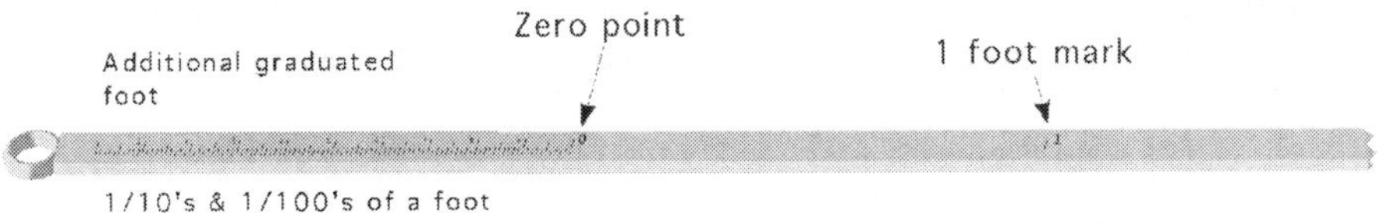

FIGURE 12.4. Typical add tape.

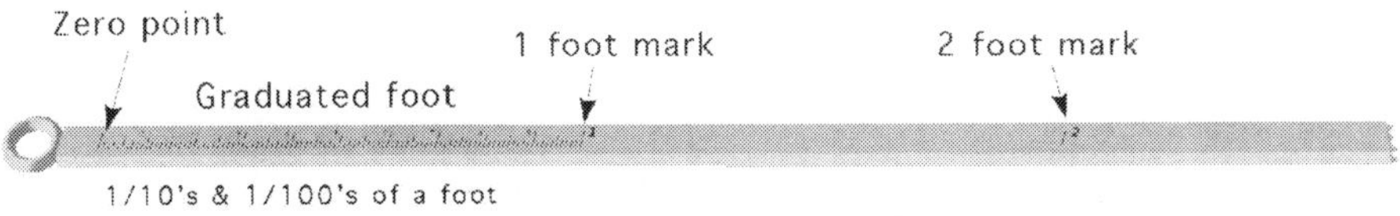

FIGURE 12.5. Segment of cut tape.

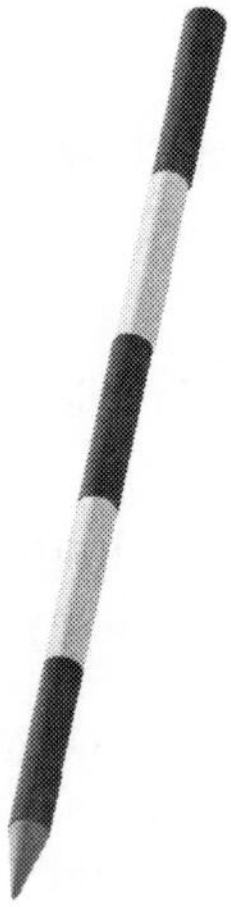

FIGURE 12.6. Range pole.

Nonmetallic tapes are usually lighter and easier to use. They require less maintenance and are not as easily damaged by moisture. Inexpensive nonmetallic tapes will stretch under tension and therefore are not as accurate as steel chains or tapes.

12.3.3.3. Chaining Pins

Chaining pins are made of heavy gauge wire and are 12 to 15 inches long, are painted red and white, and sometimes have a bright cloth attached to help locate them in tall grass. They are used to mark the end of each tape length and come with 11 pins in a set; One pin is used to mark the start of the chaining (taping), and the remaining 10 pins are used to mark lengths of tape. When all 11 pins have been used, assuming a 100 ft tape, 1,000 ft has been measured.

12.3.3.4. Range Poles

Range poles are 1-inch-diameter tubular steel or nonmetallic shafts 6 to 10 ft long with one pointed end. They are alternately painted red and white and are used for "lining in" when one is taping or measuring angles, Figure 12.6.

12.3.3.5. Plumb Bobs

Plumb bobs with 6 to 10 ft of cord attached are used when measuring horizontal distances on sloping or irregular ground to transfer the distance from the horizontally held tape to the point on the ground. They also are attached to a surveyor's level, to locate the level over a stake, when one is measuring distances by stadia or EDM, or laying out angles, Figure 12.7.

12.3.3.6. Hand Level

A hand level consists of a small sighting tube 5 to 6 inches long equipped with a spirit level, a glass tube filled with a liquid and a bubble. The image of the bubble

FIGURE 12.7. Plumb bob.

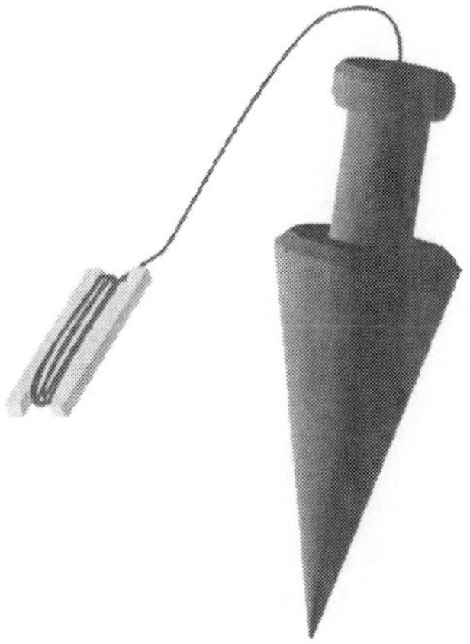

is reflected by a prism and can be observed by looking through the tube. The instrument is held to the operator's eye and is leveled by raising or lowering the end until the cross-hair intersects the image of the spirit-level bubble. This is a low-precision instrument used to make rough measurements of the slope and as an aid in keeping the surveyor's tape level. Newer models may also include stadia cross-hairs and direct-reading angle scales, Figure 12.8.

12.3.3.7. Taping Procedures

There are six basic steps involved in taping: (1) lining in, (2) applying tension, (3) plumbing, (4) marking tape lengths, (5) reading the tape, and (6) recording the distance.

Pre computer and pre calculator taping methods were measured by stations. A distance of 100 ft is called a full station and is written as 1 + 00 ft. A distance of 123 ft is written as 1 + 23 ft. The modern practice is to record all distances in decimal feet.

For many surveys the intent is to discover the true horizontal distance between two points. Three methods are commonly used: (1) tape and plumb bob, (2) tape and calculation, or (3) Electronic distance measuring (EDM). When method two is used, either the percent slope, change in elevation or vertical angle must be measured so that the horizontal distance can be calculated. When an EDM is used

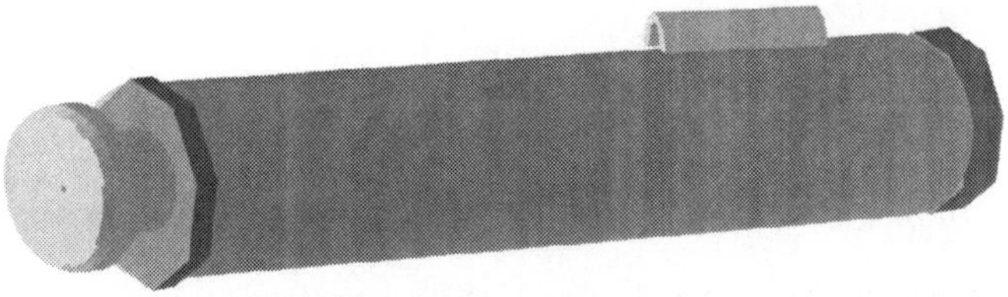

FIGURE 12.8. Hand level.

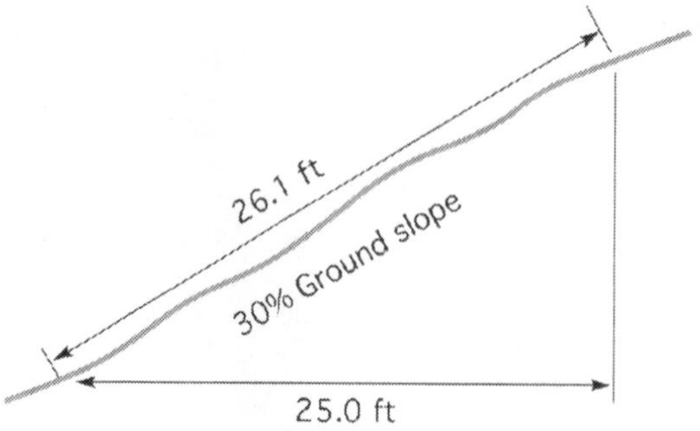

FIGURE 12.9. Effect of ground slope on true horizontal distance.

the distance is horizontal if the instrument is horizontal at the time of measurement.

The need for horizontal measurements is illustrated in Figure 12.9. When measured along the surface of the ground, the distance would be 26.1 ft, whereas the true horizontal distance is 25.00 ft.

When slope distance is measured and the desired result is horizontal distance, either the percent slope, difference in elevation or the vertical angle must be measured and the true horizontal distance calculated or determined from tables. The percent slope can be measured by a hand level or a surveying level. When the percent slope is known the horizontal distance can be determined through the use of the slope equation and the Pythagorean theorem.

Problem: Determine the horizontal distance when the slope distance is 234.5 ft and the percent slope is 3.4%.

Solution: The first step is to determine the amount of change in elevation that occurs. This can be done using the slope equation:

$$\% \text{ Slope} = \frac{\text{rise}}{\text{run}} \times 100$$

$$\text{rise} = \frac{\%\text{Slope} \times \text{run}}{100} = \frac{3.4 \times 234.5 \text{ ft}}{100} = 7.973 \text{ ft}$$

Where the rise is the change in elevation and the run is the horizontal distance, the horizontal distance can be determine by rearranging Pythagorean's theorem:

$$a^2 = b^2 + c^2$$

$$b^2 = a^2 - c^2$$

$$b = \sqrt{a^2 - c^2} = \sqrt{234.5^2 - 7.973^2} = 234.36\ldots \text{ or } 234.4 \text{ ft}$$

Table 12.1 can also be used to determine the horizontal distance for various slopes up to 30%. Note that for slopes up to 5% the correction factor is less than 0.125

TABLE 12.1. Correction factors for converting slope distance to horizontal distance.

Slope	Correction factor (ft/100 ft)	True horizontal distance (100 ft)
1	0.005	99.995
2	0.020	99.980
3	0.045	99.955
4	0.080	99.920
5	0.125	99.875
6	0.180	99.820
7	0.245	99.755
8	0.321	99.679
9	0.406	99.594
10	0.501	99.499
11	0.607	99.393
12	0.723	99.277
15	1.131	98.869
18	1.633	98.367
20	2.020	97.980
25	3.175	96.825
30	4.606	95.394

ft/100 ft (1.5 in/100 ft), but the correction factor increases more dramatically after 5%.

When Table 12.1 is used, the correction factor per 100 ft is multiplied by the slope distance and divided by 100, and the product is subtracted from the slope distance. It is important to remember that the correction factors are for 100 ft of distance. You must divide the measured slope distance by 100 first:

$$\text{HD} = \text{SD} - \left(\left(\frac{\text{SD}}{100}\right) \times \text{CF}\right)$$

where HD = Horizontal distance; SD = Slope distance; CF = Correction factor.

Problem: Use Table 12.1 to determine the horizontal distance (HD) when the slope distance (SD) is 623.82 ft and the slope is 12.0%.

Solution:

$$\begin{aligned}\text{HD} &= \text{SD} - \left(\left(\frac{\text{SD}}{100}\right) \times \text{CF}\right)\\ &= 623.82\text{ ft} - \left(\left(\frac{623.82\text{ ft}}{100}\right) \times 0.723\right)\\ &= 623.82\text{ ft} - 4.5102186 = 619.309\ldots\text{ or } 619\text{ ft}\end{aligned}$$

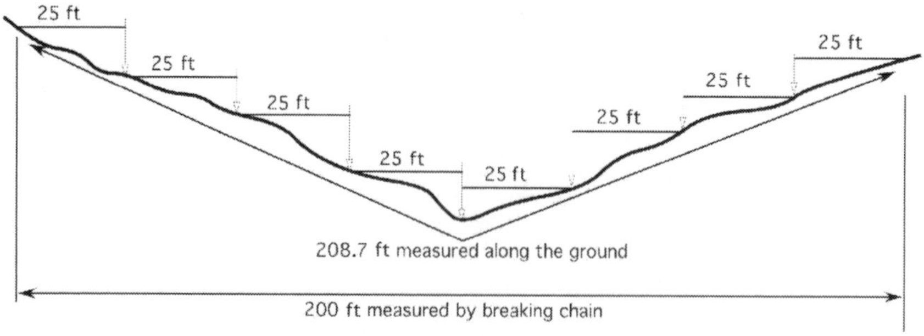

FIGURE 12.10. Measuring sloping ground by breaking chain.

When the ground slope is not uniform, the slope and the slope distance for each segment of the line must be determined and a separate correction applied to each segment.

When the horizontal tape method is used, one end of the tape is held on the ground, while the other end is raised until the tape is horizontal. The true distance is transferred to the ground from the elevated end of the tape by a plumb bob. When the slope is more that 5%, it is necessary to use a process known as "breaking chain." In this method the head chainman lays out the full length of the tape. The 100-ft length then is divided into convenient increments, usually 25 or 50 ft, with the chainman holding the tape horizontal and plumbing down to the ground at each increment. This process is illustrated in Figure 12.10. In this example, the chain was "broken" into 25-ft sections. Every 25 ft a plumb bob and line was used to set a pin.

For accurate results, a taping activity must be very carefully thought out and well organized. The following procedure for taping is recommended to ensure accurate results. A taping party consists of at least three people: the head chainman, the rear chainman, and a note keeper. An axe man also may be necessary in brushy areas. The 11 chaining pins serve as temporary markers for each station, and also help to count the number of full stations measured.

The head chainman begins by setting a pin for the starting point and then leads off with the 100 foot end of the tape. After one full station has been measured, the head chainman will have placed one pin in the ground at the beginning and one to mark the end of the first station. At this point he or she will have 9 pins left on the ring. As the chain is moved to the next station, the rear chainman does not pull the pin used to start the chaining. He only pulls the pins used at the 100 ft marks. After two full stations, the head chainman will have placed three pins in the ground and have 8 pins on the ring, and the rear chainman will have one in hand and one in the ground. If this system is carefully followed, the number of full stations measured will always be the same as the number of pins held by the rear chainman.

As described above, most surveyor's tapes are graduated in feet throughout their full length, with the first and/or last foot of the tape graduated in tenths or tenths and hundredths of a foot. A different procedure is required for measuring a distance shorter than a full tape length. When the party nears the end of the line and the remaining distance is less than 100 ft, the rear chainman holds the zero mark on the last pin and the head chainman pulls the chain taunt. The head chainman then moves the chain forward to the last foot mark if a add chain is used or backward to the next foot mark if a cut chain is being used. They both read the tape and the rear chainman's reading is subtracted from the head chainman's reading when a cut tape is used. The rear chainman's reading is added to the head chainman's reading when an add tape is used. For example, suppose that an add chain is being used and the head chainman reads 53 ft, and the rear chainman reads 0.21 ft. If the rear chainman has 6 pins, the total length of the line is 653.21 ft [(6 × 100) + (53 ft + 0.21 ft)]. If a cut chain was being used the distance would be 652.79 ft [(6 × 100 ft) + (53 ft − 0.21 ft)].

The following rules, if carefully followed, will help guarantee accuracy in taping.

1. Align the tape carefully, and keep the tape on the line being measured.
2. Keep a uniform tension of 15 lb of pull on the tape for each measurement.
3. Keep in mind the style of tape being used to avoid an error of 1 or 2 ft at each end of the tape.
4. "Break chain" on slopes as necessary to keep the tape level, or calculate the percent slope if measuring with the tape on the ground.
5. Carefully mark each station and keep an accurate count of the stations.

12.3.4. Stadia

Measuring distance by stadia relies upon a fixed angle being designed into the instrument. "Stadia" comes from an early Greek word for a unit of length. Surveying instruments equipped for stadia measurement have two additional horizontal cross-hairs, called stadia hairs. They are placed equidistant above and below the horizontal leveling cross-hair, Figure 12.11.

The distance between the stadia hairs and the horizontal cross-hair is fixed by the manufacturer to provide a constant stadia interval factor (SIF) for the instrument. The most common stadia interval factor is 100. Instruments that have a SIF of 100 will have a one foot stadia interval, that is the difference between the top stadia reading (TSR) and the bottom stadia reading (BSR), when the rod is held 100 ft away from the instrument, Figure 12.12.

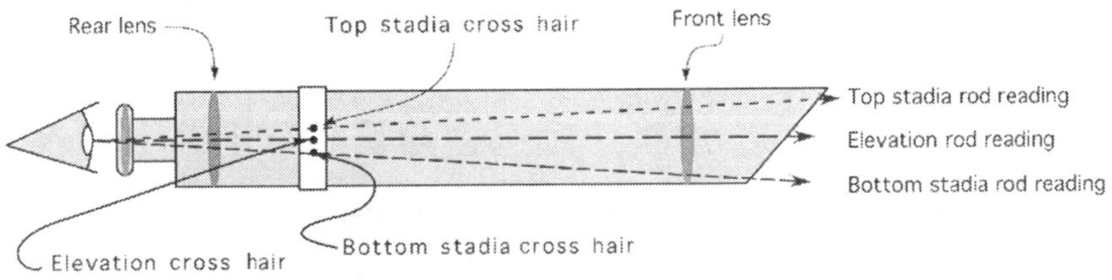

FIGURE 12.11. Principles of stadia.

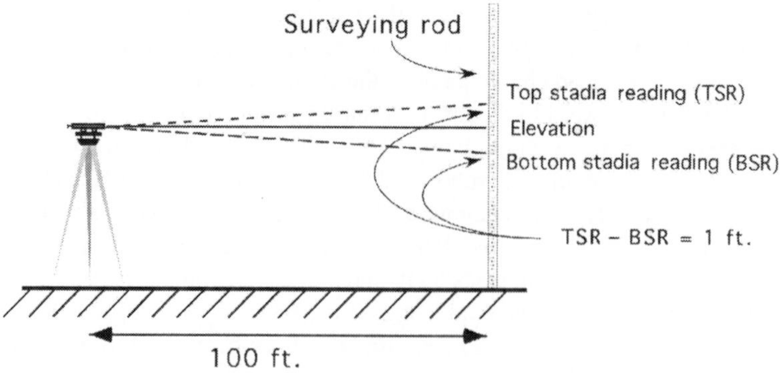

FIGURE 12.12. Measuring distance by stadia.

The angle formed by reading the top and bottom cross-hairs is constant, therefore the unit of the measure is determined by the unit of measure of the rod. When a Philadelphia style rod is used without the target the distance will be in units of feet. When a metric rod is used the distance will be in units of meters. When the stadia method is used to measure a distance, the instrument person reads the TSR and the BSR, and then multiplies the difference by the stadia interval factor. In the form of an equation:

$$D\ (\text{ft}) = (\text{TSR} - \text{BSR}) \times 100$$

Refer to Figure 12.13. The top stadia hair reading (TSR) is 6.29, and the bottom stadia hair reading (BSR) is 3.71. Thus the stadia interval is equal to 6.29 minus 3.71, or 2.58 ft. The distance from the instrument to the rod is equal to 2.58 multiplied by 100, or 258 ft.

Problem: What is the distance to the rod if the stadia readings are TSR = 6.07 and the BSR = 3.02?

Solution:

$$D\ (\text{ft}) = (\text{TSR} - \text{BSR}) \times 100 = (6.07\ \text{ft} - 3.02\ \text{ft}) \times 100 = 305\ \text{ft}$$

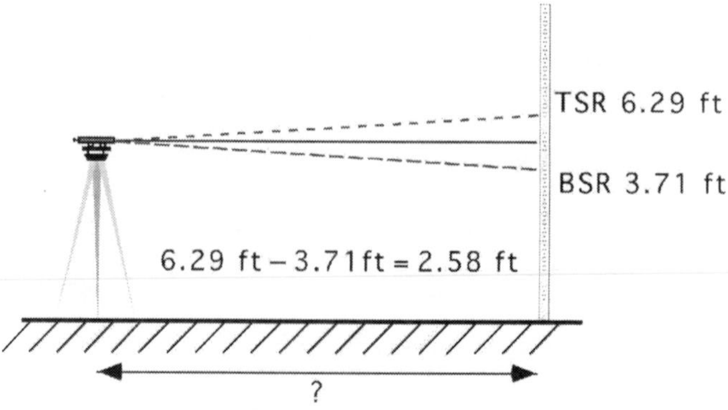

FIGURE 12.13. Example using stadia.

12.3.5. Optical Range Finders

Many different technologies are used with optical range finders. One simple type of instrument operates on the function of triangles. When the operator looks through the eyepiece, when the instrument is not set for the distance, they will see two identical images. The operator adjusts the focus until two images are in coincidence, superimposed on top of each other. The distance is read from a scale on the instrument. Optical range finders have an acceptable level of accuracy for reconnaissance or sketching purposes for short distances, but the error increases as the distance increases.

12.3.6. Electronic Distance Measurement (EDM)

EDM instruments determine lengths based on the time that it takes for an energy wave to travel from one end of a line to the other and return. They have replaced the difficult and painstaking task of taping for accurate measurement of horizontal distances. They also can measure distance over bodies of water or other inaccessible terrain as long as they have a clear line of sight. Some require a reflective surface to bounce the beam back, some are reflector less. Others use a second unit to return the signal or a reflector, Figure 12.14.

EDM's use an energy beam to send a signal to a reflector or a reflecting object and measure the time elapse from when the beam was sent and when it returns to the instrument. Distance is determined by rearranging the velocity equation.

$$V\left(\frac{\text{ft}}{\text{min}}\right) = \frac{\text{Distance (ft)}}{\text{Time (min)}}$$

$$\text{Distance} = V\left(\frac{\text{ft}}{\text{min}}\right) \times \text{Time (min)}$$

A big advantage of EDM's is that the operator does not need to walk the path, all they need is a line of sight. They can also be used when the line of sight is obscured by fog or light rain.

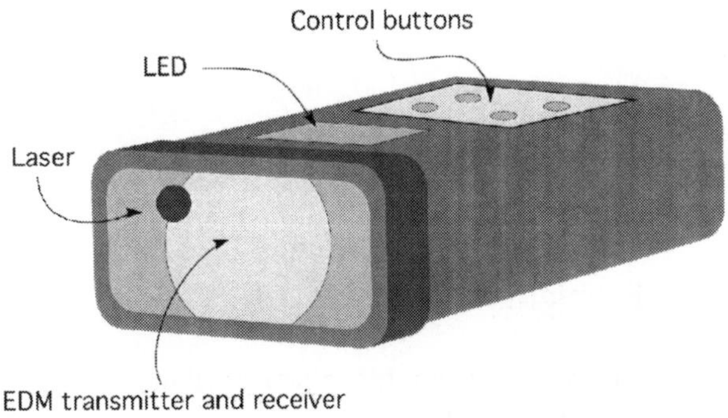

FIGURE 12.14. An example of an EDM.

12.4. Calibration Procedure

Measuring instruments are mechanical/electrical devices. Their accuracy will sometimes change with use, and is guaranteed to change with abuse. Checking the accuracy of instruments is called calibration. Instruments are calibrated by checking their performance against a standard. The measured difference in the length or angle between the averaged measurements and the standard, is error. This error can be a random error or a systematic error. Random error is unpredictable and must be controlled by following recommended techniques and procedures. Systematic errors are usually caused by damage or wear to the instrument. If the calibration process indicates a systematic error, an error that is constant with each measurement, then a correction factor can be calculated so that each measurement taken with the instrument can be corrected.

12.4.1. Determining Correction Factor

During the measurement of a known 50.0-ft distance with an odometer wheel measurements of 49.9, 49.8, and 49.7 ft were recorded. What is the correction factor?

Solution: The first step is to determine the average distance.

$$(49.8 + 49.8 + 49.7)/3 = 49.8$$

$$\begin{aligned}\text{CF}\left(\frac{\text{ft}}{\text{ft}}\right) &= \frac{\text{Known distance} - \text{Measurement}}{\text{Known distance}}\\ &= \frac{50.0\text{ ft} - 49.8\text{ ft}}{50\text{ ft}} = 0.004\ \frac{\text{ft of error}}{\text{ft of distance}}\end{aligned}$$

This instrument will have 0.004 ft of error for every foot of distance it measures. Once the correction factor is determined it can be used to adjust each of the measurements for the error in the equipment.

Problem: Determine the corrected distance for a measured distance of 150.5 ft when the instrument has an error of 0.004 ft/ft.

Solution:

$$\begin{aligned}D_c &= 150.5\text{ ft} + \left(150.5\text{ ft} \times 0.004\ \frac{\text{ft}}{\text{ft}}\right)\\ &= 150.5\text{ ft} + 0.602 = 151.102 = 151.1\text{ ft}\end{aligned}$$

The corrected distance is 151.1 ft.

12.5. Metric Problems

Problem: Determine the pace factor for an individual that counted 132, 134, and 133 paces for a distance of 100 meters.

Solution:

$$PF = \frac{\text{Distance}}{\text{Paces}} = \frac{100 \text{ m}}{\dfrac{133 + 134 + 132}{3}} = \frac{100 \text{ m}}{133 \text{ paces}} = 0.7518 \text{ or } 0.75 \; \frac{\text{m}}{\text{pace}}$$

Problem: An individual with a pace factor of 0.65 m/pace counts 380 paces for the unknown distance. What is the distance?

Solution:

$$\begin{aligned} D &= \text{PF} \times \text{P} \\ &= 0.65 \; \frac{\text{m}}{\text{pace}} \times 380 \text{ paces} = 247 \text{ m} \end{aligned}$$

Problem: A wheel with a diameter of 70 cm is used to measure a distance by rolling it along the ground. It was rolled 124 revolutions. What is the distance?

$$\begin{aligned} D &= \frac{\pi \times \text{dia}}{\text{revolution}} \times \text{revolutions} \\ &= \frac{\pi \times 70 \text{ cm}}{1 \text{ rev}} \times 124 \text{ rev} \times \frac{1 \text{ m}}{100 \text{ cm}} = 272.69 \ldots \text{or } 273 \text{ m} \end{aligned}$$

Problem: Use Table 12.2 to determine the horizontal distance (HD) when the slope distance (SD) is 350 m and the slope is 12.0%.

TABLE 12.2. Correction factors for converting slope distance to horizontal distance.

Slope	Correction factor (m/100 m)	True horizontal distance (100 m)
1	0.01	99.99
2	0.02	99.98
3	0.05	99.95
4	0.08	99.92
5	0.13	99.87
6	0.18	99.82
7	0.25	99.75
8	0.32	99.68
9	0.41	99.59
10	0.50	99.50
11	0.61	99.39
12	0.72	99.28
13	1.13	99.15
14	1.63	99.02
15	1.13	98.87
18	1.63	98.37
25	3.18	96.82
30	4.61	95.39

Solution:

$$HD = SD - \left(\left(\frac{SD}{100}\right) \times CF\right)$$

$$= 350\text{ m} - \left(\left(\frac{350\text{ m}}{100}\right) \times 0.72\right) = 350\text{ m} - 2.52 = 347.48 \text{ or } 47.5\text{ m}$$

Problem: Determine the horizontal distance when the stadia method is used with a metric rod. The top stadia reading is 1.23 m and the bottom stadia reading is 0.82 m.

Solution:

$$D\,(\text{m}) = (\text{TSR} - \text{BSR}) \times 100$$

$$= (1.23 - 0.82) \times 100 = 0.41 \times 100 = 41\text{ m}$$

Problem: Determine the corrected distance for a measured distance of 750.5 m when the instrument measured distances of 49.8, 49.8, 49.7, and 49.8 meters for a standard distance of 50 m.

Solution: The first step is to determine the correction factor. Before determining the correction factor the average measurement must be calculated.

$$\text{Distance} = \frac{49.8 + 49.8 + 49.7 + 49.8}{4} = 49.775\text{ m}$$

$$\text{CF}\left(\frac{\text{m}}{\text{m}}\right) = \frac{\text{Known distance} - \text{Measurement}}{\text{Known distance}}$$

$$= \frac{50.0\text{ m} - 49.775\text{ m}}{50\text{ m}} = 0.0045\,\frac{\text{m of error}}{\text{m of distance}}$$

The last step is to use the correction factor to determine the corrected distance.

$$\text{Distance} = \text{Measurement} + \text{Correction}$$

$$= 750.5\text{ m} + \left(750.0\text{ m} \times 0.0045\,\frac{\text{m}}{\text{m}}\right)$$

$$= 750.5\text{ m} + 3.375\text{ m} = 753.875 \text{ or } 735.9\text{ m}$$

13
Angles and Areas

13.1. Objectives

1. Be able to use three indirect methods to lay out and/or measure angles.
2. Be able to find the area of standard geometric shapes.
3. Be able to determine the area of irregular-shaped tracts of land by division into standard geometric shapes.
4. Be able to determine the area of irregular-shaped tracts of land using two trapezoidal equations.

13.2. Introduction

The layout and measurement of angles are very important parts of agricultural surveying. In this section, you will learn, using simple tools and procedures, three indirect methods that can be used to lay out a perpendicular (90°) angle and one method that can be used to lay out or measure any angle between 0° and 90°. These procedures are useful in planning buildings and fences and in determining the corner angles of irregular-shaped tracts, fields, and/or smaller areas.

13.3. Angles

Surveyors label the three parts of an angle as the backsight (base line), the vertex and the foresight. Angles can be measured or laid out using instruments, direct method, or by measuring, indirect method. There are three common indirect methods for measuring or laying out an angle. Two of these are limited to 90° angles, the chord method and the 3-4-5 method, the third, tape sine, can be used with any angle between 0° and 90°. The following discussion illustrates all three methods.

13.3.1. Chord Method

A chord is a line connecting any two points on a circle. This geometric principle can be used to lay out a line at a 90° angle to a base line. This method is very

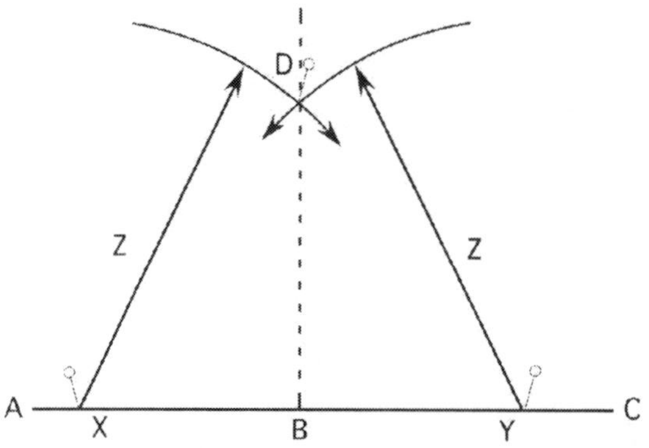

FIGURE 13.1. Laying out a right angle by the chord method.

simple and can be accomplished with two different lengths of string or even tree branches. The one disadvantage is that the base line must be extended past the turning point (B). This method is illustrated in Figure 13.1

The line BD is established at a 90° angle by completing the following steps:

1. Establish the base line AC if it is not in existence (fence or road edge, etc.).
2. Establish the vertex at B.
3. Set points X and Y equidistant from point B and on line AC.
4. Use a length greater than the distance XB or YB for Z, and scribe an arc from X and Y as shown.
5. Set a stake at the intersection of the two arcs (D).
6. The line established by this stake and B will be at 90° with the base line AC.

This method is simple in principle but not easy to complete because of the difficulty in marking the arcs. On tilled ground they can be formed by marking the surface, but when working on grass or taller vegetation it is much more difficult. An alternative is to use two tape measures. Attach one at X and the other at Y. Any point where both tapes have the same reading will be on a 90 degree angle from the base line.

13.3.2. 3-4-5 Method

The 3-4-5 method of laying out a right angle is based on the Pythagorean theorem: for any right triangle, the square of the hypotenuse is equal to the sum of the squares of the other two sides. In this method, any multiples of 3 and 4 used as the sides of a right triangle will result in the hypotenuse being a multiple of 5. To prove this, study the following equation:

$$\begin{aligned} 5 &= \sqrt{3^2 + 4^2} \\ &= \sqrt{25} \\ 5 &= 5 \end{aligned}$$

Shown graphically, Figure 13.2.

This method has three requirements: (1) the same units (feet, yards, etc.) must be used on all three sides; (2) the same multiples of 3, 4, and 5 are used for the

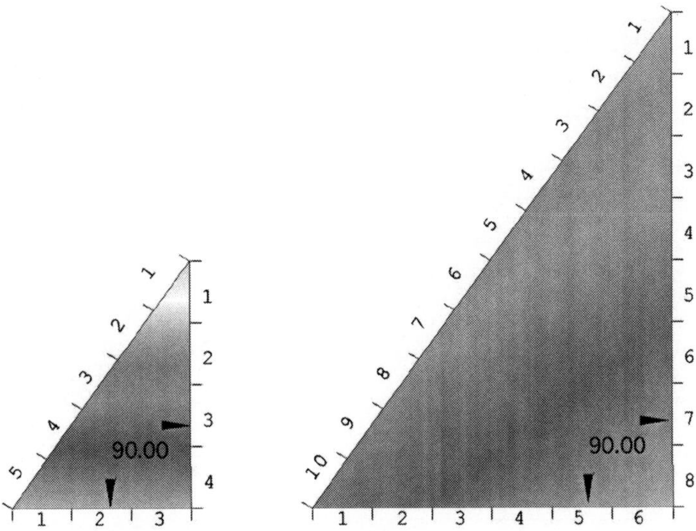

FIGURE 13.2. Principles of right triangles.

lengths of the three sides; (3) the longest length is used for the hypotenuse. When these three requirements are met, a 90° angle will be established.

When using the 3-4-5 method to lay out a right angle the easiest way is to use three people and two tape measures. When two tapes are used, either the 3 or the 4 dimension is used as the base line, and the two corners on the base line are marked. Then two people, one standing at each base line corner, hold two tape ends together at the correct dimensions, and a third person holds the remaining tape ends together at the correct dimensions and moves the third corner until the tapes are all at equal tension.

This process also may be accomplished by using a 100-ft tape. Because surveyor's steel tapes are not designed to be bent at a sharp angle, loops must be formed at two of the corners. It is recommended that at least a 5-ft loop be used. Study Figure 13.3.

By completing the following steps, a 90° angle may be laid out using the 3-4-5 method. This procedure will require three people.

1. Establish the base line (AB) and corner B.
2. Lay out the tape along the base line with the 20-ft mark (4 ft × 5 ft) at corner B and the zero mark at corner A.
3. Set corner A, and have a person hold the zero mark on the corner.
4. Form a 5-ft loop in the tape, and have a person hold the 20-ft mark over the 25-ft mark, and align these marks over corner B.
5. Lay out the remaining tape in the direction of corner C.
6. Note the position of the 40-ft mark (25 ft + 15 ft) (15 ft = 3 ft × 5), and form a 5-ft loop in the tape. Hold the 40 and 45 ft marks on corner C.

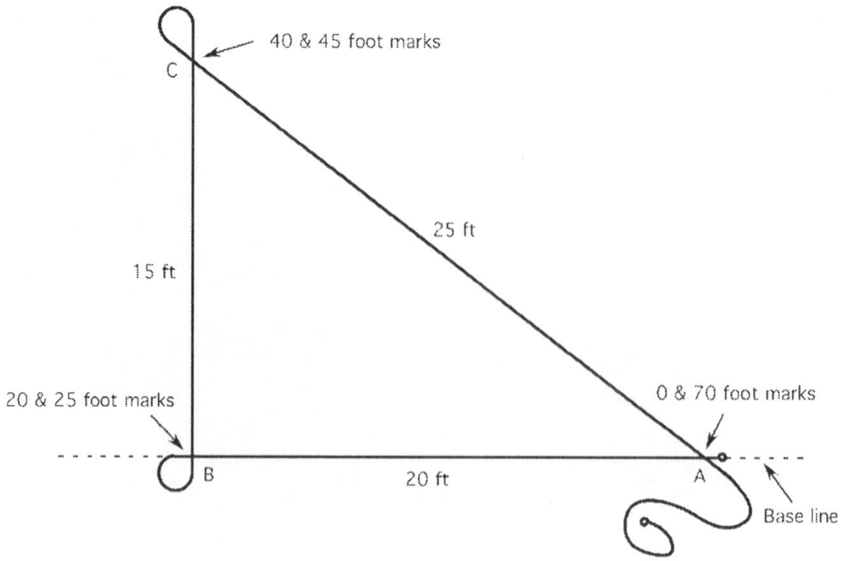

FIGURE 13.3. Laying out a right angle by the 3-4-5 method using a 100-ft tape.

7. Extend the tape back to corner A.
8. Hold the 70 ft (45 ft + 25 ft) (25 ft = 5 ft × 5) and 0 foot marks together at corner A.
9. If the individuals at A and B hold their positions carefully on the baseline while the individual at C tightens the tape in both directions, a 90° angle will be made at B.

This process will work for any combination of lengths as long as they are multiples of 3, 4, and 5. One advantage of this method is that the base line does not need to extend past the 90° corner.

13.3.3. Tape-Sine Method

The tape-sine method uses a combination of distances measured by a tape and the sine trigonometric function. This method is not limited to 90° and can also be used to lay out an angle or measure the angle between two existing lines. A review of the three commonly used trigonometric functions will make this method clear.

Trigonometric functions are based on the principle that one unique ratio exists between the lengths of any two sides for angles B and C, Figure 13.4. Because a right triangle has three sides, each angle (B and C) has six possible combinations of two sides. Each of these combinations has been given a name. The three common

FIGURE 13.4. Notation for right triangle.

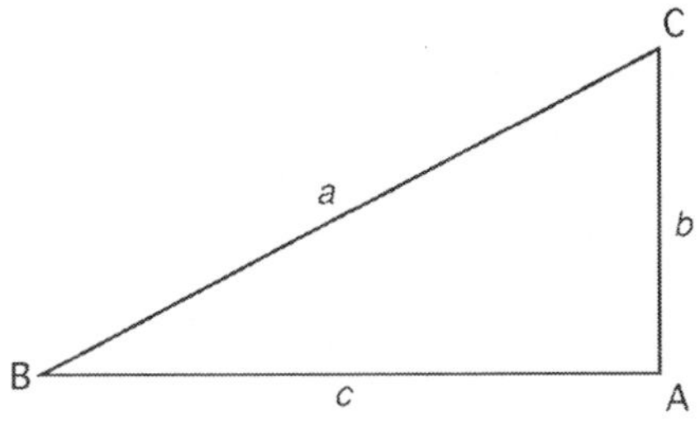

ratios are:

$$\text{Sine B} = \frac{\text{Length of opposite side } (b)}{\text{Length of hypotenuse } (a)}$$

$$\text{Cosine B} = \frac{\text{Length adjacent side } (c)}{\text{Length of hypotenuse } (a)}$$

$$\text{Tangent B} = \frac{\text{Length opposite side } (b)}{\text{Length adjacent side } (c)}$$

The same relationships are true for angle C. Each of these functions forms an equation with three variables—the function of the angle and the lengths of two sides. If any two of the variables are known, the third can be determined.

In the tape-sine method, only the sine function is used. The procedure for laying out an angle (ø) is slightly different from the procedure for measuring an existing angle.

The procedure for measuring an existing angle will be explained first. Using the example illustrated in Figure 13.5, the task is to determine the angle formed by BAC. The first step is to mark an equal distance along each side AB and AC. The

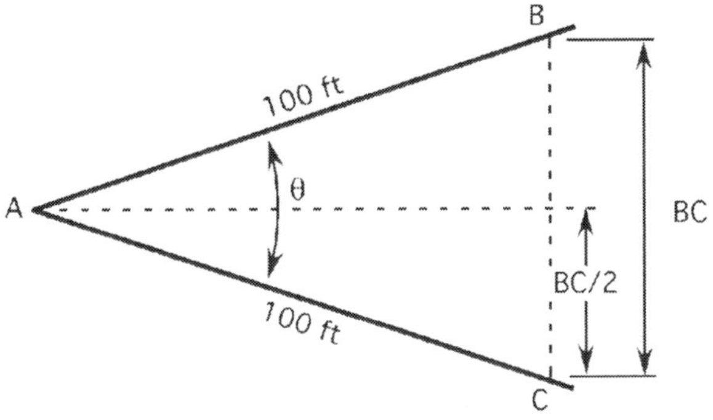

FIGURE 13.5. Example of tape-sine method.

next step is to measure the distance BC. The last step is to form two right triangles by drawing a line from corner A to the mid point of line BC.

The distances BC/2 and AC are two sides of a right triangle. For this example we will assume that the distance BC = 61.8 ft. The angle for either triangle can be found by using the Sine equation.

$$\text{Sine B} = \frac{\text{Length of opposite side } (b)}{\text{Length of hypotenuse } (a)}$$

$$= \frac{\text{Distance BC/2}}{\text{Distance AC}} = \frac{30.9\text{ ft}}{100\text{ ft}} = 0.309$$

The value 0.309 is the sine ratio of angle A. The next step is to determine the angle having a sine ratio of 0.309.

To determine the angle using a calculator, enter 0.309 and the inverse of the sine function. The angle can also be determined by consulting Table 13.1.

TABLE 13.1. Sine values.

Angle 0°	Sine of the angle	Angle 0°	Sine of the angle	Angle 0°	Sine of the angle
0	0.000	31	0.515	61	0.875
1	0.017	32	0.530	62	0.883
2	0.035	33	0.545	63	0.891
3	0.052	34	0.559	64	0.899
4	0.070	35	0.574	65	0.906
5	0.087	36	0.588	66	0.914
6	0.105	37	0.602	67	0.921
7	0.122	38	0.616	68	0.927
8	0.139	39	0.629	69	0.934
9	0.156	40	0.643	70	0.940
10	0.174	41	0.656	71	0.946
11	0.191	42	0.669	72	0.951
12	0.208	43	0.682	73	0.956
13	0.225	44	0.695	74	0.961
14	0.242	45	0.707	75	0.966
15	0.259	46	0.719	76	0.971
16	0.276	47	0.731	77	0.974
17	0.292	48	0.743	78	0.978
18	0.309	49	0.755	79	0.982
19	0.326	50	0.766	80	0.985
20	0.342	51	0.777	81	0.988
21	0.358	52	0.788	82	0.990
22	0.375	53	0.799	83	0.993
23	0.391	54	0.809	84	0.995
24	0.407	55	0.819	85	0.996
25	0.423	56	0.829	86	0.998
26	0.438	57	0.839	87	0.999
27	0.454	58	0.848	88	0.999
28	0.469	59	0.857	89	1.000
29	0.485	60	0.866	90	1.000
30	0.500				

FIGURE 13.6. Laying out an angle by the tape-sine method.

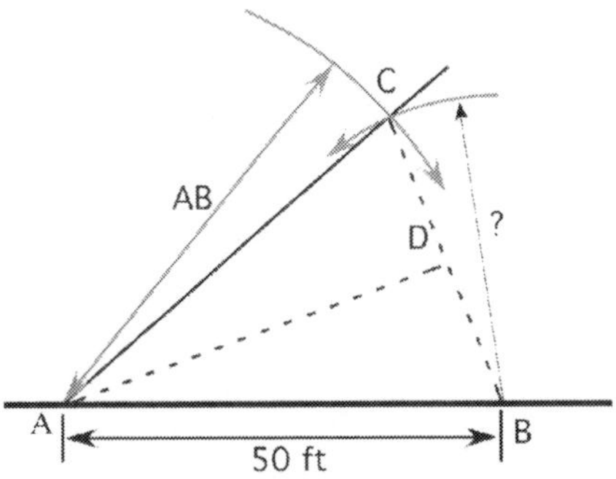

Either source should give an angle of 18°. Because we used the distance BC/2 to solve for the angle, the angle CAB is actually 18° × 2, or 36°.

The same procedures can be used to lay out an angle. For example, suppose you need to establish a fence at 40° to an existing fence. How would the tape-sine method be used? Study Figure 13.6.

The principles involved are basic trigonometry. It takes two points to establish a line and to locate a point either three dimensions or an angle and one dimension must be known. To establish the line AC so that it forms a 40° angle with the base line AB, three dimensions are used to locate point C. Two of the dimensions are the distance selected by the person laying out the angle. In this example 50 ft is used. The third dimension is the distance from point B to point C. To determine this dimension mentally split the triangle CAB into two right triangles (BAD and CAD), solve for the length of the opposite side of either, and then double this distance. The doubled distance is distance from B to C. The distance A to C is the same as A to B. Using point B and the length BS scribe a short arc in the vicinity of point C. Using point A and the length AB scribe another arc. The intersection of the two arcs is point C. A line from point A through C will establish the fence at the correct angle.

To determine the distance BC/2 the sine trig functions is rearranged to solve for the length of the opposite side:

$$\text{Sine A} = \frac{\text{Opposite}}{\text{Hypotenuse}}$$

$$\begin{aligned}\text{Opposite} &= \text{Sine A} \times \text{Hypotenuse}\\ &= \text{Sine } 20^\circ \times 50.0 \text{ ft}\\ &= 0.3420\ldots \times 50.0 \text{ ft}\\ &= 17.1010\ldots \text{ ft}\\ 17.1010\ldots \times 2 &= 34.2020\ldots \text{ or } 34.2 \text{ ft}\end{aligned}$$

To lay out the angle, mark an arc, with a radius of 34.2 ft, from point B and another arc, with a radius of 50, from point A, and set a pin at the intersection of

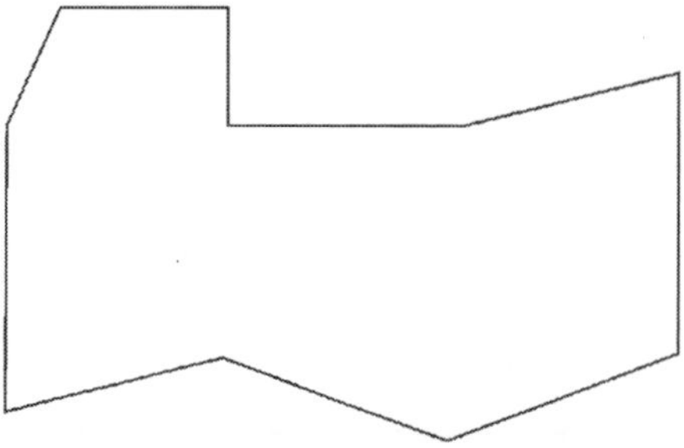

FIGURE 13.7. Irregular-shaped field.

the two arcs (point C). A line drawn from point A through point C will establish a line at a 40° angle to the base line (AB). In this method, the marking of two arcs could also be eliminated by using two tape measures.

13.4. Areas of Standard Geometric Shapes

One of the most common applications of surveying is to measure the area of a lot, field, farm, or ranch. When a high level of accuracy is required, a professional engineer or a land surveyor should be employed. If this level of accuracy is not required, many areas can be determined with nothing more elaborate than a steel tape and the application of area formulas from geometry. The area of an irregularly shaped field can be found by subdividing it into standard shapes, determining the areas of the subdivisions by calculation, and then summing the areas of the subdivisions to find the total area. For example consider Figure 13.7.

There is no right or wrong way to divide the field. The goal is to collect the required measurements with the expenditure of the least amount of resources. The best way will depend on the equipment used and the obstacles that exist in the field.

Figure 13.8 shows two different ways to divide the irregular-shaped field into standard shapes. Either method will determine the area. Method B will probably take less time and resources because it will require fewer measurements. To use this method, it is necessary to apply the correct area formula for each common geometric shape. The next section will review the common shapes and the area formula for each.

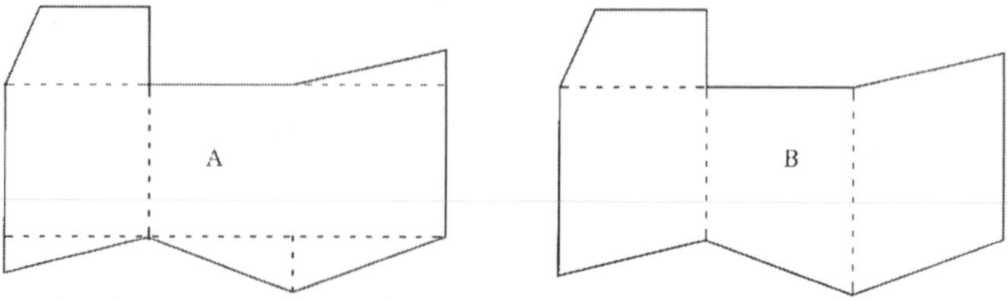

FIGURE 13.8. Two alternatives for dividing irregular-shaped field into standard shapes.

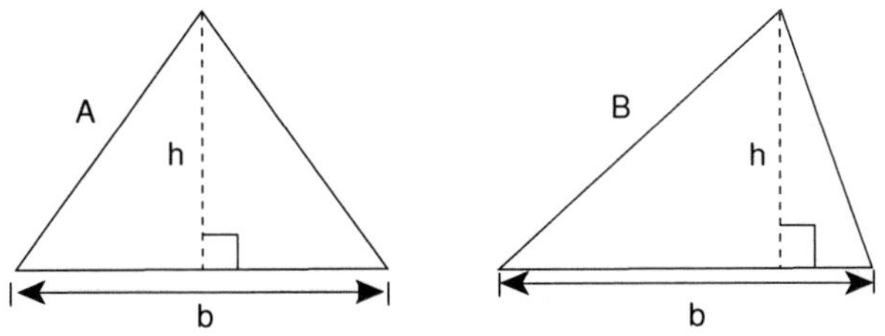

FIGURE 13.9. Triangle area with base and height known.

13.5. Triangle

Three different equations can be used to determine the area of a triangle, depending on the known dimensions of the triangle.

13.5.1. Base and Height

The first equation that can be used is when the base and height are known, see Figure 13.9.

This is the easiest equation to use when calculating the area of a triangle, but it is difficult to collect the required dimensions. The difficulty is determining the intersection of the height with the base line. If the triangle is equilateral (A), or an isosceles triangle, the problem is not as difficult because the intersection is one half of the distance of the base line. When the triangle is not an equilateral triangle, the intersection is much more difficult to determine. The equation is:

$$A = \frac{1}{2} \times b \times h$$

where A = Area; b = Length of base; h = Height perpendicular to base.

Problem: What is the area of a triangle (ac) with a base 150.0 ft long and a height of 100.0 ft.

Solution:

$$A = \frac{1}{2} \times \ 150.0 \text{ ft} \ \times \ 100.0 \text{ ft} = 7{,}500 \text{ ft}^2$$

$$\text{acre} = \frac{1 \text{ ac}}{43{,}560 \text{ ft}^2} \times \ 7{,}500 \text{ ft}^2 = 0.17217\ldots \text{ or } 0.1722 \text{ ac}$$

13.5.2. Three Sides

The second type of triangle is one where the lengths of the three sides are known, Figure 13.10. This method for determining the area of a triangle is easier to use because the only measurements are the three sides of the triangle. The shape of the triangle doesn't influence the difficulty in completing the measurements.

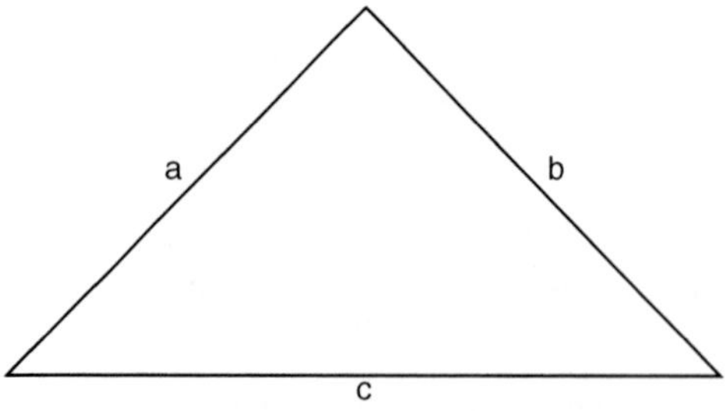

FIGURE 13.10. Triangle area with three sides known.

The primary limitation of this method is the requirement of measuring the length of all three sides. This method cannot be used if one or more of the side lengths cannot be measured. The equation for area is:

$$A = \sqrt{S\,(S - a)\,(S - b)\,(S - c)} \qquad S = \frac{a + b + c}{2}$$

where a, b, and c are the lengths of the three sides.

Problem: Determine the area (ft^2) for a triangle having sides measuring 650.0 ft, 428.0 ft, and 282.0 ft.

Solution:

$$S = \frac{a + b + c}{2} = \frac{650.0\text{ ft} + 428.0\text{ ft} + 282.0\text{ ft}}{2} = 680.0$$

$$\begin{aligned} A &= \sqrt{680.0\,(680.0 - 650.0)\,(680.0 - 428.0)\,(680.0 - 282.0)} \\ &= \sqrt{680.0\,(30.0)\,(252.0)\,(398.0)} \\ &= \sqrt{2046038400} \\ &= 45{,}233.15598 \text{ or } 45{,}230\text{ ft}^2 \end{aligned}$$

13.5.3. One Angle and Two Sides

The third type of triangle illustrated is one in which one angle and the length of the two adjacent sides is known, Figure 13.11.

The limitations on this method are that the angle must be less than 90° and the at least one angle must be known or measured. The equation for area (A) is:

$$A = \frac{1}{2} \times (b \times c \times \text{ sine } \theta)$$

where A = Area; b = Known side; c = Known side; ø = Angle between sides b and c; sine = Sine trigonometric function.

This equation is very useful in situations where one side of the triangle cannot be measured.

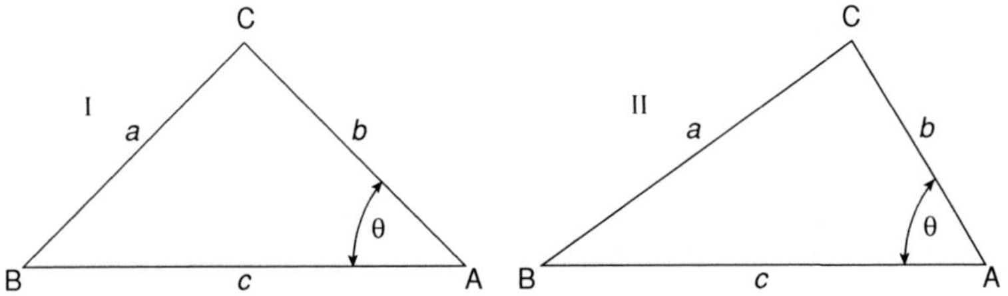

FIGURE 13.11. Triangle with one angle and the adjacent sides known.

Problem: Determine the area (ft^2) of a triangle with sides of 350.0 ft and 555.0 ft and an included angle of 45°.

Solution:

$$A = \frac{1}{2} \times a \times b \times \text{sine}\ \theta = \frac{1}{2} \times 350.0\ \text{ft} \times 555.0\ \text{ft} \times \text{sine}\ 45$$

$$= \frac{137355.4922}{2} = 68677.746\ldots\ \text{or}\ 68{,}700\ \text{ft}^2$$

13.6. Rectangle, Square, and Parallelogram

These shapes are considered together because their areas are calculated by the same equation.

Remember that the height (h) for a parallelogram is perpendicular to the base (b), see Figure 13.12. The equation for area is:

$$A = b \times h$$

where A = Area; b = Length of base; h = Height perpendicular to base.

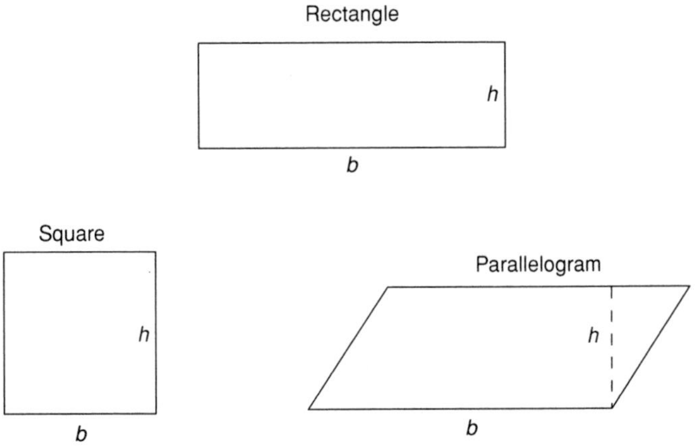

FIGURE 13.12. Base and height for rectangles, squares and parallelograms.

Problem: What is the area (ft^2) of a rectangle measuring 1320.0 ft by 660.0 ft?

Solution:

$$A = b \times h = 1{,}320.0 \text{ ft} \times 660.0 \text{ ft} = 871{,}200 \text{ ft}^2$$

Problem: What is the area (ac) of a parallelogram where the base measures 1,050.0 ft, and the height measures 750.0 ft?

Solution: The area equation can be used and then the square feet converted to acres, but the acres conversion can be added to the area equation:

$$\begin{aligned} A\,(\text{ac}) &= b \times h \times \frac{1 \text{ ac}}{43{,}560 \text{ ft}^2} \\ &= 1{,}050.0 \text{ ft} \times 750.0 \text{ ft} \times \frac{1 \text{ ac}}{43{,}560 \text{ ft}^2} \\ &= 18.078\ldots \text{ or } 18.08 \text{ ac} \end{aligned}$$

13.7. Circle or Sector

The area of a circle, in a slightly different form, was used in Chapter 5 in figuring engine displacement.

$$A = \pi r^2$$

where A = Area; $\pi = 3.14$; r = Radius of the circle.
A sector is slice of a circle. The known dimensions, the angle or the arc length, dictate the equation used to solve for the area of the sector. Study Figure 13.13.

When the angle is known, the equation for a sector is:

$$A = \frac{\pi \times r^2 \times \theta}{360}$$

where A = Area; $\pi = 3.14$; θ = Included angle of the sector; r = Radius of sector.

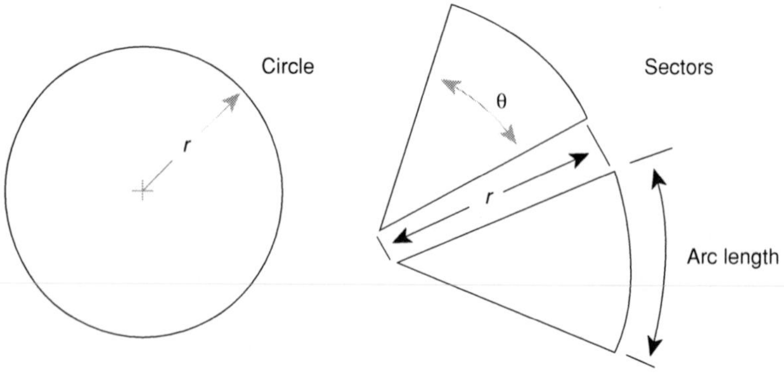

FIGURE 13.13. Circle and sectors.

When the length of the arc is known, the equation for a sector is:

$$A = \frac{r \times \text{arc length}}{2}$$

Problem: Find the area of a circle (ft^2) having a radius of 75.0 ft.

Solution:

$$A = \pi \times r^2 = 3.14 \times (75.0\text{ ft})^2 = 17662.5 \text{ or } 17{,}700\text{ ft}^2$$

Problem: Find the area of a sector (ft^2) having a radius of 135.0 ft and an angle of 60.0°.

Solution:

$$A = \frac{\pi \times r^2 \times \theta}{360} = \frac{3.14 \times (135.0\text{ ft})^2 \times 60.0^\circ}{360}$$
$$= \frac{3{,}433{,}590}{360} = 9{,}537.75 \text{ or } 9{,}540\text{ft}^2$$

Problem: What is the area of a sector (ac) if the radius is 100.0 ft and the arc length is 210.0 ft?

Solution: (Adding the conversion value from square feet to acres):

$$A = \frac{r \times \text{ arc length}}{2} \times \frac{1\text{ ac}}{43{,}560\text{ ft}^2}$$
$$= \frac{100.0\text{ ft} \times 210.0\text{ ft}}{2} \times \frac{1\text{ ac}}{43{,}560\text{ ft}^2}$$
$$= \frac{21{,}000}{87{,}120} = 0.24104\ldots \text{ or } 0.241\text{ ac}$$

13.8. Trapezoid

A trapezoid is a four-sided figure with only two parallel sides. Note that the height must be measured perpendicular to the parallel sides, Figure 13.14.

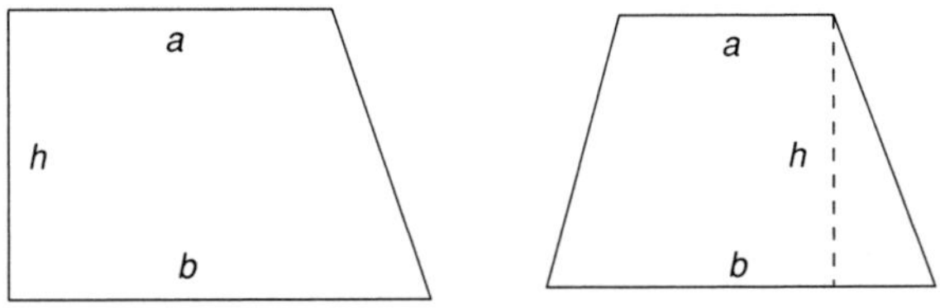

FIGURE 13.14. Two types of trapezoids.

The area of a trapezoid is determined by the equation:

$$A = h \times \left(\frac{a+b}{2}\right)$$

where A = Area; h = Height (distance between the parallel sides); a = Length of one parallel side; b = Length of the second parallel side.

Problem: What is the area (ac) of a trapezoid with parallel sides of 300.0 ft and 450.0 ft, and with a height of 120.0 ft?

Solution: (using conversion to acres):

$$\begin{aligned} A &= h \times \left(\frac{a+b}{2}\right) \times \frac{1\text{ ac}}{43{,}560\text{ ft}^2} \\ &= 120.0\text{ ft} \times \left(\frac{300.00\text{ ft} + 450.00\text{ ft}}{2}\right) \times \frac{1\text{ ac}}{43{,}560\text{ ft}^2} \\ &= 45{,}000\text{ ft}^2 \times \frac{1\text{ ac}}{43{,}560\text{ ft}^2} \\ &= 1.03305\ldots \text{ or } 1.033\text{ ac} \end{aligned}$$

13.9. Determining Areas of Irregular-Shaped Fields Using Standard Geometric Shapes

Very few agricultural fields or lawns are standard geometric shapes. They may not be as complicated as the field illustrated in Figure 13.15, but the principles

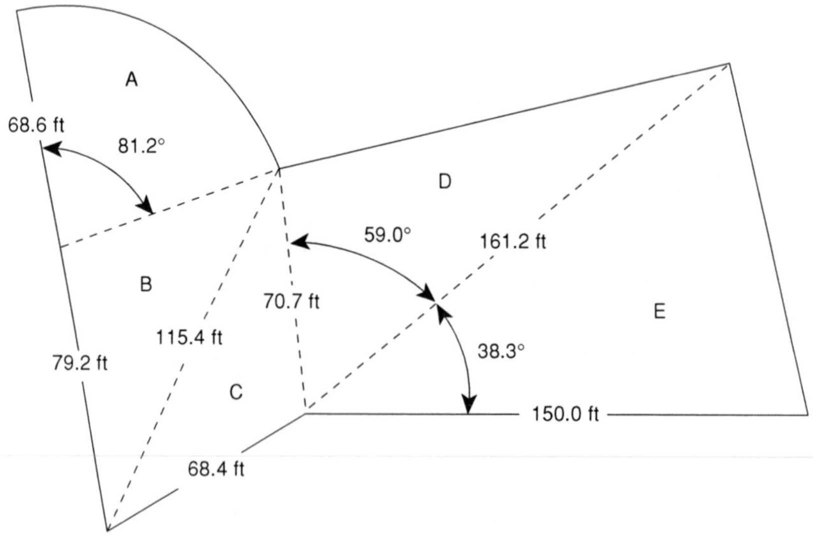

FIGURE 13.15. Example of determining the area of an irregular shape.

for determining areas are the same. The total area is the sum of each individual area. The key to efficient calculations is dividing the irregular shape into the fewest possible number of standard shapes.

Shape A is a sector of a circle with a known radius and angle. The area of this shape can be found by using the sector equation.

$$A_{\mathrm{A}} = \frac{\pi \times r^2 \times \phi}{360} = \frac{\pi \times \ 68.6 \text{ ft } \times \ 81.2}{360}$$
$$= \frac{12004778}{360} = 3334.66\ldots \text{ or } 3{,}330 \text{ ft}^2$$

Shape B is a triangle with the length of the sides known.

$$A_{\mathrm{B}} = \sqrt{S(S-a)(S-b)(S-c)} \quad S = \frac{a+b+c}{2}$$
$$S = \frac{a+b+c}{2} = \frac{68.6 \text{ ft} + 79.2 \text{ ft} + 115.4 \text{ ft}}{2} = \frac{263.2}{2} = 131.6$$
$$A_{\mathrm{B}} = \sqrt{S(S-a)(S-b)(S-c)}$$
$$= \sqrt{131.6(131.6-68.6)(131.6-79.2)(131.6-115.4)}$$
$$= \sqrt{131.6(63)(52.4)(16.2)} = \sqrt{7037894.304} = 2652.90\ldots \text{ or } 2{,}650 \text{ ft}^2$$

Shape C is another triangle with the lengths of all three sides known.

$$A_{\mathrm{B}} = \sqrt{S(S-a)(S-b)(S-c)} \quad S = \frac{a+b+c}{2}$$
$$S = \frac{a+b+c}{2} = \frac{68.4 \text{ ft} + 70.7 \text{ ft} + 115.4 \text{ ft}}{2} = \frac{254.5}{2} = 127.25$$
$$A_{\mathrm{B}} = \sqrt{S(S-a)(S-b)(S-c)}$$
$$= \sqrt{127.25(127.25-68.4)(127.25-70.7)(127.25-115.4)}$$
$$= \sqrt{127.25(58.85)(56.55)(11.85)} = \sqrt{5018283.793}$$
$$= 2240.15\ldots \text{ or } 2{,}240 \text{ ft}^2$$

Area D is a triangle with one angle known and the length of the adjacent sides. The area (A_{D}) is:

$$A = \frac{1}{2} \times a \times b \times \text{ sine } \theta = \frac{1}{2} \times 70.7 \text{ ft} \times 161.2 \text{ ft} \times \text{sine } 59.0$$
$$= \frac{11396.84 \times 0.8571\ldots}{2} = \frac{9768.99\ldots}{2} = 4884.49\ldots \text{ or } 4{,}880 \text{ ft}^2$$

Shape E is the same as shape D. The area of shape E is:

$$A = \frac{1}{2} \times a \times b \times \text{sine}\,\theta = \frac{1}{2} \times 161.2\ \text{ft} \times 150.0\ \text{ft} \times \text{sine}\ 38.3$$
$$= \frac{24180 \times 0.61977\ldots}{2} = \frac{14986.25\ldots}{2} = 7493.12\ldots \text{ or } 7{,}490\ \text{ft}^2$$

Now that the area for each shape is known, the total area (A_T) is calculated by combining the areas for the individual shapes:

$$A_T = A_A + A_B + A_C + A_D + A_E$$
$$= 3330\ \text{ft}^2 + 2650\ \text{ft}^2 + 2240\ \text{ft}^2 + 4880\ \text{ft}^2 + 7490\ \text{ft}^2$$
$$= 20590 \text{ or } 20{,}600\ \text{ft}^2$$

13.10. Determining Areas of Irregular-Shaped Fields Using Trapezoidal Equations

Occasionally it becomes necessary to determine the area of a field or other property that has one irregular-shaped boundary. This is a very common problem when one of the boundaries is formed by surface water. In this situation, the area can be determined by dividing the shape into a series of trapezoids and summing the individual areas. In situations where the trapezoids can be laid out so the heights are the same, the trapezoidal summation equation can be used.

Problem: What is the area (acres) for the field illustrated in Figure 13.16?

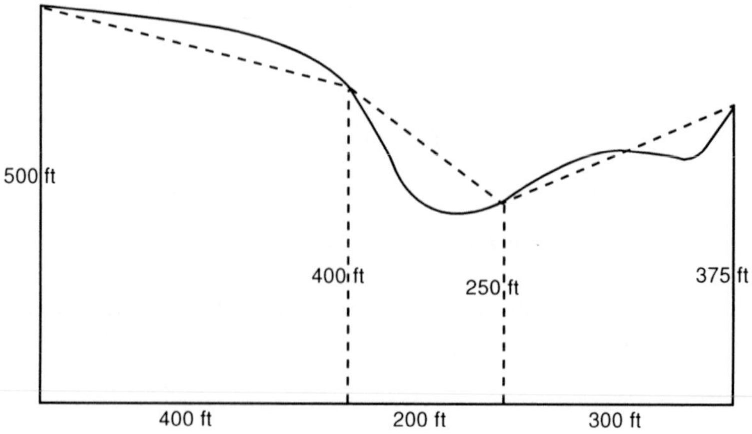

FIGURE 13.16. Example of using trapezoids for irregular shape.

Solution: The total area (A_T) is the sum of each trapezoidal shape (dashed lines).

$$A = h \times \frac{a+b}{2}$$

$$A_T = \left(h_1 \times \frac{a_1+b_1}{2}\right) + \left(h_2 \times \frac{a_2+b_2}{2}\right) + \left(h_3 \times \frac{a_3+b_3}{2}\right)$$

$$= \left(400\text{ ft} \times \frac{500\text{ ft}+400\text{ ft}}{2}\right) + \left(200\text{ ft} \times \frac{400\text{ ft}+250\text{ ft}}{2}\right)$$

$$+ \left(300\text{ ft} \times \frac{250\text{ ft}+375\text{ ft}}{2}\right)$$

$$= (400 \times 450) + (200 \times 325) + (300 \times 312.5)$$

$$= 180000 + 65000 + 93750 = 338{,}750\text{ ft}^2$$

$$A(\text{ac}) = \frac{1\text{ ac}}{43560\text{ ft}^2} \times 33{,}8750\text{ ft}^2 = 7.776 \text{ or } 7.78\text{ ac}$$

The field illustrated in Figure 13.16 contains approximately 7.78 acres. Why is it approximate? The dashed lines used to define the trapezoidal shapes only approximate the actual boundary lines of the field, Figure 13.17.

When the irregular boundary is uniform enough to allow a series of equal distances (d) along the base line to be established, the summation trapezoidal equation can be used. Study Figure 13.16 and the following equation for an illustration of this method.

$$A = d \times \left(\frac{h_o}{2} + \sum h + \frac{h_n}{2}\right)$$

where A = Area; d = Distance between offsets (must be equal); h_o and h_n = End offsets measured from base line AB; $\sum h$ = Summation of all interior offsets (excluding the two end offsets).

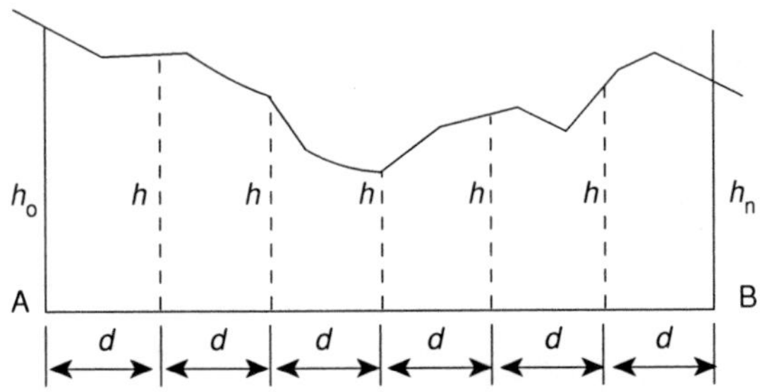

FIGURE 13.17. Sketch for summation trapezoidal equation.

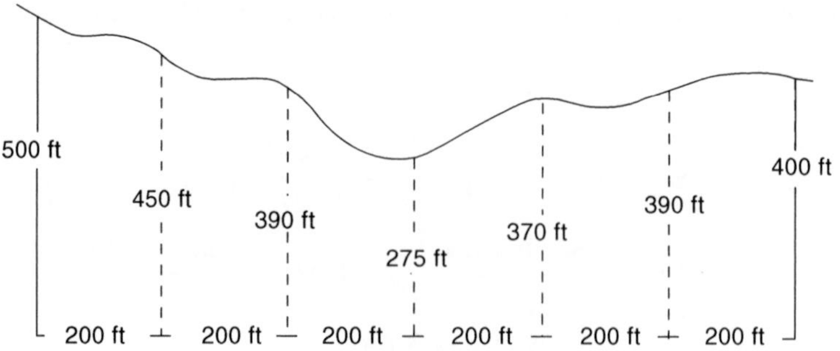

FIGURE 13.18. Summation trapezoidal equation applied to a field.

To use the summation equation the field length, AB, is divided into a number of equal distances (d) and the offsets (perpendicular distances from line AB to the curved edge of the field), h, are measured and recorded.

Problem: Determine the area (ac) of the irregular-shaped field illustrated in Figure 13.18.

Solution:

$$A = d \times \left(\frac{h_o}{2} + \sum h + \frac{h_n}{2}\right)$$

$$= 200 \text{ ft} \times \left(\frac{500 \text{ ft}}{2} + (450 \text{ ft} + 390 \text{ ft} + 275 \text{ ft} + 370 \text{ ft} + 390 \text{ ft}) + \frac{400 \text{ ft}}{2}\right)$$

$$= 200 \text{ ft} \times (250 \text{ ft} + 450 \text{ ft} + 390 \text{ ft} + 275 \text{ ft} + 370 \text{ ft} + 390 \text{ ft} + 200 \text{ ft})$$

$$= 200 \text{ ft} \times 2325 \text{ ft} = 465{,}000 \text{ ft}^2$$

$$A \text{ (ac)} = \frac{1 \text{ ac}}{43{,}560 \text{ ft}^2} \times 465{,}000 \text{ ft}^2 = 10.67\ldots \text{ or } 10.7 \text{ ac}$$

13.11. Metric Problems

Problem: What is the area of a triangle (ha) with a base 150.0 m long and a height of 100.0 m?

Solution:

$$A = \frac{1}{2} \times 150.0 \text{ m} \times 100.0 \text{ m} = 7{,}500 \text{ m}^2$$

$$\text{ha} = \frac{1 \text{ ha}}{10{,}000 \text{ m}^2} \times 7{,}500 \text{ m}^2 = 0.75 \text{ ha}$$

Problem: Determine the area (m^2) for a triangle having sides measuring 650.0 m, 428.0 m, and 282.0 m.

Solution:

$$S = \frac{a+b+c}{2} = \frac{650.0\text{ m} + 428.0\text{ m} + 282.0\text{ m}}{2} = 680.0\text{ m}$$

$$\begin{aligned} A &= \sqrt{680.0\,(680.0 - 650.0)\,(680.0 - 428.0)\,(680.0 - 282.0)} \\ &= \sqrt{680.0\,(30.0)\,(252.0)\,(398.0)} \\ &= \sqrt{2046038400} \\ &= 45{,}233.15598 \text{ or } 45{,}230\text{ m}^2 \end{aligned}$$

Problem: Determine the area (m^2) of a triangle with sides of 350.0 m and 555.0 m and an included angle of 45°.

Solution:

$$\begin{aligned} A &= \frac{1}{2} \times a \times b \times \text{ sine } \theta = \frac{1}{2} \times 350.0\text{ m} \times 555.0\text{ m} \times \text{ sine } 45 \\ &= \frac{194250}{2} \times 0.7071 = 68677.74\ldots \text{ or } 68{,}700\text{ m}^2 \end{aligned}$$

Problem: What is the area (m^2) of a rectangle measuring 1320.0 m by 660.0 m?

Solution:

$$A = b \times h = 1{,}320.0\text{ m} \times 660.0\text{ m} = 871{,}200\text{ m}^2$$

Problem: What is the area (ha) of a parallelogram where the base measures 1,050.0 m, and the height measures 750.0 m?

Solution:

$$\begin{aligned} A\,(\text{ha}) &= b \times h \times \frac{1\text{ ha}}{10{,}000\text{ m}^2} = 1{,}050.0\text{ m} \times 750.0\text{ m} \times \frac{1\text{ ha}}{10{,}000\text{ m}^2} \\ &= \frac{787500}{10000} = 78.75 \text{ or } 78.8\text{ ha} \end{aligned}$$

Problem: Find the area of a circle (m^2) having a radius of 75.0 m.

Solution:

$$A\left(\text{m}^2\right) = \pi \times r^2 = 3.14\ldots \times 75.0^2 = 17671.45\ldots \text{or } 17{,}700\text{ m}^2$$

Problem: Find the area of a sector (m^2) having a radius of 135.0 m and an angle of 60.0°.

Solution:

$$\begin{aligned} A &= \frac{\pi \times r^2 \times \theta}{360} = \frac{3.14 \times (135.0\text{ m})^2 \times 60.0^o}{360} \\ &= \frac{3{,}433{,}590}{360} = 9{,}537.75 \text{ or } 9{,}540\text{ m}^2 \end{aligned}$$

Problem: What is the area of a sector (ha) if the radius is 100.0 m and the arc length is 210.0 m?

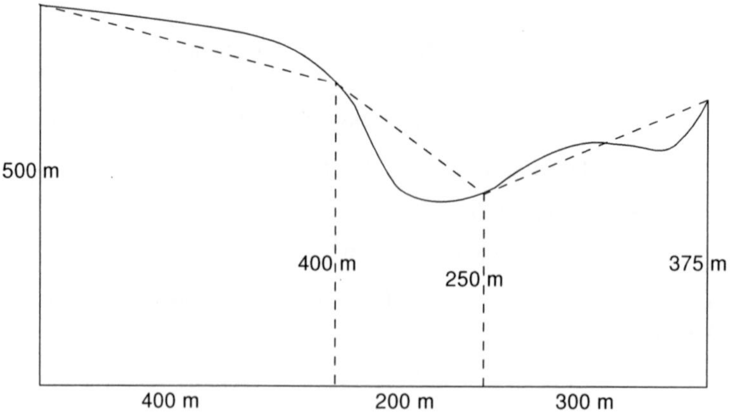

FIGURE 13.19. Metric example of using trapezoids for irregular shape.

Solution:

$$A\ (\text{ha}) = \frac{r \times \text{arc length}}{2} \times \frac{1\ \text{ha}}{10{,}000\ \text{m}^2}$$

$$= \frac{100.0\ \text{m} \times 210.0\ \text{m}}{2} \times \frac{1\ \text{ha}}{10{,}000\ \text{m}^2} = \frac{21{,}000}{10{,}000} = 2.1\ \text{ha}$$

Problem: What is the area (ha) of a trapezoid with parallel sides of 300.0 m and 450.0 m, and with a height of 120.0 m?

Solution:

$$A = h \times \left(\frac{a+b}{2}\right) \times \frac{1\ \text{ac}}{43{,}560\ \text{ft}^2}$$

$$= 120.0\ \text{ft} \times \left(\frac{300.00\ \text{ft} + 450.00\ \text{ft}}{2}\right) \times \frac{1\ \text{ac}}{43{,}560\ \text{ft}^2}$$

$$= 45{,}000\ \text{ft}^2 \times \frac{1\ \text{ac}}{43{,}560\ \text{ft}^2}$$

$$= 1.03305\ldots \text{ or } 1.033\ \text{ac}$$

Problem: What is the area (acres) for the field illustrated in Figure 13.19?

Solution: The total area (A_T) is the sum of each trapezoidal shape (dashed lines).

$$A = h \times \frac{a+b}{2}$$

$$A_T = \left(h_1 \times \frac{a_1+b_1}{2}\right) + \left(h_2 \times \frac{a_2+b_2}{2}\right) + \left(h_3 \times \frac{a_3+b_3}{2}\right)$$

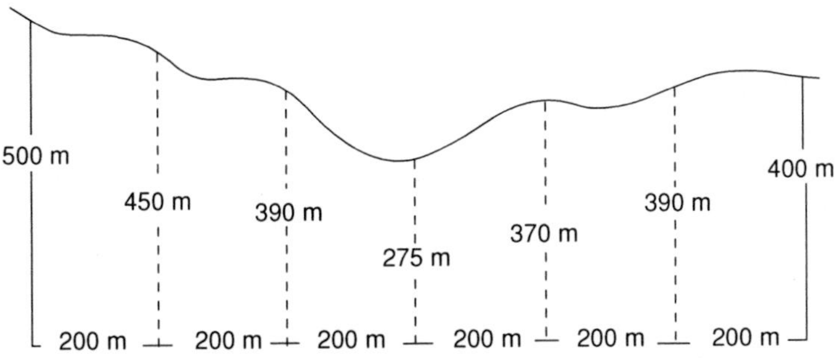

FIGURE 13.20. Summation trapezoidal equation applied to a metric measured field.

$$= \left(400\text{ m} \times \frac{500\text{ m} + 400\text{ m}}{2}\right) + \left(200\text{ m} \times \frac{400\text{ m} + 250\text{ m}}{2}\right)$$

$$+ \left(300\text{ m} \times \frac{250\text{ m} + 375\text{ m}}{2}\right)$$

$$= (400 \times 450) + (200 \times 325) + (300 \times 312.5)$$

$$= 180000 + 65000 + 93750 = 338{,}750\text{ m}^2$$

$$A\text{ (ha)} = \frac{1\text{ ha}}{10{,}000\text{ m}^2} \times 33{,}8750\text{ ft}^2 = 33.875\text{ or }33.9\text{ ha}$$

Problem: Determine the area (ac) of the irregular-shaped field illustrated Figure 13.20.

Solution:

$$A = d \times \left(\frac{h_o}{2} + \Sigma h + \frac{h_n}{2}\right)$$

$$= 200\text{ m} \times \left(\frac{500\text{ m}}{2} + (450\text{ m} + 390\text{ m} + 275\text{ m} + 370\text{ m} + 390\text{ m}) + \frac{400\text{ m}}{2}\right)$$

$$= 200\text{ m} \times (250\text{ m} + 450\text{ m} + 390\text{ m} + 275\text{ m} + 370\text{ m} + 390\text{ m} + 200\text{ m})$$

$$= 200\text{ m} \times 2325\text{ m} = 465{,}000\text{ m}^2$$

$$\text{A (ha)} = \frac{1\text{ ha}}{10{,}000\text{ m}^2} \times 465{,}000\text{ m}^2 = 46.5\text{ ha}$$

14
Land Description

14.1. Objectives

1. Understand the three common methods of describing land.
2. Be able to use the rectangular system to read and write legal descriptions.
3. Be able to use the rectangular system to determine the number of acres from a written description.

14.2. Introduction

From earliest times there was a need to mark and describe the boundaries of parcels of land. It is recorded that in about 1400 B.C. Egypt developed a system to reestablish land boundaries after each flood. The need to accurately establish and describe land boundaries has persisted to this day. To transfer ownership one must be able to describe what land is being considered, and where it is located. In the United States, a unique description for each parcel of land is required to collect property taxes. Three primary systems are used in the United States to define land boundaries and locations. These are metes and bounds, block and lot, and the Public Land Survey System (PLSS) commonly called the rectangular systems.

14.3. Metes and Bounds

The metes and bounds system is based on a verbal or written description of the boundary using land marks visible at the time. The description includes a point of beginning (POB), such as a rock, tree, stake, or post, and lengths and bearings of successive lines from this point. Lengths may be paces or in units of chains, poles, or rods. As the technology available for surveying improved, these units were replaced by feet and inches, decimal feet or metric measurements. The bearings may be magnetic, true, or grid. True or grid bearings are the preferred type. To read or write land descriptions with the metes and bounds system, the units used and the type of bearing used must be known.

A typical description might read as follows: "beginning at the large rock about 2 min walk northly from the creek, thence easterly 100 paces to the red oak tree, thence southerly 250 paces to a large partially buried rock, thence westerly to the creek and follow the creek northly to the point of beginning." The dilemma that anyone faces trying to trace the boundaries many years later should be obvious. Creeks change course, trees are removed, rocks can be dug out or covered when land leveling.

14.4. Block and Lot

This system is very common in cities where land has been arranged into lots. City or county recorders' offices have plat books giving the location and dimensions of all the blocks and lots. In most areas, developers are required to provide such plans and have them approved before construction begins. A typical description might read: "Country Club plat, block 4, lot 23, book 543, page 201." In addition, each parcel can be identified by street number.

14.5. Rectangular System

The PLSS or rectangular system of public land survey is used in 30 states. The system was adopted in 1785 by the Continental Congress to subdivide new lands northwest of the Ohio River in a logical and systematic manner. In general, the system establishes a grid by dividing the land by north and south line that follows a true meridian, with a base line that follows a latitude, see Figure 14.1.

Additional lines parallel to the base line, called standard parallels and lines perpendicular to the base lines, called guide meridians, are added to form quadrangles 24 miles square. A few of the original PLSS surveys laid out the lines 30 miles apart. A spacing of 24 miles became the later standard. In this discussion, we will

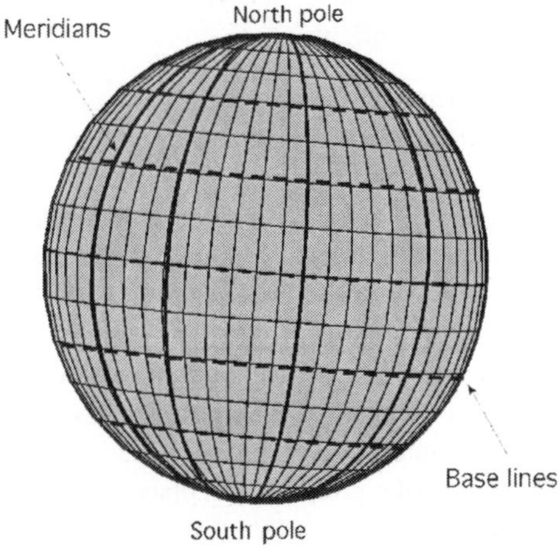

FIGURE 14.1. Meridians and base lines.

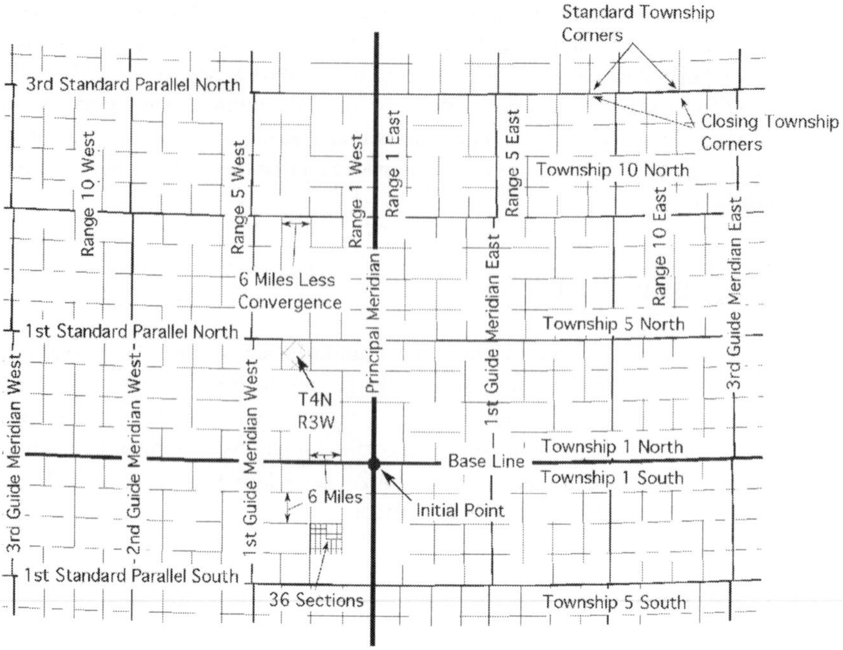

FIGURE 14.2. Rectangular system of public surveying.

consider only the 24-mile rectangular surveys. The quadrangles are divided into townships that are 6 miles square. The townships are further divided into 36 sections of approximately 640 acres each and each section is still further subdivided into fractional parts.

Because meridians converge, at the poles, it is mathematically impossible to have a true grid. A system of subdivisions was used to adjust for the converging of the meridians. This is the reason the guide meridians, Figure 14.2, are not continuous, but have an adjustment at each standard parallel. A description of the subdivisions follows:

1. *Quadrangles*: These square tracts are approximately 24 miles on each side.
2. *Townships*: Each quadrangle contains 16 townships, each approximately 6 miles on a side.
3. *Sections*: Each township is divided into 36 sections, each approximately one mile on a side and containing 640 acres.
4. *Quarter sections*: Each section is divided into quarter sections approximately one-half mile on a side and containing 160 acres. Quarter sections may be divided into fractional areas, the individual tracts containing 80, 40, 20, 10, or 5 acres or combinations of these sizes.

The intent was to produce sections of land one mile to a side. Any variation because of convergence is placed in the western column and the top row of each township.

To begin an original survey, an initial point was established in each new area of land by astronomical observations. Thirty seven initial points were eventually established. A base line, a true parallel of latitude that extends in an east–west direction, and a principal meridian, a true north–south line, passes through each initial point. The principal meridian may be designated by name or by number. The Oklahoma Territory was surveyed from two different meridians, the Indian Meridian and the Cimarron Meridian. The territory that became the states of Kansas, Colorado, and Nebraska was also surveyed from the Indian Meridian, but in these states it was called the 6th Prime Meridian. Figure 14.2 illustrates the rectangular system.

In Figure 14.2, find the initial point, base line, and principal meridian. Note that quadrangles are bounded on the north and the south by true parallels of latitude, called standard parallels, which are 24 miles apart and are numbered consecutively north and south of the base line. For example, the First Standard Parallel North is 24 miles north of the base line, and the Fourth Standard Parallel North is 96 miles north of the base line. The east and west boundaries of quadrangles follow true meridians but because of the adjustment for convergence they are not continuous, therefore they are called guide meridians. They are 24 miles apart and are numbered consecutively east and west of the principal meridian. Thus, the First Guide Meridian West is 24 miles west of the principal meridian, and the Third Guide Meridian East is 72 miles east of the principal meridian.

Because the guide meridians are true meridians, they converge as they approach the North Pole and the South Pole. This causes the north side of each quadrangle to be slightly less than 24 miles. To adjust for convergence, a closing corner is set at the intersection of each guide meridian and each standard parallel or base line. The distance between the closing corner and the standard corner causes an offset in the meridian.

Townships are bounded on the north and the south by township lines and on the east and west by range lines. Range lines are true meridians and thus converge. The south boundary of each township is 6 miles in length, but the north boundary is slightly less. Closing corners are established for townships in the same way that they are established for quadrangles. Township lines are parallel to the base line and the standard parallel.

A township is identified by a unique description based upon the principal meridian governing it. A north–south column of townships is called a range. Ranges are numbered in consecutive order east and west of the principal meridian. An east–west row of townships is called a tier. Tiers are numbered in order north and south of the base line. By common practice, the word "tier" usually is replaced by "township" or just "T" in designating the rows.

An individual township is identified by its number and direction north or south of the base line, followed by the number and direction east or west of the principal meridian. The township designation usually is abbreviated, as for example, "T5N, R3E" of the prime meridian. This township would be located between 24 and 30 miles north and 12 and 18 miles east of the initial point. Figure 14.3 shows a quadrangle divided into townships and one township divided into sections.

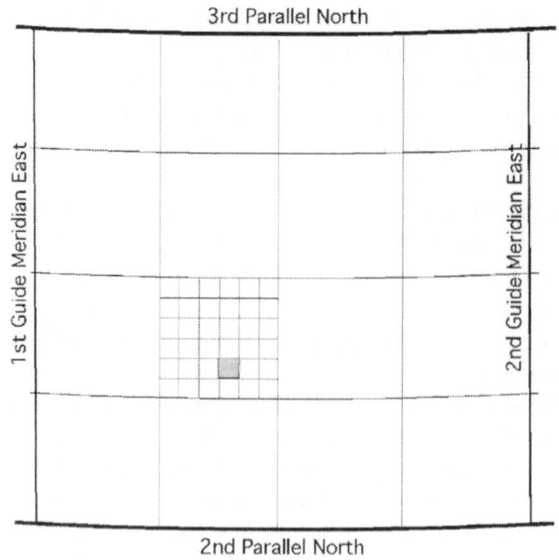

FIGURE 14.3. Division of quadrangle and township.

Based on the earlier discussion, the township divided into sections in Figure 14.3 would be identified as, T10N, R6E.

Figure 14.4 shows the method of subdividing a township into 36 sections, one mile square (640 acres). Sections are numbered by starting in the northeast corner of the township and continuing in a serpentine fashion. If the survey is error free, which is extremely improbable, all sections are one mile square except those along the west and north boundary of the township. These sections are less than one mile in width because of convergence of the range lines and any error in measurements. The description of the highlighted section in Figure 14.4 is "S21," and the township identified in Figure 14.3 is T10N, R6W, Principal Meridian (or PM). "The complete description would be written as, "S21, T10N, R6W, PM."

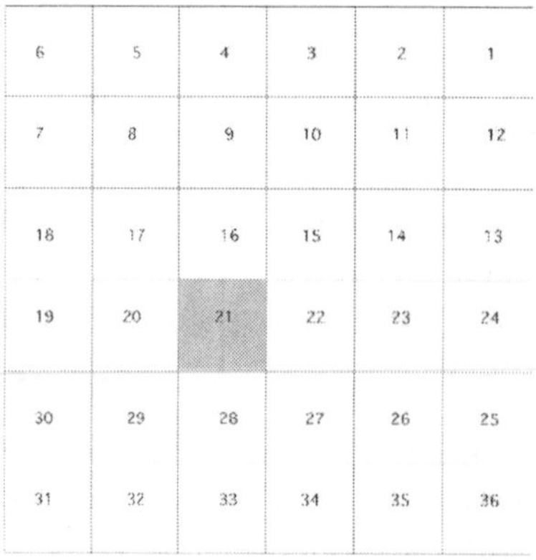

FIGURE 14.4. Section numbers in township.

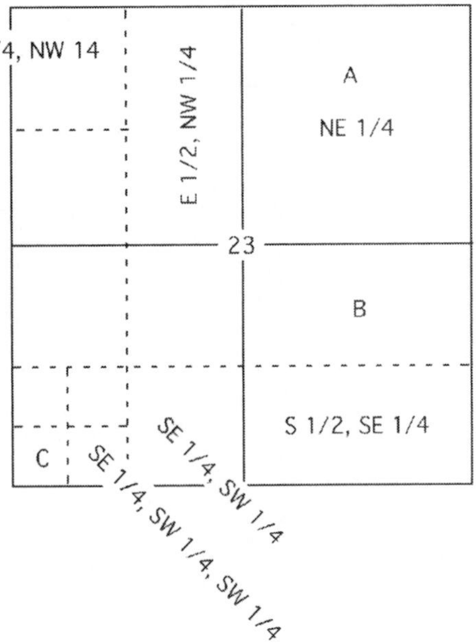

FIGURE 14.5. Subdivision of a section.

As shown in Figure 14.5, each section may be further subdivided into smaller tracts, with north being to the top of the page and east to the right.

The fractions refer to the fraction of the section subdivision being considered and only 1/4 and 1/2 are used unless using lot numbers. The rectangular description of the section subdivisions begins with the smallest unit of area.

Problem: What are the descriptions of the subdivisions labeled A, B, and C in Figure 14.5?

Solution:

A = NE 1/4, S23, T4N, R7E, Principal Meridian.
B = N 1/2, SE 1/4, S23, T4N, R7E, Principal Meridian.
C = SW 1/4, SW 1/4, SW 1/4, S23, T4N, R7E, Principal Meridian.

The description for subdivision A is read as: "the northeast one quarter of section 23, located in township 4 north and range seven east of the Principal Meridian."

It is not unusual for a field to be contained in more that one fraction of a section. In these situations a "&" can be used to link the parts.

Problem: What is the description of the field in Figure 14.6?

Solution: NE1/4, SE1/4, NW1/4 & SW1/4,N, E1/4 see Figure 14.7.

Land descriptions also can be used to determine the acres contained in each subdivision. To determine the acres, divided the area of a section, 640 acres, by each of the denominators in the fractions.

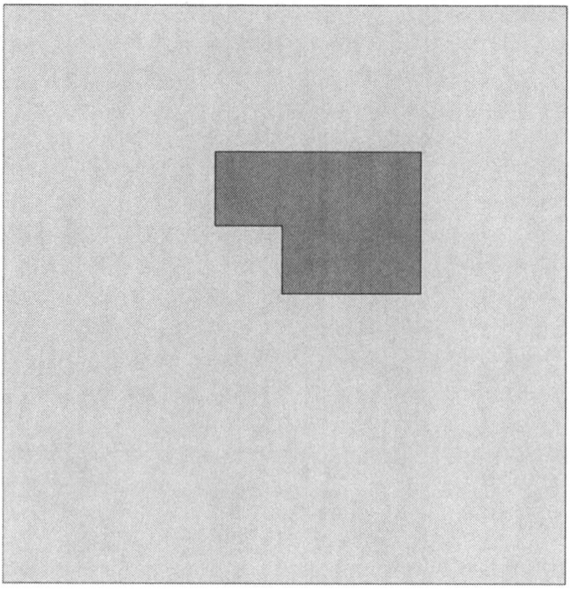

FIGURE 14.6. Fractions of a section using 1/2.

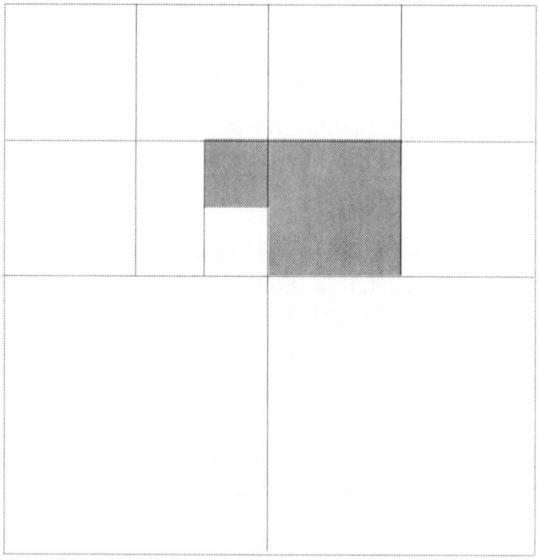

FIGURE 14.7. Solution of fractions of section using an &.

Problem: What is the area (acres) for each of the subdivisions labeled A, B, and C in Figure 14.5?

Solution: Because a section is nominally 640 acres, the areas are:

$$\text{A: } 640 \times \frac{1}{4} = 160 \text{ acres,}$$

$$\text{B: } 640 \times \frac{1}{2} \times \frac{1}{4} = 80 \text{ acres,}$$

$$\text{C: } 640 \times \frac{1}{4} \times \frac{1}{4} \times \frac{1}{4} = 10 \text{ acres.}$$

14.6. Metric Problems

The PLSS is unique to the United States and there is no comparable system in the SI units.

15
Differential and Profile Leveling

15.1. Objectives

1. Be able to describe the equipment used in differential and profile surveying.
2. Understand the terms used in differential and profile leveling.
3. Be able to read a surveying rod.
4. Be able to identify and control the common sources of error during leveling.
5. Be able to describe the processes of differential and profile leveling.
6. Be able to record a set of notes for differential and profile leveling.

15.2. Introduction

Leveling is the process of determining the elevation of points on, above or below the surface of the earth. Many different types of surveys can be used depending on the desired results. Differential and profile leveling are two surveying methods that are very useful for agricultural and horticultural projects. They are both useful for planning and layout of projects. For planning purposes they are used to provide the information needed to develop the maps, charts, and drawings necessary to lay out buildings, roads, drains, etc. They can also be used for layout. Layout is used to establish the boundaries, lines, and elevation for the construction of those structures. Differential and profile leveling rank next to the measurement of distance in importance as a surveying technique. The following chapter will discuss the terms, equipment, and procedures for two types of leveling.

15.3. Leveling Terms

15.3.1. Bench Mark

A bench mark (BM) is an object whose elevation above mean sea level is known or assumed to be known. A bench mark must be an object that is dimensionally stable because it is the reference point for all of the elevations for a survey. If the bench mark elevation is accidentally changed, all surveys that used it must be redone. Bench marks allow a survey to be repeated at a later date and permit a surveyor to tie elevations from the current survey to elevations established in previous surveys. A network of bench marks can thus be established over a large area, all tied to the same reference elevation. Bench marks may vary in character and permanency according to the survey for which they were established. The U.S. Geological Survey (USGS) has established a nationwide network of bench marks, all referenced to mean sea level. These marks consist of bronze disks set in concrete monuments, similar to right-of-way markers, which have been firmly set in the ground. The date of the survey and the elevation and bench mark number usually are stamped in each bronze disk. The adoption of Global Positioning Satellites (GPS) for surveying has provided another means of establishing benchmarks.

In many situations it is not necessary to know the exact elevation above sea level. For such surveys, a local bench mark is used. Frequently, this bench mark is given the elevation of 100.00 ft. If the terrain is hilly, the surveyor should choose a larger number for the starting elevation, as it is not standard practice to use negative numbers for elevations in common surveys.

When using a local bench mark, the survey crew must select and establish its location. Two rules should be followed. The object selected should: (1) be reasonably permanent for as long as it will be needed, and not easily moved or otherwise destroyed; and (2) be capable of being described in such a way that it can be easily relocated. A typical local bench mark might be an "X" chipped in a concrete curb or a bridge abutment, an iron pin driven firmly into the ground, or the rim of an electrical or sewer access hole. It is the job of the note keeper to describe accurately the name, number, type, elevation, and location of each bench mark, and to record this information on the right-hand page in the surveying notebook.

15.3.2. Backsight

A backsight (BS) is a rod reading taken on a point of known or assumed elevation. It is the vertical distance between the line of sight through the instrument and the point of known or assumed elevation on which the rod is set. When a Philadelphia or similar type rod is used, the vertical measurement will be in units of decimal feet.

The backsight reading is used to establish height of the instrument, see Figure 15.1. A backsight will always be taken on a bench mark or a turning point.

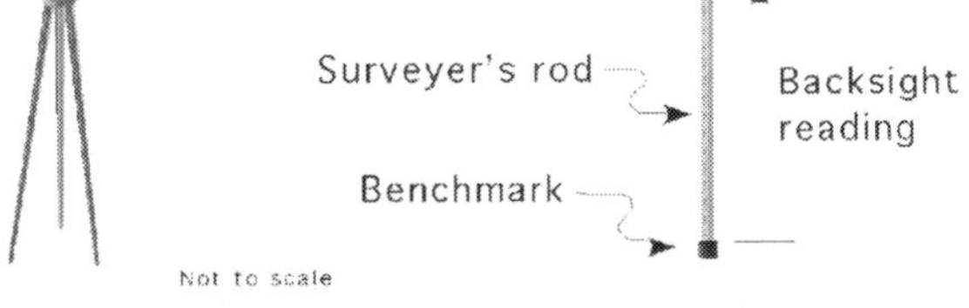

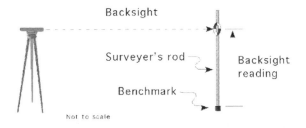

FIGURE 15.1. Backsight rod reading.

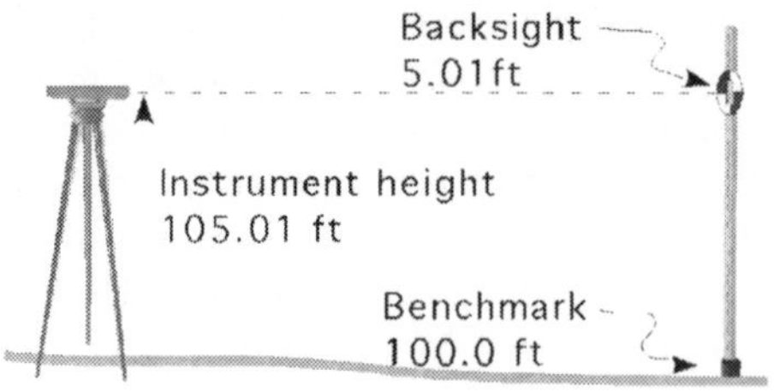

FIGURE 15.2. Determining instrument height.

A backsight on a turning point is used to continue a survey beyond the starting instrument position. The word "backsight" has nothing to do with the direction in which the instrument is pointed. It is important to remember that there is only one backsight for each setup of the instrument.

15.3.3. Height of Instrument

The height of instrument (HI), also called instrument height (IH), is the elevation of the line of sight through the instrument when it is mounted on a tripod and leveled.

It is found by adding the backsight rod reading to the elevation of the point on which the backsight was taken. In Figure 15.2 the instrument height is 105.1 ft because the line of sight through the instrument is 5.01 ft higher than the benchmark.

15.3.4. Foresight

A foresight (FS) is a rod reading taken on any point of unknown elevation. In differential leveling, there is only one foresight for each instrument setup, whereas profile leveling may have several foresights per instrument setup.

The FS is subtracted from the HI to find the elevation of an unknown point. In Figure 15.3, the foresight rod reading, 3.21 ft, is subtracted from the instrument height, 105.01 ft to determine the elevation at the unknown point, 101.8 ft.

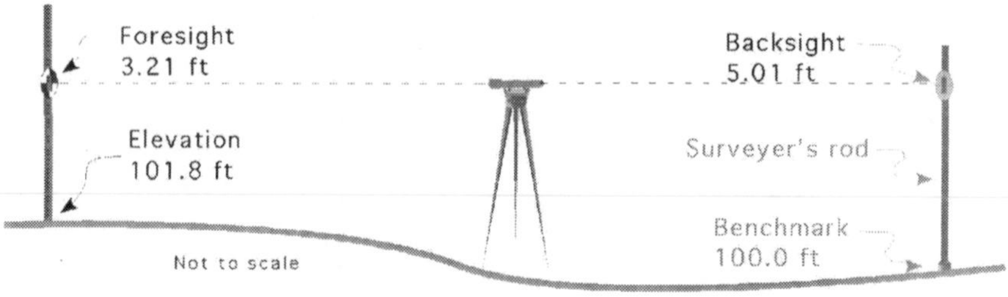

FIGURE 15.3. Foresight rod reading.

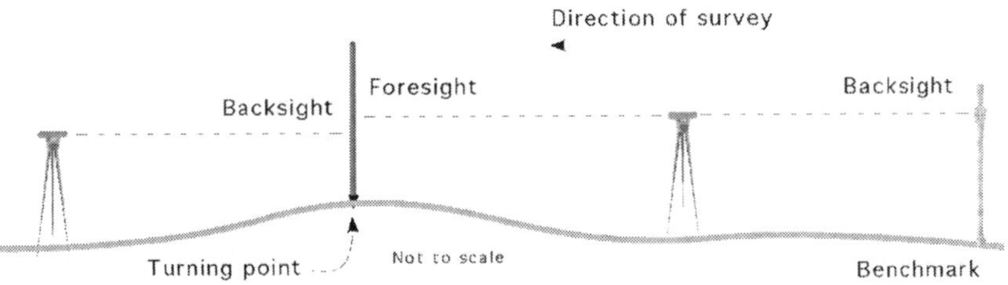

FIGURE 15.4. Turning point.

15.3.5. Turning Point

A turning point (TP) is a temporary bench mark that is used to extend the survey a greater distance. They are used anytime it is necessary to move the instrument. Whenever the instrument is moved to a different location, a backsight is taken on a TP to establish the new instrument height. Turning points are usually used wherever the starting station and ending station can not be seen from one instrument position.

The number of turning points used during a survey is controlled by the maximum distance for the instrument, the distance traveled and the variations in the topography, see Figure 15.4. They should be selected with care and must not be moved until the survey is complete. The structure used for a turning point should be dimensionally stable just like a bench mark. A stake, or permanent structure such as a curb must be used. Do not set the rod on the ground when making a turning point; doing so could result in a serious error.

15.4. Surveying Equipment

The equipment used in leveling consists of a leveling instrument and a leveling rod. The most common type of level is called an engineer's or a dumpy level. The leveling instrument is a telescope containing both vertical and horizontal cross-hairs and one or more spirit levels (bubble in a liquid-filled tube) to indicate when the instrument base is horizontal. The entire assembly, consisting of the frame, telescope and spirit level, can be "leveled" by turning the three or four leveling screws that hold the frame in position above the tripod head.

Another common level is the hand level. It usually has a spirit level for holding it horizontal and one set of cross-hairs. More sophisticated models may also have stadia hairs and direct reading angle scales.

The traditional leveling rod is a wooden scale about 1 inch by 2 inches in cross section and about 14 ft long, graduated in feet, tenths and hundredths of feet. Rods measuring in feet and inches, and meters and centimeters are also available. Rods are used to measure the vertical distance between the line of sight through the telescope and the object on which the leveling rod is resting.

15.5. Rocking a Rod

A rod is used to measure the vertical distance between the line of sight, established by the instrument, and the object on which the rod is resting. To provide an accurate reading it must be held in an upright position. Failure to hold the rod in a vertical position is a common error in surveying. The easiest way to ensure that the rod is vertical is to use a rod level. If a rod level is not available, the instrument person (person on the instrument) can determine if the rod is vertical to the left or the right by instructing the rod person to move the rod until it is aligned with the vertical cross-hair of the instrument. The fore and aft alignment must be controlled by the rod person. When a rod level is not available, the fore and aft vertical position of the rod can be determined by rocking the rod. When rocking the rod the rod person slowly rocks the rod forward and backward past center. The instrument person records the shortest reading. This is illustrated in Figure 15.5.

The shortest reading will occur when the rod is vertical. The distance D_1 will always be less than distance D_2.

15.6. Reading a Philadelphia Rod

The most common type of rod is known as the Philadelphia type. This rod has two sections, each approximately 7 ft long that can be extended to give continuous readings from zero at the base to 13.00 ft at the top. The graduations consist of black marks painted on a white background. The black graduations are 0.01 ft thick and are spaced 0.01 ft apart. The size of the graduations allows the rod to be read to the nearest 0.01 ft, directly for distances up to 250 ft. The tenths of a foot are indicated by black numerals, and each foot is indicated by a larger rod numeral. The foot interval usually is also indicated by a small rod numeral between the whole foot marks. Figure 15.6 is read as 7.02 ft.

When the sight distance is greater than 250 ft, or if the rod cannot be read directly for any reason, the target should be used. The target, Figure 15.7 is a round or oval disk divided into quadrants, which are alternately red and white. When the rod

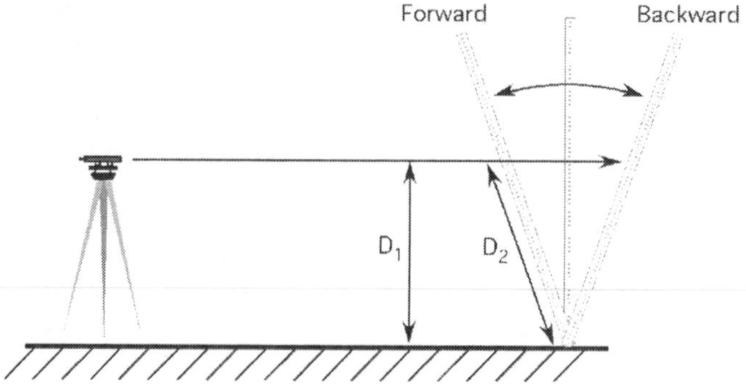

FIGURE 15.5. Rocking the rod.

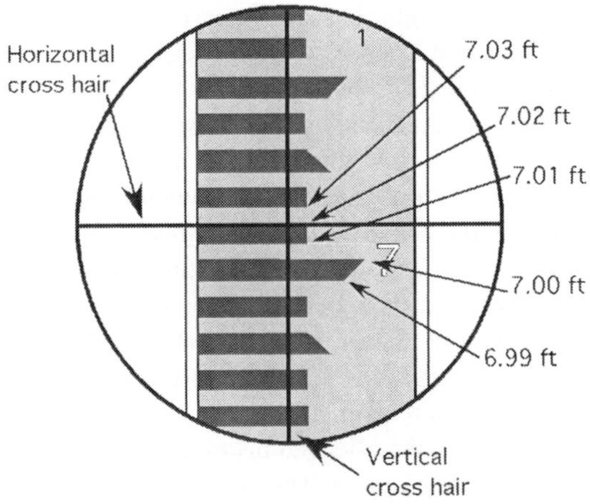

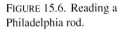

FIGURE 15.6. Reading a Philadelphia rod.

cannot be read directly, the instrument person can signal the rod person to raise and lower the target until it is aligned with the horizontal cross-hair. The note taker can then record the reading from the rod. The target also has a vernier scale that allows readings to be made to 0.001 ft. The reading for the rod in Figure 15.7 is 5.753 ft.

15.7. Setting up a Surveying Level

Many different types and variations of levels are used for surveying. The following procedure for setting up an engineer's level will also work with most other types. When not in use, the instrument is housed in a case for protection. To be used, it must be removed from this case and carefully threaded on the tripod head. The tripod legs are set about 4 ft apart and pushed firmly into the ground, thereby

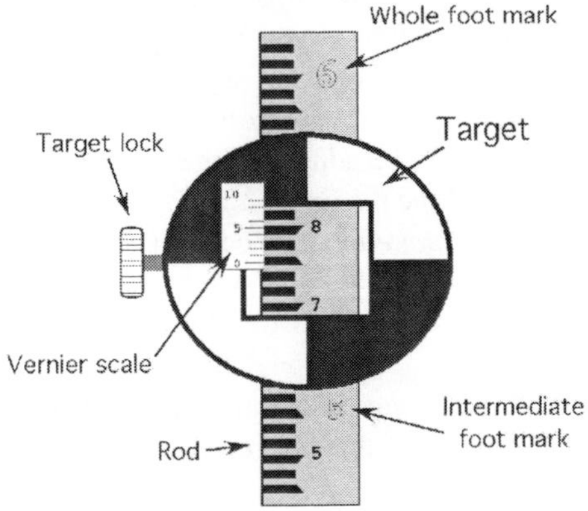

FIGURE 15.7. Reading the rod using a target.

providing a stable base for the instrument and helping to ensure an accurate survey. If the tripod is set up with the head nearly horizontal, leveling of the instrument will be much easier and faster.

To level a four-legged (four leveling screws) instrument, the telescope is lined up over one pair of leveling screws, and the bubble is centered. The screws are turned in opposite directions. The leveling process is expedited when the left thumb rule is used, the bubble in the level moves in the direction of the left thumb. Then the telescope is lined up over the other pair of leveling screws and centered again. The procedure is repeated until the bubble remains centered for any position of the telescope. The leveling screws should not be over tightened, or they will bind and put excessive strain on the instrument frame, or left loose, the instrument will rock and not remain level.

To level a three-legged instrument, the telescope is aligned over any one of the three leveling screws. The instrument is leveled with the adjusting screws and the spirit level. Once leveled, the telescope is should be slowly rotated to check the accuracy of the leveling.

15.8. Common Sources of Error

The accuracy of leveling can be greatly improved if several common errors are controlled:

1. *Instrument out of adjustment*: Anytime that the instrument is "bumped" or otherwise moved, it must be releveled. It is a good practice to check the leveling bubble both before and after reading the rod. Note: resetting the level may change the instrument height. It must be checked, and if different or all or some, of the measurements will need to be repeated.
2. *Rod not plumb*: The rod must always be held plumb using a rod level, or rocked back and forth as explained in the section on reading the rod. A good method of plumbing the rod is to stand behind the rod and balance it carefully on the stake by holding your hands lightly on each side.
3. *Parallax*: If the cross-hairs appear to move over the object as the eye is shifted slightly, parallax exists. When parallax exists, the line of sight of the eye may not be parallel to the line of sight of the instrument. This source of error is eliminated by adjusting the eyepiece until the cross-hairs are the darkest. Because the eyepiece will need to be adjusted for each person, standard practice is for just one person taking all the readings.
4. *Sights not equal*: When field conditions permit, the distance of backsights and foresights should be as nearly equal as possible. Thus, any errors due to the instrument's being out of adjustment are minimized because they tend to cancel out.
5. *Reading the rod incorrectly*: The person reading the rod must be very careful to ensure that the correct foot mark is used, and that the target is used correctly.

15.9. Recording Field Notes

Engineers and surveyors have developed a standard practice for recording surveying information because it is important that the notes be clear, complete and easy to read. Professional surveyors use electronic instruments when every possible. These instruments record the data and down load it to a data collector or computer for post processing and/or use. On electronic instruments the manufacturers of the equipment establish the process and form for collecting and recording the data.

For small projects and when mechanical equipment is used, the traditional field book is still used. The field book is divided into right- and left-hand pages. The left-hand page contains the title, the location, and the data. The right-hand page of the notes is used to record information about the weather, survey party, equipment and describe the location of the bench marks, turning points, creeks, fence or property lines, or other conditions that might influence the design of the structure for which the survey is being made. The location of the benchmark, or starting point, is also described in the notes. A sketch of the general area, showing the location of the beginning and ending stations of the survey, bench marks, ditches, roads, and other landmarks, also is helpful, Figure 15.8.

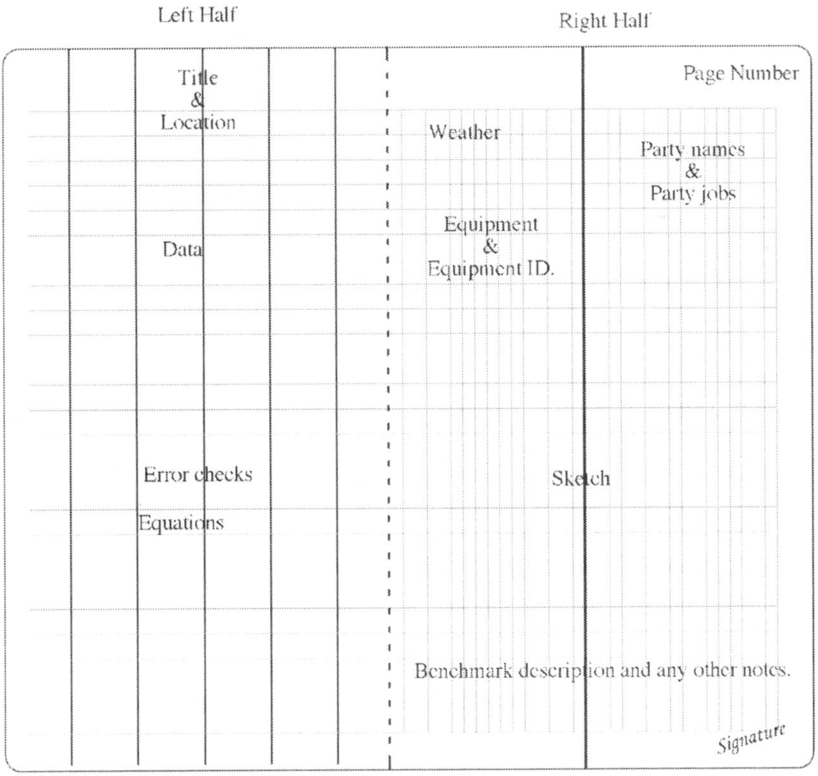

FIGURE 15.8. Standard from for field notes.

Good surveying practices dictate that the data be checked if at all possible. In differential leveling, the accuracy is checked by "closing the loop." Closing the loop, or just closure, means surveying or measuring back to the beginning. When completing a differential leveling activity, surveying from BM2 to BM1 should find the same difference in elevation as the first leg of the survey from BM1 to BM2. This is rarely the case, however, because some errors are always present in leveling. The amount by which the original BM1 elevation and the BM1 elevation calculated from loop closing fail to agree is called the error of closure. The closing of the survey is conducted following the same procedure as the first leg of the survey. The instrument is picked up and moved, leveled and a backsight is taken on BM2. The survey is completed using the necessary foresights and backsights to survey back to BM1.

15.10. Differential Leveling

Differential leveling is the process of finding the difference in elevation between two or more points. When the two points are within the sight limits of the instrument, two readings are taken. The difference in rod readings represents the difference in elevation between the two points. When one or more of the points are beyond the range of the instrument turning points are used.

One of the most common applications of differential leveling is to run a circuit of sights to determine the elevations of one or more bench marks relative to a previously established bench mark. The procedure for differential leveling will be described using this type of circuit, illustrated schematically in Figure 15.9. The diagram shows that three instrument setups were made in traveling from BM1 to BM2. Also note that a "return check" was made between BM2 and BM1, and that three more setups were made in this phase of the survey.

The survey begins with the instrument person going forward a convenient distance (not exceeding the limits of the instrument) and setting up the level, following the procedure previously described. The instrument person sights on the rod while it is held on the top of BM1 by the rod person, and notes a center cross-hair reading of 3.03 ft. This is the backsight, so the 3.03 ft rod reading is added to the BM1 elevation (assumed 100.00 ft), resulting in a height of instrument (HI) of 103.03 ft. The rod person then moves forward past the instrument and selects a turning point, TP1. The FS rod reading of this TP1 is 3.86 ft. The foresight is subtracted from the HI, and the elevation of TP1 is found to be 99.17 ft. The instrument can now be moved forward and set up at a new position. A backsight rod reading of 2.60 ft is observed on TP1 and added to the TP1 elevation of 99.17 ft, giving an instrument height of 101.77 ft. Again a new turning point, TP2, is selected, and a FS rod reading of 4.53 ft is recorded. This rod reading is subtracted from the HI of 101.77 ft, giving the elevation of 97.24 ft for TP2. This process is repeated a third time, and the elevation of BM2 is found to be 95.30 ft. We now know that

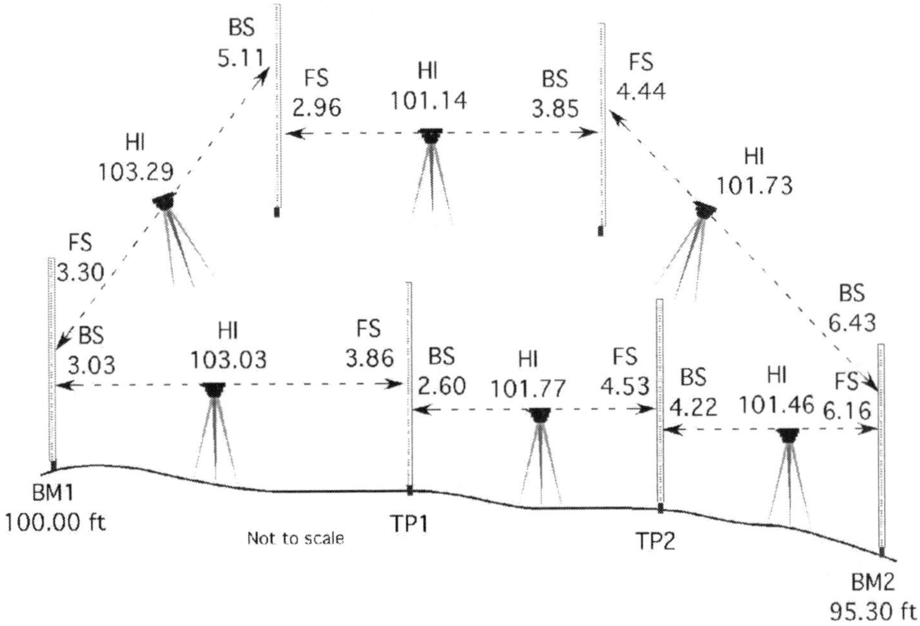

FIGURE 15.9. Differential leveling.

the difference in elevation between BM1 and BM2 is 100.00 ft minus 95.30 ft, or 4.7 ft, assuming no errors were made.

In summary, the procedure for differential leveling is as follows:

1. Set up the instrument.
2. Take the BS reading on BM1.
3. Establish the TP, and take the FS reading.
4. Move the instrument, and set up again.
5. Take the BS on the TP.
6. Establish the next TP, and take the FS reading.
7. Move the instrument, and set up again.
8. Repeat steps 5 to 7 until a foresight is taken on the last station.

The data for the differential leveling survey in Figure 15.9 are recorded in the data table, Table 15.1. The left-hand page of a standard surveying field notebook contains five columns. These columns are needed for the column headings of station (STA), backsight (BS), height of instrument (HI), foresight (FS), and elevation (ELEV). Additional columns may be added as needed to record additional information, for example, the distance for each sight.

TABLE 15.1. Data for differential leveling example.

STA	BS	HI	FS	ELEV
BM1	3.03			100.00
		103.03		
TP1	2.60		3.86	99.17
		101.77		
TP2	4.22		4.53	97.24
		101.46		
BM2	6.43		6.16	95.30
		101.73		
TP3	3.85		4.44	97.29
		101.14		
TP4	5.11		2.96	98.18
		103.29		
BM1			3.30	99.99

15.10.1. Error Control

In surveying, it is important to control as many errors as possible. For differential leveling surveys three error checks should be completed: close the loop, note check, and calculation of the allowable error of closure. The survey is closed to provide the information for the other two checks. The note check is conducted to catch any mathematical errors in the notes. For checking notes, the absolute value of the sum of the foresights minus the absolute value of the sum of the backsights, should equal the absolute value of the difference (Δ) in elevation for BM1 (beginning and closure elevation). Expressed mathematically:

$$|\Sigma \text{FS} - \Sigma \text{BS}| = |\Delta \text{ Elevation BM1}| .$$

Problem: Are the field notes in Table 15.1 accurate?

Solution:

ΣBS	ΣFS
3.86	3.03
4.53	2.60
6.16	4.22
4.44	6.43
2.96	3.85
3.30	5.11
25.25	25.24

$$\begin{aligned} |\Sigma \text{FS} - \Sigma \text{BS}| &= |\Delta \text{ Elevation BM1}| \\ |\Delta \text{ Elevation BM1}| &= \left|\text{BM1}_{\text{beginning}} - \text{BM1}_{\text{ending}}\right| \\ |25.25 - 25.24| &= |100.00 - 99.99| \\ 0.01 &= 0.01. \end{aligned}$$

The note check equation is true, this means the 0.01 ft difference in the elevation of BM1 is not caused by a math error in the data table. An error that is discovered after the fact cannot be corrected; therefore the next step is to determine if the error is acceptable.

The third check is for the error of closure. The allowable error of closure procedure was established because surveyors realized that it was impossible to survey without some error. When perfection is not possible, limits of acceptability must be established. The equation for allowable error of closure sets the acceptability limits for surveys.

$$\mathrm{AE} = K\sqrt{M}$$

where AE is the allowable error, K is a constant ranging from 0.01 to 1.0, and M is the distance traveled (mi).

The constant K is determined by the level of survey. A high order survey, one with very little allowable error, would use a K value of 0.01. A very low order survey, one where more error is acceptable, might use a K value of 1.0. The order of survey must be established before the survey is started because it will determine the quality of the equipment and the procedures that must be followed to collect the data. A K value of 0.10 or 0.05 is acceptable for most general construction and agricultural surveys.

Problem: Is the closure error of 0.01 ft for the differential leveling survey in Figure 15.9 acceptable if the total distanced surveyed, out and back, was 3,600 ft and a K value of 0.10 is acceptable?

Solution:

$$\mathrm{AE} = K\sqrt{M} = 0.10 \times \sqrt{3{,}600\ \text{ft} \times \frac{1\ \text{mi}}{5{,}280\ \text{ft}}} = 0.10 \times \sqrt{0.6818\ldots} = 0.08.$$

$0.01 < 0.08$ The closure error is acceptable.

15.11. Profile Leveling

Profile leveling is used to establish changes in elevation along a line. Common lines requiring surveying are drains, roads, fences, and retaining walls. When this information is plotted on a graph, it will give a profile of the line and will enable one to establish grades, find high or low spots, and make estimates of depths of cuts and many other decisions. The following sections will illustrate the procedure for profile leveling and the preferred way to record the data.

Before a profile can be made, the surveying crew establishes the stations by setting a stake or flag where the rod readings are to be taken. When the terrain is uniform, the stakes can be set a fixed distance apart (25, 50, or 100 ft). When the terrain is not uniform, or when there are addition stations that need to be recorded, such as a sidewalk, the survey crew reconnoiters the line and establishes a station

TABLE 15.2. Data for profile leveling.

STA	BS	HI	FS	ELEV
BM1	6.02	106.02		100.00
0.00			7.34	98.68
100			5.76	100.26
225			3.67	102.35
290	1.72	107.40	(0.34)	105.68
340			4.03	103.37
400			6.65	100.75
460	4.00	107.28	(4.12)	103.28
TP2	3.34	105.27	(5.35)	101.93
BM1			(5.25)	100.02
ΣBS	15.08	ΣFS	15.06	

$|\Sigma FS - \Sigma BS| = 0.02$

$|\Delta \text{Elevation}| = 100.02 - 100.00 = 0.02$

$0.02 = 0.02$ Notes check

Acceptable error $= 0.10$ {EQ\r(920 ÷ 5280)} $= 0.04$

$0.02 < 0.04$ Closure error acceptable

at each important point. Because the purpose of the profile is to show the true slope of the ground, the irregularity of the terrain will largely determine where the stations should be established. When there is a definite change in the slope of the ground, the crew should set a stake to determine the elevation, even if it does not fall at a uniform distance.

The major difference between profile and differential surveying is the addition of additional foresights that are used to define the profile of the terrain. These additional foresights are called intermediate foresights and they are treated differently during the error checks on the notes. Table 15.2 shows that a common way to record the intermediate foresights is to identity the true foresights with parenthesis. An additional method is to add a column to the table for the intermediate foresights.

Once the centerline of the ditch, terrace outlet, channel, road, or other line to be profiled is established, the distance from the starting point to each station is accurately measured. For higher level surveys, 2 × 2-inch stakes may be driven flush with the ground surface, and rod readings taken on the tops of the stakes. On less important surveys, the foresights may be taken with the rod set directly on the ground. Next, the level is set up, readings are taken, and elevations are established for each staked point along the line. Turning points are used as needed to complete the survey. Finally, a closing circuit of readings must be made to check the accuracy of the survey. This is done by running a line of differential levels back to the bench mark where the survey began. If no turning points are used, a sight is taken at the benchmark used to establish the height of instrument, and compared with the original backsight. Figure 15.10 illustrates the steps for conducting a profile leveling survey. The data for this survey is in Table 15.2.

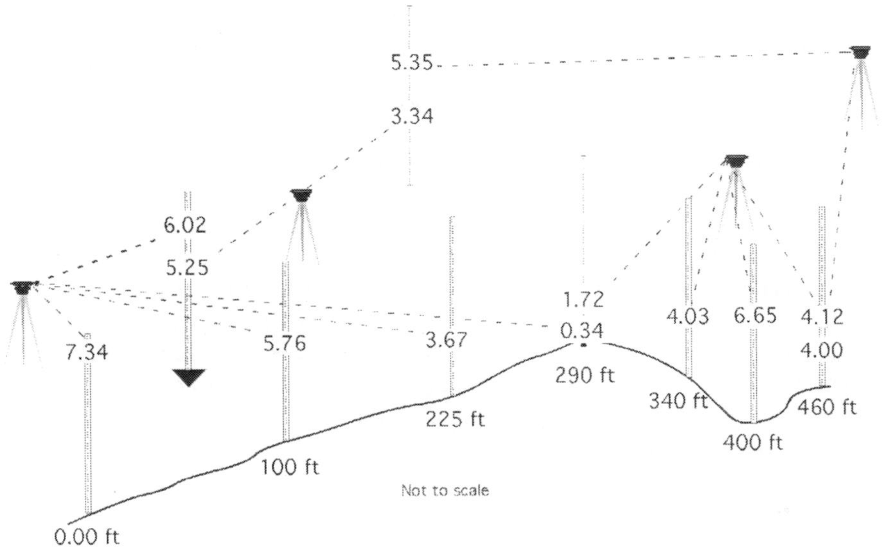

FIGURE 15.10. Profile example.

With the level set up near the line to be profiled, the rod is held on the bench mark, and a reading of 6.02 ft is recorded. This establishes the height of the instrument as 106.02 ft (100.00 + 6.02). Next the rod is held at stations 000, 100, 225, and 290, where rod readings of 7.34, 5.76, 3.67, and 0.34 ft are recorded. At station 290 a stake is set flush with the surface so it can be used to record the elevation and also be used as a turning point. The elevation of each point is calculated by subtracting the rod reading from the height of instrument. Notice that the same HI is used for all stations up to and including 290, as all readings are made from the same instrument setup.

Because the surveying crew anticipated in advance the need for a turning point, the profile leveling exercise is continued by moving the instrument to a new location to provide a view of the remaining stations. A backsight of 1.72 ft is observed on station 290, establishing the new height of instrument as 107.40 ft. The survey continues by recording the rod readings for stations 340, 400, and 460. These readings are 4.03, 6.65, and 4.12 ft.

The profile leveling survey is completed by closing the circuit. The instrument is moved, and a backsight of 4.00 ft is observed on station 460. A stake is set at TP2, and a foresight of 5.35 ft is observed. The procedure is repeated to complete the survey from TP2 to BM1.

Table 15.2 contains a set of notes for this survey. Note the similarities and differences between these notes and Table 15.1 (data for differential leveling). At the completion of the survey, it is noted that an error of 0.02 ft has accumulated during the survey.

15.11.1. Error Control

Because profile leveling notes usually have more foresights than backsights, the method of checking for arithmetic mistakes is slightly different from that for differential leveling. The only foresights to be included when calculating the sum of the foresights (ΣFS) are those taken on turning points, including bench marks if they were used as TPs. In the sample notes, Table 15.2, the foresight readings that are to be used for checking are shown in parentheses. The data in Table 15.2 indicate there were no arithmetic errors in the survey, and the closure error was acceptable. *Note*: This procedure provides a check on the turning points only—not the intermediate foresights. Any errors in the rod readings for the intermediate foresights will not be detected using the standard checks for error. Additional methods such as multiple reading or plunging the telescope must be used. Extreme care must be used to prevent mistakes in reading and recording the rod readings and in calculating the elevations at all stations.

15.12. Using Profile Data

The results of profile leveling surveys are most useful when they are plotted in a graph. The principal purposes for plotting a profile are: (1) to aid in the selection of the most economical grade, location, and depth of irrigation canals, drainage ditches, drain tile lines, sewer lines, roads, etc.; and (2) to determine the amount of cut or fill required for these installations. The graph is plotted with the elevation on the ordinate (vertical scale) and the stations on the abscissa (horizontal scale).

The use of computer spreadsheets makes it much easier to plot and complete "What if?" examples using surveying data.

Problem: What will be the maximum depth of cut for the data in Table 15.2 to install a drain at 2.0 ft below the elevation of station 0.0 and with a slope of 0.5 %?

Solution: This problem could be answered by calculating the difference in elevation between each station and the proposed drain, but that would require a calculation for almost every station. Plotting the data will provide the visual information to determine where the greatest depth will be, Figure 15.11.

The plot shows that the maximum elevation of the profile is 105.7 ft and at that station the elevation of the drain will be at 98.20 ft.

$$105.7\text{ ft} - 98.20\text{ ft} = 7.5\text{ ft}.$$

The maximum depth of cut to install the drain is 7.5 ft.

The problem could also be solved by calculating the difference in elevation

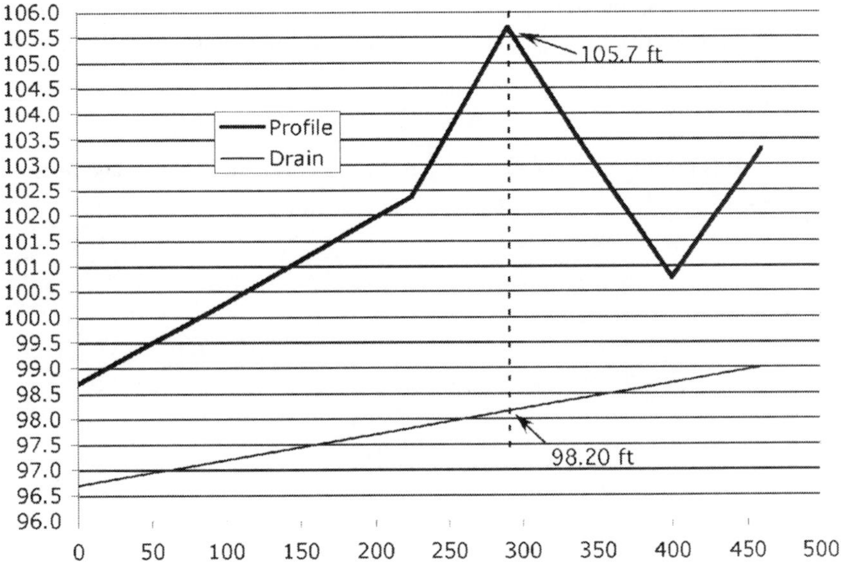

FIGURE 15.11. Profile plot of example data with drain.

between the profile and the drain for each station surveyed using the equation:

$$\text{Elevation}_{\text{drain}} = \text{Elevation}_{\text{starting}} + (\text{Station distance} \times \text{Slope})\,.$$

With a slope for the drain of 0.5%, for station 290 in the sample data the elevation of the drain is

$$\begin{aligned}\text{Elevation}_{\text{drain}} &= \text{Elevation}_{\text{starting}} + (\text{Station distance} \times \text{Slope})\\ &= 96.7\text{ ft} + (290\text{ ft} \times 0.005)\\ &= 96.7\text{ ft} + 1.45\text{ ft}\\ &= 99.15\text{ or } 98.2\text{ ft.}\end{aligned}$$

The elevation of the profile at station 290 (105.7 ft) minus the elevation of the drain at station 290 (98.2 ft) leaves a depth of cut of 105.7 – 98.2 or 7.5 ft.

15.13. Metric Problems

The procedures and techniques for completing differential and profile surveys using metric equipment are the same. The only difference is that a metric rod would be used and the allowable error of closure equation is modified for metric use. A common error of closure equation for metric differential and profile surveys is:

$$\mathrm{AE} = K\sqrt{M},$$

where AE is the allowable error of closure, K is a constant, 3–6 mm, and M is the distance surveyed, in kilometers.

16
Weather

16.1. Objectives

1. Understand the differences between high pressure and low pressure areas and their effect on the weather.
2. Be able to identify the five common air masses.
3. Understand the hydrologic cycle.
4. Be able to define rainfall intensity, duration, and recurrence interval.
5. Be able to determine rainfall intensity when given the duration and return interval.

16.2. Introduction

The weather is the largest uncontrolled variable, as well as the most unpredictable variable, in the production of agricultural products. Agricultural production is based on the growth of plants, and those plants require an optimum environment for maximum production. However, maximum production is seldom realized in the natural environment because the weather seldom matches the needs of plants. Whenever the real environment is significantly different from the optimum, plants are stressed, and production is reduced. The major limiting factor is water. An understanding of the mechanisms of the weather will improve the decision-making process for activities such as cutting hay, irrigating, harvesting, and tillage. Better decisions mean increased productivity for any agricultural enterprise. This chapter will discuss the pressure systems that influence the weather, the hydrologic cycle, and a few of the characteristics of rainfall and runoff.

16.3. Areas of High and Low Pressure

Weather is greatly influenced by atmospheric pressure systems. High pressure and low pressure centers are indicators of the type of weather to be expected. Winds blow in a counterclockwise direction around a low pressure center and in a clockwise direction around a high pressure center. Closely spaced systems indicate a steep pressure gradient and high wind speeds; when the systems are farther apart, the wind speeds are lower. A trough line may develop between two low pressure areas, and a ridgeline may develop between two high pressure areas. In general, cloudy or rainy weather is associated with a low pressure center, and clear, sunny weather accompanies areas of high pressure.

16.4. Air Masses

Differences in pressure are caused by air masses moving across the country. An air mass is a large body of air that has a more or less uniform temperature and moisture content throughout the mass. Air masses that have been over water for a period of time may contain large amounts of moisture, whereas those originating over land areas are usually dry.

The weather in the United States is influenced by five air masses, Figure 16.1. The following section lists and briefly describes these air masses.

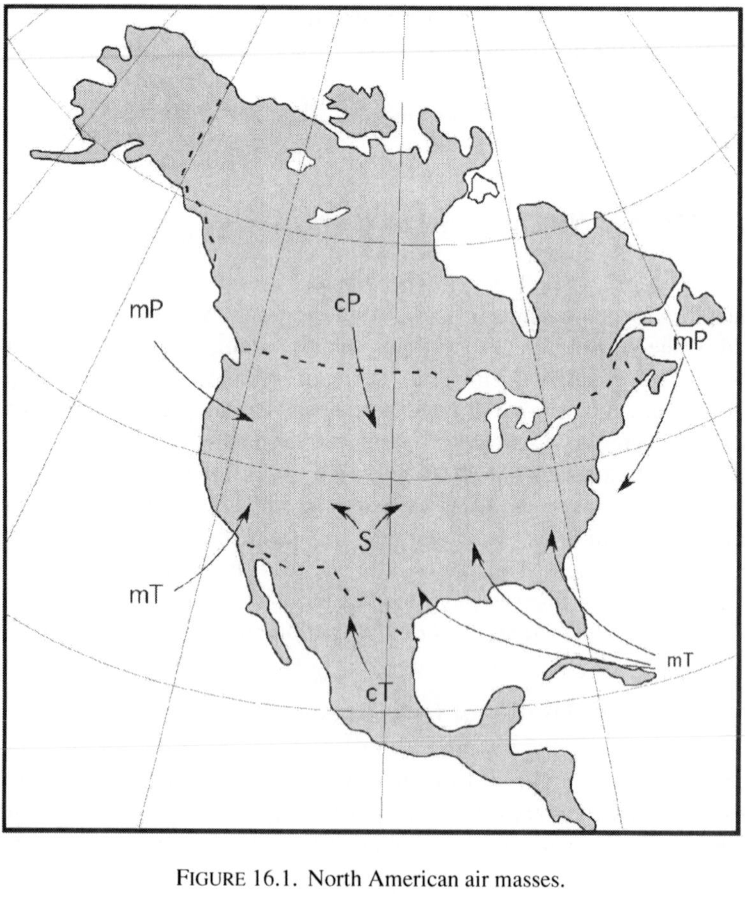

FIGURE 16.1. North American air masses.

1. *Tropical maritime* (mT): This air mass forms off the Gulf of Mexico where it is subject to tremendous heating by the sun. This heating causes evaporation from the ocean, resulting in a warm, moist air mass. This mass generally moves north and west into the central region of the country. This mass contributes the largest amount of moisture to the central and eastern regions of the United States. Tropical maritime air masses also form seasonally of the southern coast of California.
2. *Tropical continental* (cT): The tropical continental forms over the Mexican countryside. The cT is subject to tremendous heating by the sun; but because it forms over land, it is dry. It tends to move north into the central plains. When a cT moves into a region dominated by a mT, a dry line can develop at the boundary line between the two air masses. This condition is famous for the development of severe thunderstorms.
3. *Polar maritime* (mP): This air mass forms over the polar regions of the Pacific and Atlantic Oceans. They are very cool and usually saturated. Consequently any additional cooling can produce precipitation. They are notorious for producing fog, drizzle, cloudy weather and long lasting, light to moderate rain. The Pacific mP occurs more frequently and tends to move south and east into the central plains. As it is forced up over the Rocky Mountains it loses most of its moisture and takes on the characteristics of a continual air mass. The Atlantic mP occurs less frequently and tends to follow the east coast.
4. *Polar continental* (cP): This mass forms over the central plains of Canada. This area of Canada has long winter nights and a large amount of radiation cooling resulting in a very cold, dry air mass. This air mass also tends to move very slowly, and consequently usually is very cold. Consequently this air mass does not produce very much precipitation. The exception is along the boundary line between the cP and a mT. In this situation the rapid cooling of the mT caused by the cP can product localized intense thunderstorms. Its normal movement is south into the Great Plains.
5. *Superior* (S): The superior air mass is unique because it forms at high altitudes over the southwestern desert and occasionally descends to the surface. It is usually very hot and dry.

16.5. Storms

Storms are the result of conflict between warm air masses and cold air masses. The zone of contact between the contrasting air masses is called a front. Fronts are classified as cold fronts or warm fronts depending on which air mass is dominant. When the cold air mass is dominant, the change occurs quickly because cold air masses usually move rapidly. Cold air is heavier than warm air; therefore, the warm air is forced upward, as shown in the cross section in Figure 16.2.

As it rises, warm air cools rapidly. Extreme turbulence and heavy rainfall may occur over small areas. From a soil and water conservation standpoint, cold fronts may cause local rainstorms of high intensity that may result in serious soil erosion and local flooding.

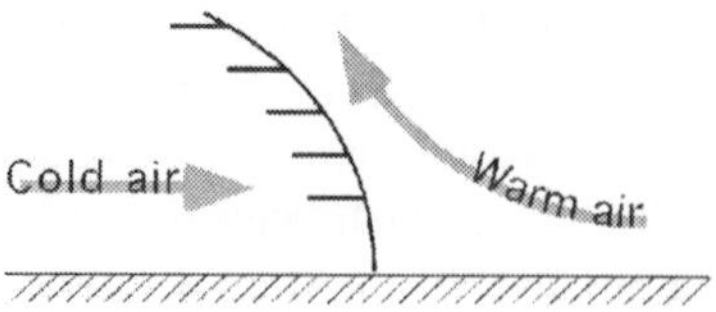

FIGURE 16.2. Cross section of a cold front.

A cross-sectional view of a warm front is shown in Figure 16.3. A warm front occurs when the warm air mass is dominant and overtakes cooler air. The rate of cooling is much less than for a cold front, and resulting rains are gentler. The rains usually cover very large areas. Thus the potential for erosion is reduced, but the rains may cause widespread flooding if they persist for a long period of time.

16.6. Hydrologic Cycle

Climatological concerns involve more than the effects of pressure and air masses on the weather. The conservation of soil and water is based on an understanding of the hydrologic cycle. Figure 16.4 illustrates the movement of water in, on, and above the earth.

The earth is a closed system in which all of the water circulates from one form to another. A study of the cycle can begin at any point, but we will start with precipitation.

Precipitation is caused by the condensation of water vapor in the atmosphere. Water vapor collects in the atmosphere as the sun evaporates water from the oceans, rivers, lakes, and plants. It falls to earth in the form of rain, hail, or snow, or forms on the surfaces of objects as dew or frost. Not all precipitation reaches the earth's surface; some evaporates as it falls, some reaches the surface but does not move through the cycle because it is held in the form of snow and ice and some is intercepted by plants. The moisture stored as snow and ice may be stored for a long period of time in glaciers and the polar ice caps. Precipitation follows different paths before it eventually returns to the atmosphere in the form of vapor.

A portion of the precipitation will infiltrate the soil. It is not unusual for the infiltration rate (inches/hour) of the soil to be less than the rainfall intensity (inches/hour). When this occurs, the excessive precipitation becomes runoff, which is one of two causes of erosion. Not all runoff reaches the ocean; some evaporates, some is collected in different size impoundments and then infiltrates into the soil,

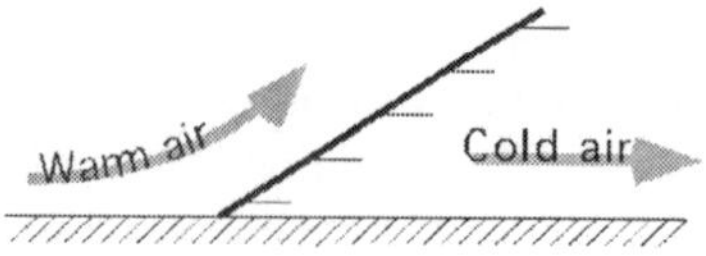

FIGURE 16.3. Cross section of a warm front.

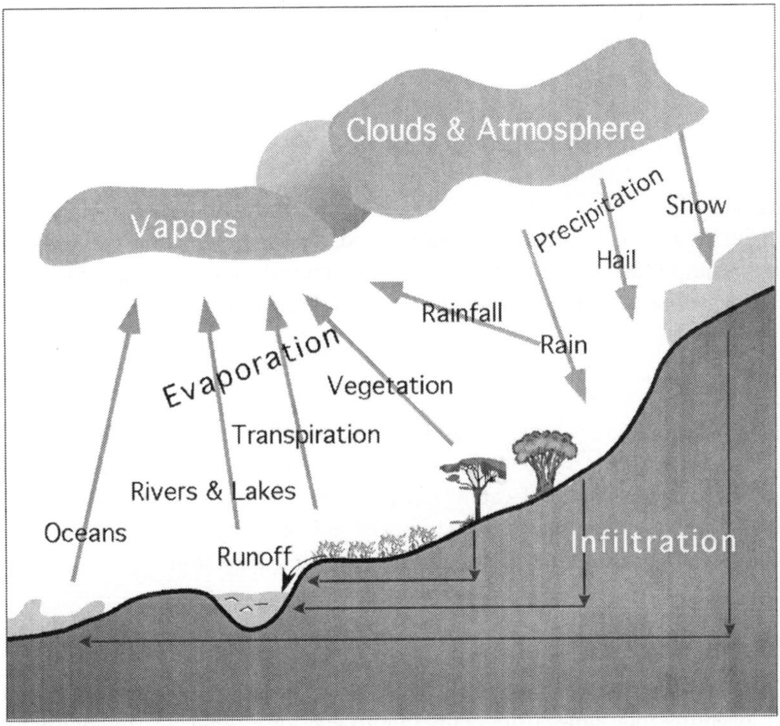

FIGURE 16.4. Hydrologic cycle.

and some is used by vegetation. The water used by vegetation may be evaporated quickly or may become part of the plant processes. Water can be tied up in plants for extended periods, but it eventually returns to the cycle.

The precipitation that infiltrates into the soil also takes different paths. Some may fall on areas of deep soil and percolate deeply into the earth. Other infiltration may reach an impervious layer close to the surface and start to move horizontally quickly. The underground horizontal movement may end at some type of surface water, or may flow out of the ground as a spring or an artesian well. The water that infiltrates deeper into the soil may collect in large underground basins such the Ogallala aquifer, and may be pumped out of the ground for domestic, industrial, or agricultural use, or it eventually may reach the ocean. Horizontal movement through the soil may be as little as a few inches per year. Once the water reaches the ocean, the sun causes it to evaporate, and the cycle begins again.

The hydrologic cycle can be used to explain the importance of water conservation. Activities such as pumping, dam building, and so on, change the amount and direction of the flow of water. For example, any water captured by a dam, not only reduces the amount of water available to down stream users, it also changes the amount of water that infiltrates into the soil and evaporates into the atmosphere. Contamination, in the form of chemicals, silt, and so forth, added to the water at one point in the cycle may remain in the water and cause problems for the next user.

16.7. Rainfall Intensity, Duration, and Recurrence

Engineers, conservationists, ecologists, and agricultural producers are interested in rainfall because of its impact on erosion, floods, and water available for crops. There are four important characteristics of rainfall; intensity, duration, total amount, and recurrence interval (for storms). These are all discussed in the following sections.

16.7.1. Intensity

Rainfall intensity is the rate of rainfall. It is usually expressed in units of inches per hour or millimeters per hour. The intensity is an important characteristic of rainfall because, other things being equal, more erosion is caused by one rainstorm of high intensity than by several storms of low intensity. A high intensity storm will have a greater volume of runoff than a low intensity storm.

16.7.2. Duration

Duration is the period of time that rain falls at a constant rate or intensity. It does not mean the total time, from beginning to end, of rainfall. During any rain fall event (storm) the rainfall intensity will not be constant. It may vary from quite high to very low; so it is necessary to think in terms of how long a specific rainfall intensity lasts (min or hr) at a certain rate (in/min or mL/min). The example of the rainfall intensity-duration for a rainfall event in Figure 16.5 has eight (8) different durations during a rainfall event that lasted for 94 min. The total amount of rain

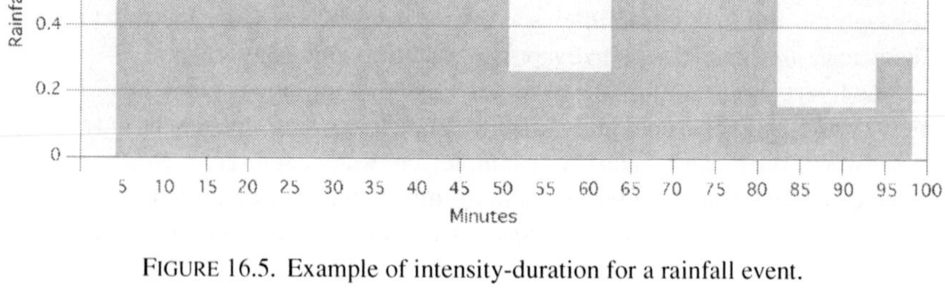

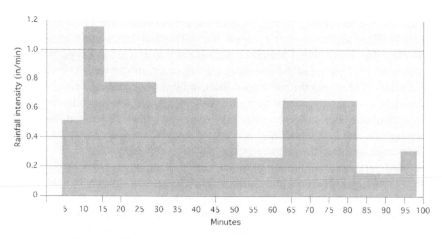

FIGURE 16.5. Example of intensity-duration for a rainfall event.

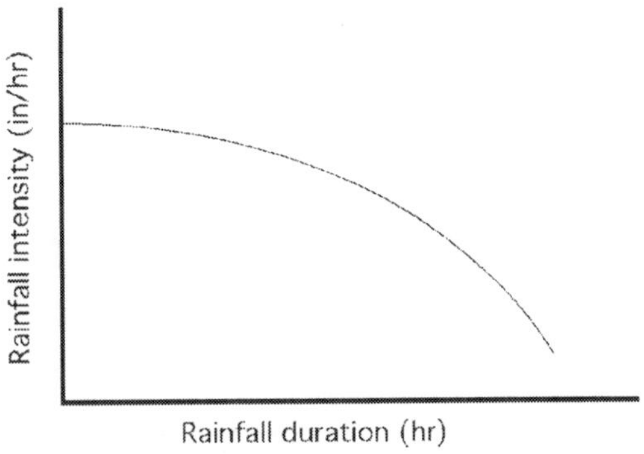

FIGURE 16.6. Typical rainfall intensity—duration curve.

(inches or millimeters) for a rainfall event is the sum of each rainfall intensity times the duration of that intensity.

The average rainfall intensity for a storm is between the highest and the lowest rainfall intensity. As a general rule, the high-intensity portion of a storm has a shorter duration than the low-intensity portions. A typical relationship between rainfall intensity and duration is illustrated in Figure 16.6.

16.7.3. Recurrence Interval

Another important aspect of rain is how often a storm of a specified intensity and duration may be expected to occur. The recurrence interval is defined as the number of years (on the average) before a storm of a given intensity and duration can be expected to recur. A storm that would be expected on an average of once in 25 years is said to have a 25-year recurrence interval, or is called a 25-year storm. It is important to remember that this is all based on the laws of probability, or chance, and that these estimates are based on averages only. There is nothing to prevent a 25-year storm from happening in two successive years, or even more than twice in one year, although the odds against such frequent occurrence are great.

The National Weather Service has studied the rainfall records of major storms for many years. The results of these studies have been published in the form of Rainfall Intensity Recurrence Interval Charts. Such a chart would look like Figure 16.7.

Thus, a storm is considered to have a recurrence interval of 2, 5, 10, 25, 50, or 100 year, or longer, depending on the average number of years expected to pass before a storm of similar intensity and duration occurs again. For a given duration we would expect the intensity of a 100-year recurrence interval storm to be greater than that of a 10-year storm. A curve can be plotted showing the relationship between rainfall intensity and the expected duration for each recurrence interval.

16.7.4. Intensity–Duration–Recurrence Interval

Figure 16.8 shows typical intensity-duration-recurrence interval curves for a specific location. The lines in the chart represent the recurrence intervals of 2, 5, 10, 25, 50, and 100 years.

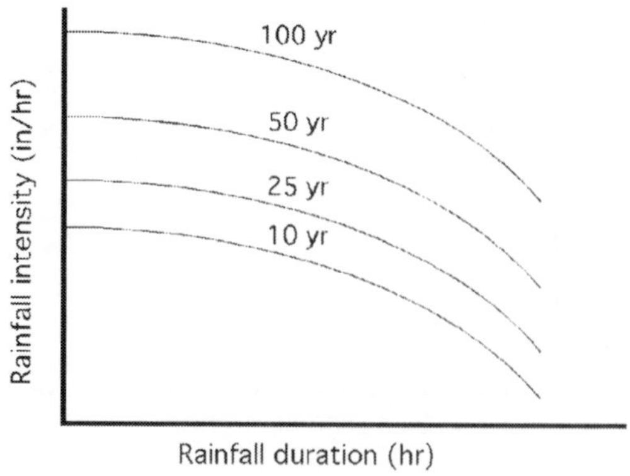

FIGURE 16.7. Typical rainfall intensity–duration–recurrence interval curves.

Note that both axes in the graph are logarithmic scales and must be carefully read. The time in the left half of the graph is in minutes. The time in the right half of the graph is in hours. To design water conservation structures, a graph appropriate for the area under study should be used.

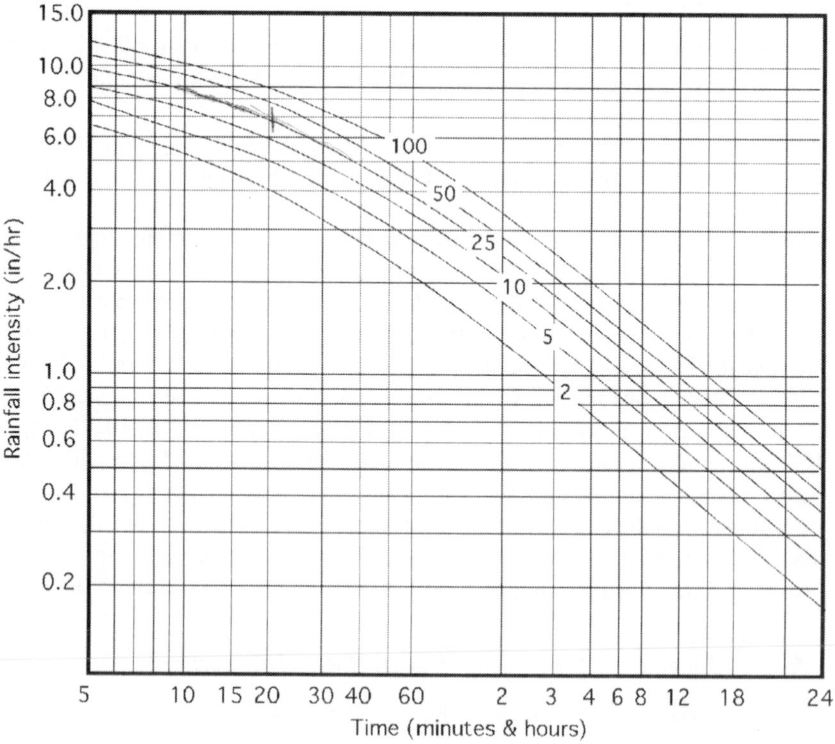

FIGURE 16.8. Example of rainfall intensity–duration–recurrence interval.

Problem: What is the expected rainfall intensity for a duration of 40 min and a 10 year recurrence interval.

Solution: Enter the table at 40 min along the horizontal axis. Move vertically until intersecting with the ten year curve. Move horizontally to the left until intersecting with the rainfall intensity axis. Read the rainfall intensity = 3.5 in/hr, Figure 16.9.

16.8. Metric Problems

The hydraulic cycle is same regardless of the units being used. The only difference is in the units used to measure rainfall. The rainfall-intensity-duration graph included in this section with SI units looks different because it is not a logarithmic graph, Figure 16.10.

Problem: Determine the rainfall intensity for a 5 year storm and duration of 20 min using SI units

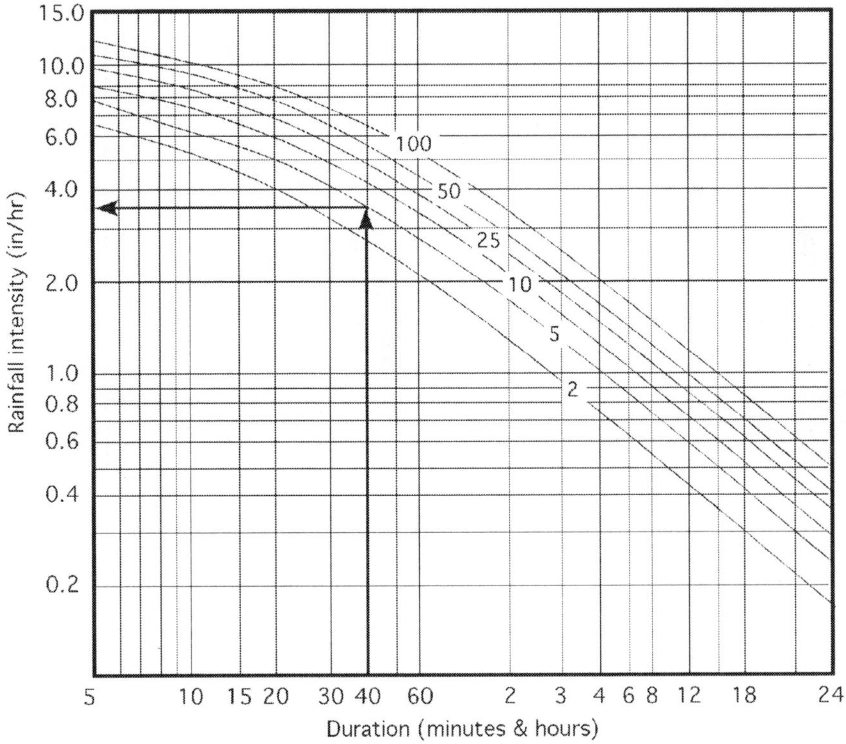

FIGURE 16.9. Rainfall intensity for sample problem.

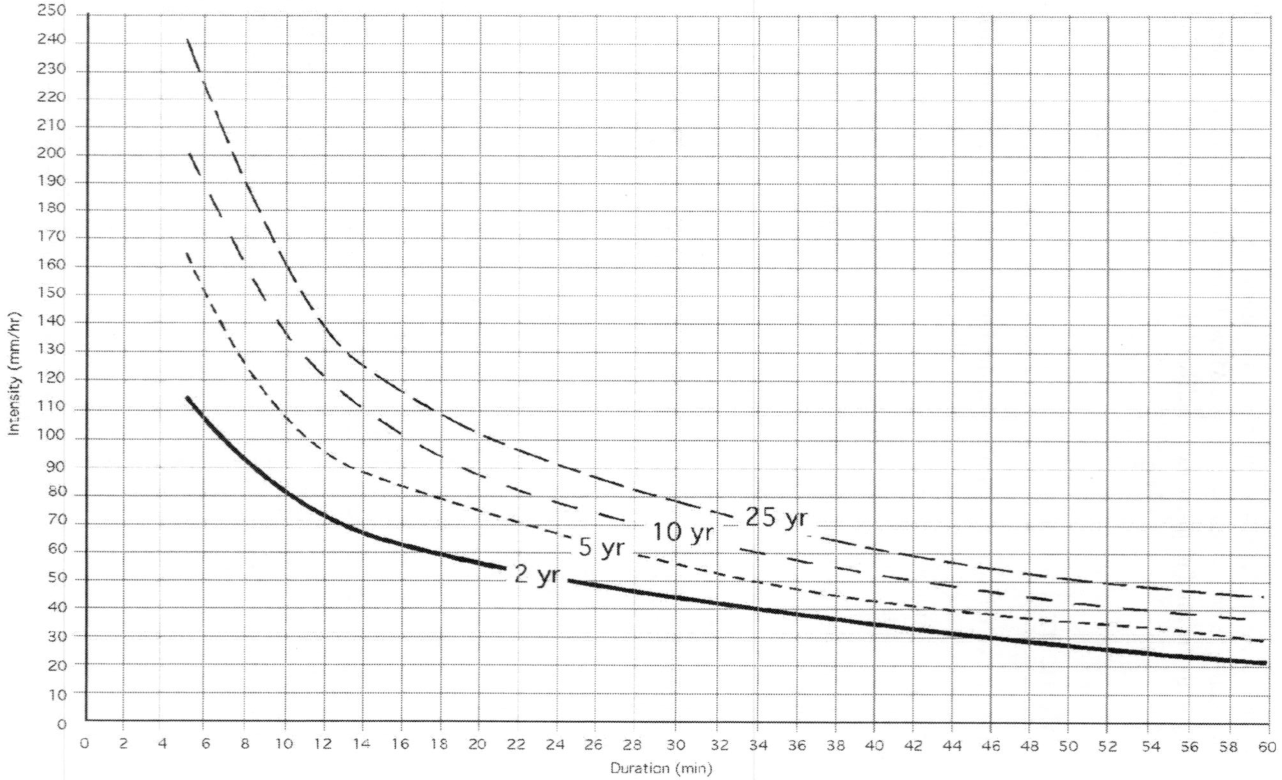

FIGURE 16.10. Rainfall Intensity—duration–reoccurrence interval graph with SI units.

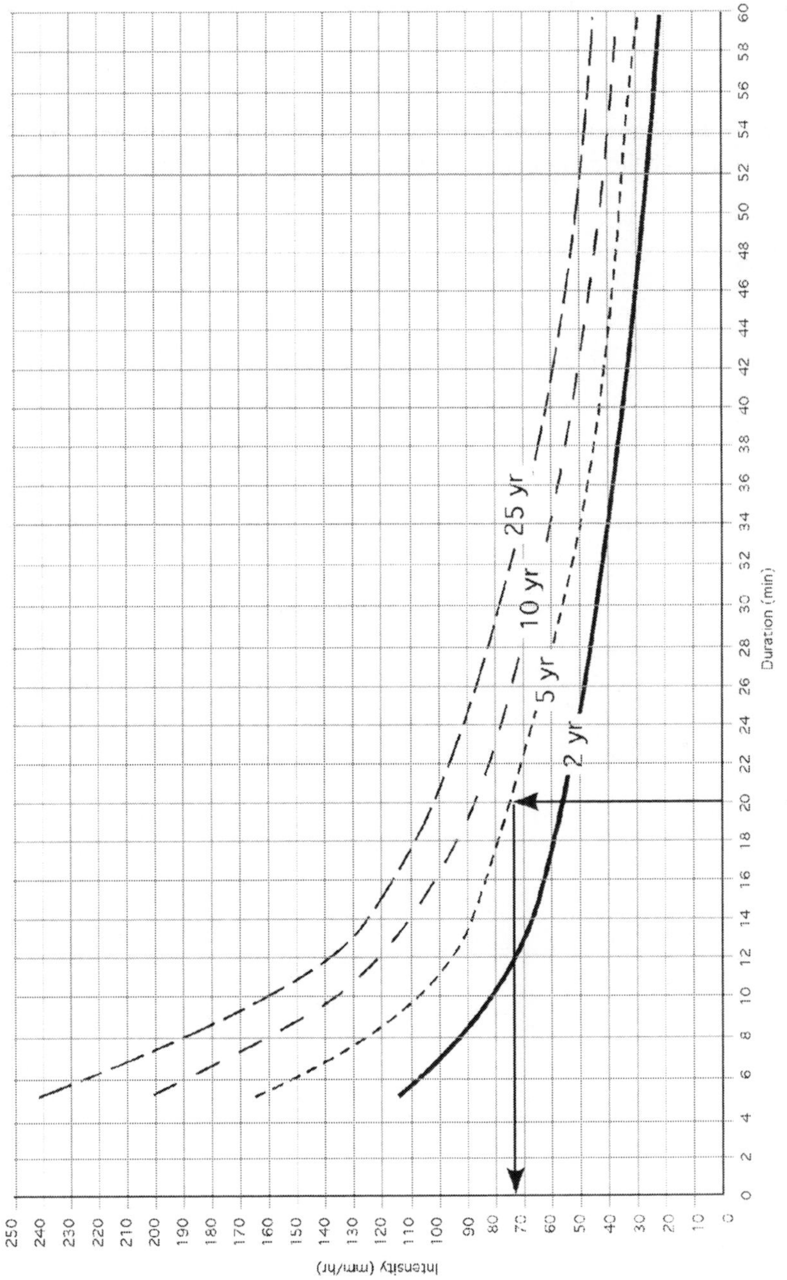

FIGURE 16.11. Rainfall intensity for sample problem.

Solution: Enter Figure 16.11 along the horizontal axis at the 20 min mark and move vertically to the 5 year storm curve and then move horizontally left to the rainfall intensity scale. Read rainfall intensity = 72 mm/hr.

Problem: What is the expected recurrence interval for a storm with a duration of 28 min and an intensity of 70 mm/min?

Solution: Move upward from 28 min duration and to the right from 70 mm/hr, Figure 16.12. At the point were they converge read 10-year interval.

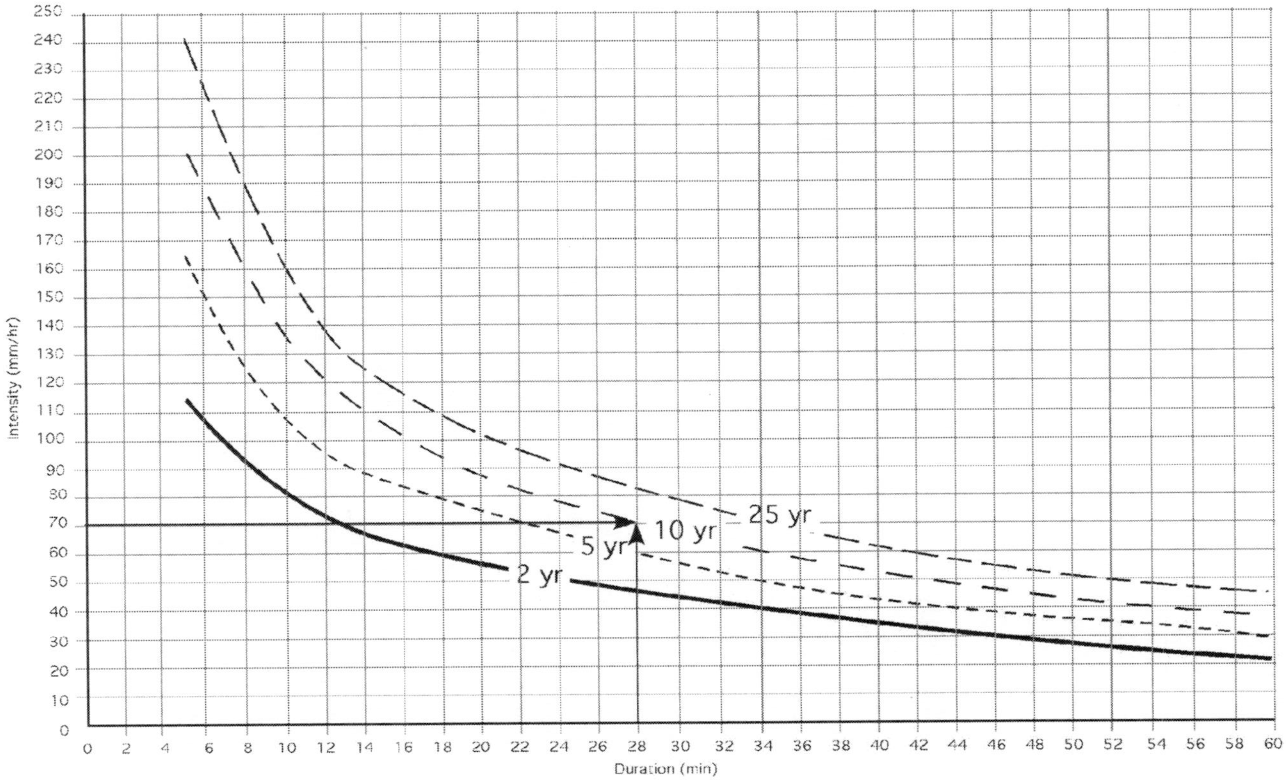

FIGURE 16.12. Reoccurrence interval for sample problem.

17
Water Runoff

17.1. Objectives

1. Be able to discuss the watershed characteristics that influence water runoff
2. Be able to use the Rational method for determining peak runoff rates
3. Be able to utilize a rainfall intensity-duration-reoccurrence interval graph
4. Given the characteristics of a mixed watershed, be able to determine the peak runoff rate for a specified storm.

17.2. Introduction

In the previous chapter it was determined that runoff occurs whenever the rainfall intensity exceeds the infiltration rate. In many situations it is important to be able to determine the volume of water that will run off an area and/or the maximum rate of runoff. The volume of runoff is used to size drainage structures and water impoundments. One of the uses of the peak rate is to size drainage ways, culverts, and bridges. The peak rate can be easily visualized if the flow rate is plotted in the form of a hydrograph. A hydrograph plots the runoff rate for a watershed versus time. A watershed is a drainage basin where all of the water that runs off passes through one point, Figure 17.1.

This point may be the outlet into another stream or an impoundment like a pond or lake.

Figure 17.2 is a typical hydrograph and it shows that runoff gradually increases to a peak rate and then drops off after the rainfall stops and the watershed drains. Flow rates in streams are generally measured in units of cubic feet per second (cfs) or cubic meters per second (cms). This chapter will discuss the Rational method, which is one method that is used to calculate the peak rate of runoff for areas less than 20 acres.

17.3. Peak Rate of Runoff

The rate of runoff is not constant for any rainfall event. Depending on the characteristics of the watershed there will be some delay from the start of the rain until

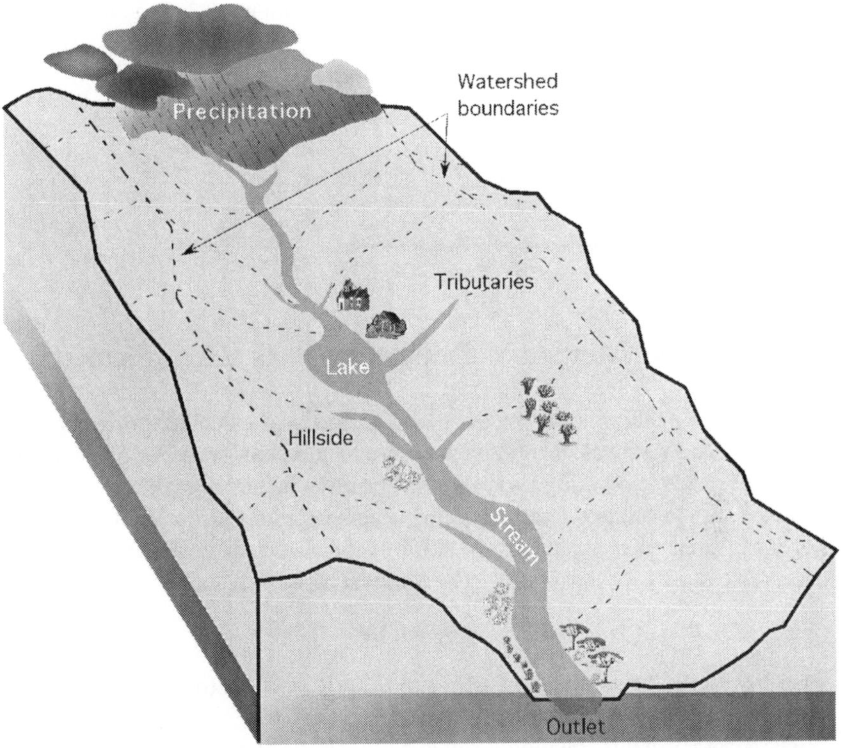

FIGURE 17.1. Model of watershed.

runoff starts to occur. Once runoff occurs each watershed will have a different time lapse until the peak rate occurs. The effect of these differences can be seen by evaluating the hydrograph for each watershed.

The shape of the curve will be different for each rainfall event and for each watershed. For example, if the rainfall event has a longer duration, the hydrograph could look more like Figure 17.3.

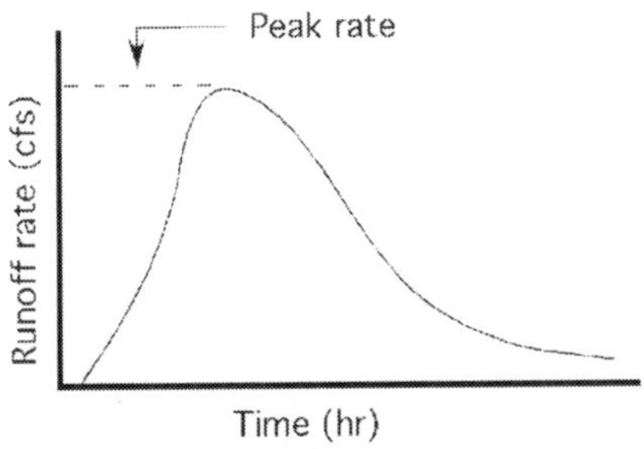

FIGURE 17.2. Typical runoff hydrograph.

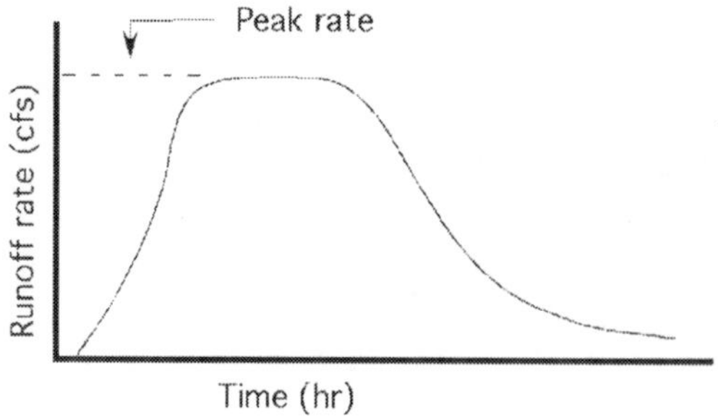

FIGURE 17.3. Typical hydrograph with longer duration.

17.4. Rational Method of Calculating Peak Rate of Runoff

The Rational method is one of the oldest methods of calculating the peak rate of runoff. The Rational method is considered simplistic, but it is still useful for small watersheds and using it explores the primary factors that affect runoff. The confidence in the output decreases as the watershed size and variability increases. For small areas that are relatively uniform, the Rational method will produce acceptable results for most needs. The Rational method equation is:

$$Q = \mathrm{CIA}$$

where Q = Peak rate of runoff (ft^3/sec); C = Runoff coefficient; I = Rainfall intensity (in/hr); A = Drainage area, watershed (ac).

The accuracy of this equation as a predictor of the peak runoff rate is only as good as the accuracy of the numbers used for each of the variables. The following tables contain generalized values and should be used only as examples. When the calculations are critical, the actual values for a particular watershed should be obtained from the National Resource Conservation Service or the appropriate Web site.

17.4.1. Runoff Coefficient (C)

The runoff coefficient (C) is defined as the ratio of the peak runoff rate to the rainfall intensity. The runoff coefficient mathematically indicates for a watershed whether the runoff is likely to be high or low. The value of C depends on the type and characteristics of the watershed. If the watershed is composed of very tight soil, steep slopes, or cultivated land (or all three of these), the runoff rate will be high. If the soil is sandy, with flat slopes, and covered with good vegetation, the runoff will be low.

The runoff coefficient for watersheds with various topographic, soil, and cover conditions can be estimated by using the values given in Table 17.1.

17.4.2. Rainfall Intensity (I)

The rainfall intensity used in the rational method is based on a specific rainfall duration and recurrence interval. The recurrence used depends on the importance of

TABLE 17.1. Table of runoff coefficients (C).

Topography, vegetation and slope	Soil texture: Sandy loam	Clay and silt loam	Tight clay
Woodland			
Flat 0–5%	0.10	0.30	0.40
Rolling 5–10%	0.25	0.35	0.50
Hilly 10–30%	0.30	0.50	0.60
Pasture			
Flat 0–5%	0.10	0.30	0.40
Rolling 5–10%	0.16	0.36	0.55
Hilly 10–30%	0.22	0.42	0.60
Cultivated			
Flat 0–5%	0.30	0.50	0.60
Rolling 5–10%	0.40	0.60	0.70
Hilly 10–30%	0.52	0.72	0.82

the project. Terraces and waterways are designed for a 10-year recurrence, whereas spillways for dams may require a design based on a recurrence interval of 100 years or more. The duration of rainfall used in the rational method is determined by the time of concentration of the watershed.

17.4.3. Time of Concentration

The time of concentration for a watershed is defined as the time required for water to flow from the most remote point of the watershed to the outlet. The peak rate will occur when the entire watershed contributes to the runoff.

The peak runoff flow is influenced by many factors. For the purposes of this text, the only factors that will be considered are slope and length of the drainage way. Obviously, if the drainage way is short and steep, the water will arrive at the outlet quickly and the time of concentration will be short. A flat drainage way gradient has the opposite effect. The time of concentration for small watersheds with various lengths and drainage way gradients is shown in Table 17.2.

TABLE 17.2. Time of concentration for small watersheds (min).

Maximum length of flow (ft)	Drainage way gradient (slope), %: 0.05	0.10	0.50	1.00	2.00	5.00
500	18	13	7	6	4	3
1,000	30	23	11	9	7	5
2,000	51	39	20	16	12	9
4,000	86	66	33	27	21	15
6,000	119	91	46	37	29	20
8,000	149	114	57	47	36	25
10,000	175	134	67	55	42	30

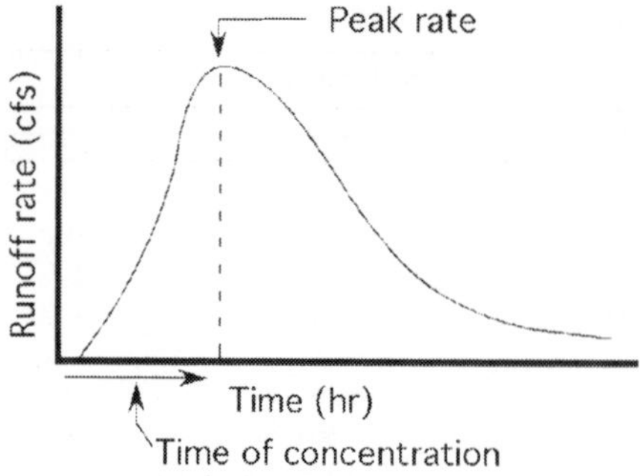

FIGURE 17.4. Determining time of concentration.

The Rational method equation has three variables, C, I, and A. A value must be determined for each one before the equation can be used. The value of C is determined using Table 17.1. The value of I is more complex. To determine I the rainfall intensity–duration–recurrence interval, Figure 16.10, is used. To find rainfall intensity from this graph the duration and recurrence interval must be known. In the Rational method the duration is determined from the time of concentration. The time of concentration is the amount of time that occurs from when the rainfall starts until the peak rate or runoff occurs, Figure 17.4.

Once the time of concentration is determine from Table 17.1, and the desired recurrence interval is known, Figure 16.10 can be used to determine the rainfall intensity.

Problem: Determine the peak runoff for a watershed consisting of 90.0 acres of pasture with tight clay soil and an average slope of 4.0%. The drainage way for the watershed is approximately 4,000 ft with a gradient of 0.50%. Assume a recurrence interval of 10 years.

TABLE 17.3. "C" value for sample problem.

	Soil texture		
Topography, vegetation and slope	Sandy loam	Clay and silt loam	Tight clay
Woodland			
Flat 0–5%	0.10	0.30	0.40
Rolling 5–10%	0.25	0.35	0.50
Hilly 10–30%	0.30	0.50	0.60
Pasture			
Flat 0–5%	0.10	0.30	0.40
Rolling 5–10%	0.16	0.36	0.55
Hilly 10–30%	0.22	0.42	0.60
Cultivated			
Flat 0–5%	0.30	0.50	0.60
Rolling 5–10%	0.40	0.60	0.70
Hilly 10–30%	0.52	0.72	0.82

The correct value for "C" is 0.40.

TABLE 17.4. Time of concentration for sample problem.

	Drainage way gradient (slope), %					
Maximum length of flow (ft)	0.05	0.10	0.50	1.00	2.00	5.00
500	18	13	7	6	4	3
1,000	30	23	11	9	7	5
2,000	51	39	20	16	12	9
4,000	86	66	33	27	21	15
6,000	119	91	46	37	29	20
8,000	149	114	57	47	36	25
10,000	175	134	67	55	42	30

The time of concentration is 33 min.

Solution: To determine the solution a value must be determined for each on the three variables in the rational equation using Table 17.1, Table 17.2, and Figure 16.10. The correct numbers to use for this example problem are shown in Table 17.3, Table 17.4, and Figure 17.5.

The rainfall intensity (I) is 4.4 in/hr.

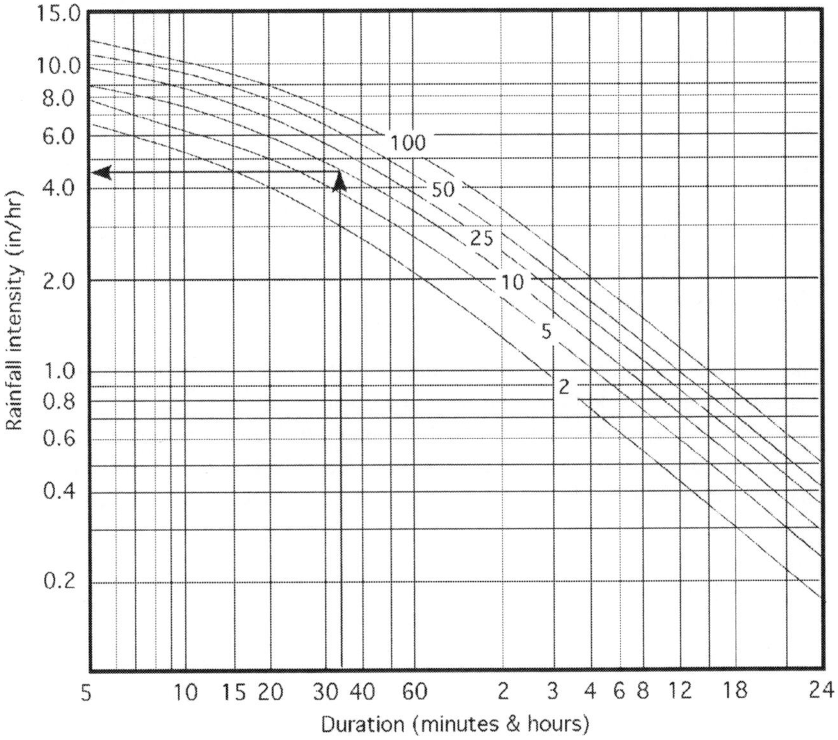

FIGURE 17.5. Rainfalll intensity for sample problem.

With $C = 0.40$, $I = 4.5$ in/hr, and $A = 90$ ac the peak rate of runoff is:

$$Q = \text{CIA} = 0.40 \times 4.5\,\frac{\text{in}}{\text{hr}} \times 90\text{ ac} = 162\frac{\text{ft}^3}{\text{sec}}$$

In this example the best estimate of the peak runoff rate is 162 cfs. Note that the units do not cancel in the equation. This is because the C value is determined so that the answer is in cubic feet per second.

17.5. Effect of Varying Recurrence Interval

The choice of recurrence interval will have a great influence on the peak rate of runoff to be expected from a watershed. This is illustrated by the following example.

Problem: Given the same watershed as the previous example, calculate the peak runoff rate for recurrence intervals of 2, 5, 10, 25, 50, and 100 years.

Solution: Table 17.5

TABLE 17.5. Effect of varying recurrence interval.

Recurrence interval (yr)	Time of concentration (min)	Rainfall intensity (in/hr)	Runoff coefficient (C)	Watershed area (ac)	Peak runoff rate (cfs)
2	33	3.0	0.40	90	108
5	33	3.8	0.40	90	137
10	33	4.5	0.40	90	162
25	33	5.5	0.40	90	198
50	33	6.3	0.40	90	227
100	33	7.0	0.40	90	252

17.6. Mixed Watersheds

The previous examples were the simplest kind because they had the same slope, vegetation, and soil conditions throughout the watershed. In nature, this only occurs for very small areas. Watersheds usually contain different slopes, vegetation, soil types, and farming practices. All of these variables affect the value used for C. Any changes in the value used for C will change the calculated runoff. To cope with a variable watershed it is necessary to calculate the weighted runoff coefficient (C_w) for the watershed.

The weighted runoff coefficient is determined by finding the appropriate value of C for each field or portion of the watershed that is different, multiplying that value of C by the appropriate area (ac), adding up these products for all of the different areas in the watershed, and then dividing their sum by the total watershed area.

Study Figure 17.6 and the following equation:

$$C_w = \frac{(A_1C_1) + (A_2C_2) + \cdots + (A_nC_n)}{\sum A}$$

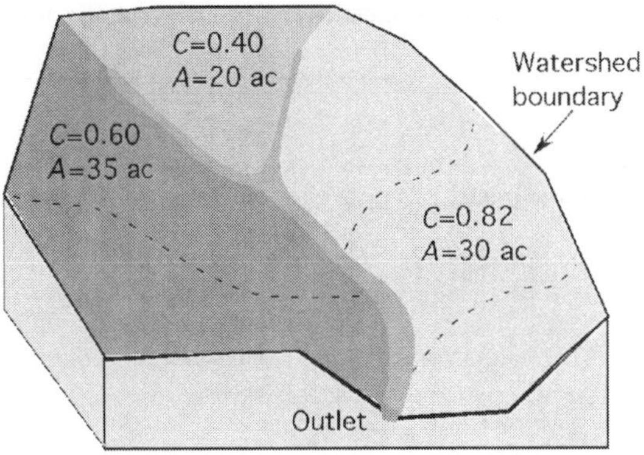

FIGURE 17.6. Example of a mixed watershed.

Problem: Determine the weighted C for the watershed in Figure 17.6.

Solution:

$$\begin{aligned} C_w &= \frac{(A_1C_1) + (A_2C_2) + (A_3C_3)}{\Sigma A} \\ &= \frac{(35 \times 0.60) + (20 \times 0.40) + (30 \times 0.82)}{35 + 20 + 30} \\ &= \frac{21 + 8 + 24.6}{85} \\ &= \frac{53.6}{85} \\ &= 0.6305 \ldots \text{ or } 0.63 \end{aligned}$$

The weighted C value for this watershed is 0.63.

17.7. Metric Problems

The Rational method was developed during the 1880s in the United States using standard units. For use with metric units the user has two options. The first one is to use the standard data and convert the answer from cubic feet per second to cubic meters per second by dividing by 360. This results in the equation:

$$Q\left(\frac{\text{m}^3}{\text{sec}}\right) = \frac{\text{CIA}}{360}$$

where Q = peak flow (m^3/s); A = drainage area (ha); C = runoff coefficient (weighted); I = rainfall intensity (mm/hr); 360 = Conversion constant.

The second method is to use tables and graphs that have metric units. When this method is used the coefficient of runoff is taken from Table 17.6.

The time of concentration is found in Table 17.7.

The rainfall intensity is taken from Figure 17.7.

TABLE 17.6. Recommended coefficient of runoff values for various selected land uses, SI units.

Land use	Description	Hydrologic soils group*			
		A	B	C	D
Cultivated Land	without conservation treatment	0.49	0.67	0.81	0.88
	with conservation treatment	0.27	0.43	0.67	0.67
Pasture or Range Land Meadow	poor condition	0.38	0.63	0.78	0.84
	good condition	—	0.25	0.51	0.65
	good condition	—	—	0.41	0.61
Wood or Forest Land	thin stand, poor cover, no mulch	—	0.34	0.59	0.70
	good cover			0.45	0.59
Open Spaces, Lawns, Parks, Golf Courses, Cemeteries					
Good Condition	grass cover on 75% or more	—	0.25	0.51	0.65
Fair Condition	grass cover on 50% to 75%	—	0.45	0.63	0.74
Commercial; and Business Area	85% impervious	0.84	0.90	0.93	0.96
Industrial Districts	72% impervious	0.67	0.81	0.88	0.92
Residential	average % impervious				
Average Lot Size (acres)					
1/8	65	0.59	0.76	0.86	0.90
1/4	38	0.29	0.55	0.70	0.80
1/3	30	—	0.49	0.67	0.78
1/2	25	—	0.45	0.65	0.76
1	20	—	0.41	0.63	0.74
Paved Areas	parking lots, roofs, driveways, etc	0.99	0.99	0.99	0.99
Streets and Roads	paved with curbs and storm sewers	0.99	0.99	0.99	0.99
	gravel	0.57	0.76	0.84	0.88
	dirt	0.49	0.69	0.80	0.84

Note: Values are based on NRCS (formerly the SCS) definitions and are average values.
Source: Technical Manual for Land Use Regulation Program, Bureau of Inland and Coastal Regulations, Stream Encroachment Permits, New Jersey Department of Environmental Protection
* Group A Soils: High infiltration (low runoff). Sand, loamy sand, or sandy loam. Infiltration rate > 0.3 inch/hr when wet.
Group B Soils: Moderate infiltration (moderate runoff). Silt loam or loam. Infiltration rate 0.15 to 0.3 inch/hr when wet.
Group C Soils: Low infiltration (moderate to high runoff). Sandy clay loam. Infiltration rate 0.05 to 0.15 inch/hr when wet.
Group D Soils: Very low infiltration (high runoff). Clay loam, silty clay loam, sandy clay, silty clay, or clay. Infiltration rate 0 to 0.05 inch/hr when wet.

TABLE 17.7. Time of concentration (minutes) SI units.

Drainage way change in elevation (m)	Drainage way length (m)							
	100	150	200	250	500	1,000	1,500	2,000
1	4	6	8	11	23	55	85	120
2	3	4.5	6.5	8	18	40	65	95
5	2	3	4.5	6	13	29	44	65
10	1.5	2.4	3.5	4.5	9	22	36	50
15	1.25	2	3	4	8	20	33	42
20	0.25	1.75	2.5	3.5	7	17	27	40

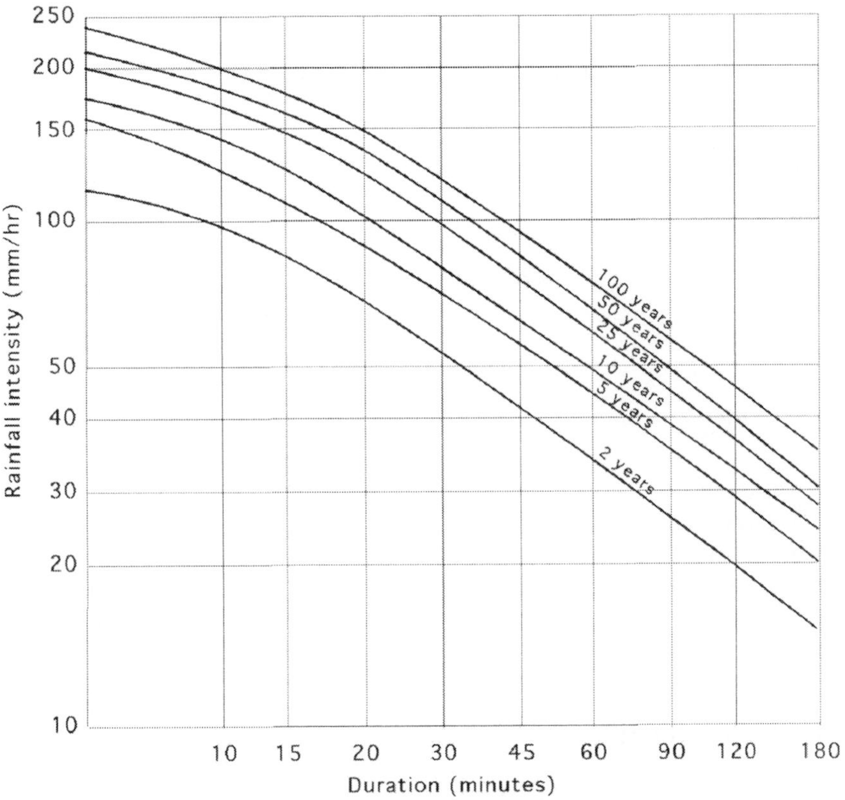

FIGURE 17.7. Rainfall intensity, duration, and frequency for metric units.

Problem: Determine the peak rate of runoff for a 50 hectare cultivated field with conservation practices and soil group B. The drainage way has a change of elevation of 2 m and a length of 1,000 m. The design calls for a 25-year storm.

Solution:

$$Q\left(\frac{\mathrm{m}^3}{\mathrm{sec}}\right) = \mathrm{CIA}$$

$$= 0.43 \times 120 \times 50$$

$$= 2{,}580\ \frac{\mathrm{m}^3}{\mathrm{sec}}.$$

18
Erosion and Erosion Control

18.1. Objectives

1. Understand the development of soil.
2. Understand the causes of erosion.
3. Be able to identify the two types of erosion.
4. Determine the rate of erosion using the Universal Soil Loss Equation.
5. Be able to explain common erosion control practices.

18.2. Introduction

Soil is a complex, constantly changing resource. It affects the life of plants and thereby all animals, and it is affected by plants and animals. Soil is the primary medium of growth for plants. It supplies nutrients, water, and a place to anchor roots. Soil has an economic value and therefore any that is lost from production is an economic loss. The topsoil has the most value because it is the primary source of nutrients, chemicals, seeds, water-holding capacity, and organic matter. Soil is formed and destroyed by erosion. In this chapter soil development, destruction, and conservation will be discussed.

18.3. Soil Development

Soil is developed and destroyed by erosion. Soil develops from the breakdown, or geological erosion, of sedimentary or volcanic parent material. As the forces of nature: wind, water, heat, and cold, erode the parent material, the particle size gradually decreases, and nutrients, water, and organic matter are mixed in. The exact rate of breakdown is debatable, but one estimate is that it takes 100 to as long as 1,000 years to develop 1 inch (25.4 mm) of topsoil.

18.4. Causes of Erosion

Destructive erosion is a natural process. Rivers have always flooded and eroded the banks, intense storms have caused excessive runoff which has eroded soil and left it in a water way or pond. The problem is that many of mankind activities greatly accelerate natural erosion. The productivity of soil is destroyed by man-made or man-accelerated erosion that occurs from poor agricultural and construction practices. Land with permanent vegetation is very stable, but as soon as vegetation is removed, the soil is exposed to rainfall and other forces of nature and erosion accelerates. Activities such as the construction of roads and buildings, cultivation of fields, and timber harvesting all remove the natural protective cover. The two agents of erosion, water and wind, will be discussed in the next section, and then a method for estimating the amount of soil that is lost by water erosion will be illustrated.

18.5. Two Types of Erosion

18.5.1. Water

The amount of erosion caused by water is dependent on four factors: climate, soil, vegetation, and topography. The impact of climate is related to the amount and the intensity of rain. The greater the annual rainfall—the greater the potential for water erosion; and as the frequency of intense storms increases—the greater the potential for water erosion.

The contribution of soil to erosion relates to the size of the soil particles and the moisture content of the soil. Sandy and organic soils have the greatest potential for water erosion. These soils are not bound together, especially when wet, and the soil particles are easily moved.

Vegetation plays an important role in water erosion; it reduces the energy of the raindrops striking the soil, in turn reducing the displacement of the soil. In cultivated fields, crop residues in and on the soil provide the same function as natural vegetation. The amount of crop residues can be measured by randomly laying out a 100-ft tape across a field and counting the number of one foot marks that are over or touching any residue. This number is the residue cover as a percent. If four foot marks are over a piece of residue, then the percent is 4%. The recommended minimum percentage is based on the location and the soil type.

The influence of topography is associated with the slope of the land. The greater the slope, the greater the potential for water erosion.

These four factors predict what type of water erosion occurs and how much occurs. Water erosion usually is divided into three stages: raindrop, sheet and rill, and gully. Raindrop erosion is the soil splash resulting from the impact of water on soil particles. If the soil is covered with vegetation, raindrop erosion is almost zero. When the soil is a bare cultivated field, raindrop erosion can be significant. It is estimated that raindrops can displace soil particles 2 ft vertically and 5 ft horizontally. The effect of raindrop erosion increases as the slope increases because there is greater movement down slope than there is up slope.

Sheet and rill erosion are the next two stages. They are combined because many experts believe sheet erosion only exists in theory. As the rainfall intensity exceeds the infiltration rate of the soil, water starts to move across the soil's surface in a thin sheet. Almost as soon as the movement begins, small but well-defined channels develop, which are called rills. Rills can be easily farmed over and are easy to overlook, but they account for most of the soil loss to water erosion.

Gully erosion is the next stage. Gullies are an advanced stage of rills. Small ones may still be farmed over, but if they are not checked, they become too large and then must be farmed around.

18.5.2. Wind

Movement of the soil due to wind erosion is not down the slope as in water; instead the direction is determined by the wind. When the soil is completely covered with vegetation, very little wind erosion occurs because the vegetation reduces the velocity of the wind at the surface of the soil. When the wind has access to the surface, erosion may occur. The velocity of the wind, the soil moisture, and the soil particle size determine the amount of erosion and type of erosion. The greater the velocity—the greater the volume and the larger the size of particle that the wind can pick up. Fine sandy soils are affected by wind more than clay soils. Dry soils are easier to erode than moist soils.

Wind erosion is not divided into types; it only varies by degree (stages). There are three stages of movement. All three stages may occur simultaneously. The first stage, suspension, occurs when soil particles are fine enough and the velocity is high enough to keep the particles suspended in the air. Very fine particles can be transported for hundreds of miles. The second stage is skipping and bouncing, or saltation. This state of wind erosion accounts for the largest volume of soil movement. In saltation, the wind is strong enough to pick up the soil particle, but not strong enough to hold them in suspension. The wind lifts up the particles and carries them for a short distance, but then they are dropped. The third stage is rolling or creep. In the creep stage the wind moves or rolls the soil particles across the surface of the soil, but it does not pick them up.

The five factors used to estimate wind erosion are soil erodibility, climate, soil roughness, field length, and vegetation. To estimate the amount of wind erosion for a specific location, contact the National Resource Conservation Service (NRCS).

18.6. Estimating Soil Loss

Equations have been developed to estimate the amount of soil loss to water and wind erosion. These are called the Universal Soil Loss Equation (USLE) for water and the wind erosion equation (WEQ) for wind. The USLE for water has been updated and the new version is called the Revised Universal Soil Loss Equation

(RUSLE). Another version, RUSLE2, is a computer program. The WEQ is a computer program that can be utilized through a NRCS office. The USLE for water is:

$$A = R \times K \times L \times S \times C \times P$$

where A = Predicted average annual soil loss (T/ac/yr); R = Rainfall factor; K = Soil erodibility factor; L = Slope length factor; S = Slope gradient factor; C = Cropping management factor; P = Erosion control practice factor.

For this discussion, several of these factors will be combined to give the following equation:

$$A = R \times K \times LS \times CP$$

The USLE assigns numerical values to the factors that influence water erosion. Therefore, the accuracy of the calculated soil loss is only as good as the numerical values representing these factors. The tables and graph included here are generalized to show the influence of the different factors and to aide in the understanding of the equation. These values should not be used for the design of soil conservation structures or for estimating soil loss for government programs. More accurate values for specific areas can be obtained from the NRCS or similar agency.

18.6.1. Rainfall Factor (R)

The rainfall factor is a measure of the intensity and duration of the expected rainfall events. Common values for R range from 100 to 350. A higher number should be used in an area of high annual rainfall and high rainfall intensities. Numbers are available for specific locations from government sources.

18.6.2. Soil Erodibility (K)

The K value is a measure of the soil's susceptibility to erosion. The soil's texture, organic matter content, and permeability influence the K value. The values are based on a cultivated, continuous fallow arbitrary field with a slope length of 72.6 ft and a slope of 9%. Typical values for some soils and organic matter contents can be found in Table 18.1.

18.6.3. Topographic Factor (LS)

The topographic factor is a measure of the expected soil loss for the field compared to the arbitrary field with a slope length of 72.6 ft and a slope of 9%. It is a combination of the slope length and the percent of slope. An estimate for LS can be found in Figure 18.1.

TABLE 18.1. Soil erodibility factor (K) (ton/ac).

	Organic matter content %		
Textural class	0.5	2.0	4.0
Fine sand	0.16	0.14	0.10
Very fine sand	0.42	0.36	0.28
Loamy sand	0.12	0.10	0.08
Loamy very fine sand	0.44	0.38	0.30
Sandy loam	0.27	0.24	0.19
Very fine sandy loam	0.47	0.41	0.33
Silt loam	0.48	0.42	0.33
Clay loam	0.28	0.25	0.21
Silty clay loam	0.37	0.32	0.26
Silty clay	0.25	0.23	0.19

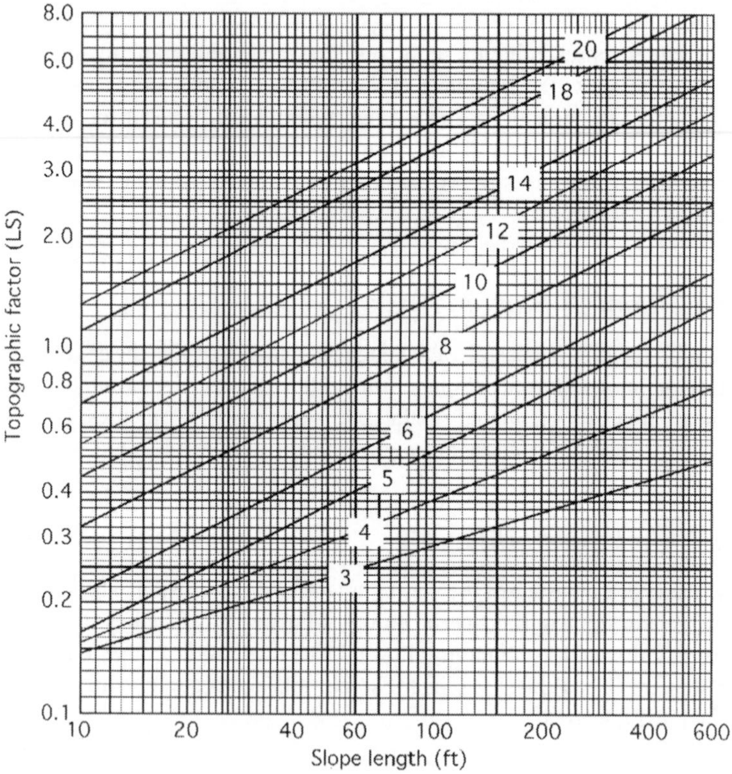

FIGURE 18.1. Typical topographic factors (LS) for slopes of 3–20% and slope lengths of 10–600 ft.

TABLE 18.2. Typical cropping and management factors (*CP*).

Management factors	Cropping practices: Up and down the slope	Terraces and field boundary	On the contour
Continuous small grain MRU (6/20)	0.29	0.21	0.15
Continuous small grain HRU (6/20)	0.22	0.16	0.11
Continuous small grain MRU (8/1)	0.22	0.16	0.11
Continuous small grain HRU (8/1)	0.18	0.13	0.09
Continuous small grain ROS	0.12	0.09	0.06
Continuous cotton MF no WC	0.59	0.42	0.30
RC Continuous grain sorghum (25–30 bu)	0.48	0.34	0.24
RC Continuous grain sorghum (35–45 bu)	0.42	0.30	0.21
Continuous peanuts with WC	0.43	0.30	0.22
Continuous peanuts no WC	0.54	0.38	0.27
Alfalfa 5 yr/small grain 2 yr	0.05	0.05	0.05

MRU = Moderate Residue Turned Under; HRU = Heavy Residue Turned Under; RC = Row Crop; ROS = Residue on Surface at Seeding Time; MF = Moderate Fertilizer; WC = Winter Cover.

18.6.4. Cropping and Management Factor (CP)

The original USLE separated the cropping factor (C) and management factor (P) into two different terms. In this discussion, they have been combined. The cropping factor is used to estimate the relative effectiveness of the practice to retard soil loss when compared to a continuously tilled field. The management factor (P) is an estimation of the effect of conservation practices in reducing soil loss. These factors combined are a function of the crop raised, the management practice, the planting date, the amount of residue on the surface, and the tillage practice. Values representing these factors can be found in Table 18.2.

Problem: Estimate the annual erosion for a field having a rainfall factor of 220, consisting mainly of loamy sand with 2% organic matter, and averaging a 4% slope with a slope length of 400 ft, which has been in continuous small grain, normally planted in June, with moderate residue worked under and farmed up and down the slope.

Solution: Using the USLE and the appropriate tables and figure:

$$A = R \times K \times LS \times CP$$

where $R = 220$; $K = 0.10$, Table 18.3.
$LS = 0.73$, Figure 18.2.
$CP = 0.29$, Table 18.4.

$$A = R \times K \times LS \times CP = 220 \times 0.10 \times 0.73 \times 0.29$$
$$= 4.6574 \text{ or } 4.6 \text{ T/ac/yr}$$

TABLE 18.3. *K* factor for sample problem.

	Organic matter content %		
Textural class	0.5	2.0	4.0
Fine sand	0.16	0.14	0.10
Very fine sand	0.42	0.36	0.28
Loamy sand	0.12	0.10	0.08
Loamy very fine sand	0.44	0.38	0.30
Sandy loam	0.27	0.24	0.19
Very fine sandy loam	0.47	0.41	0.33
Silt loam	0.48	0.42	0.33
Clay loam	0.28	0.25	0.21
Silty clay loam	0.37	0.32	0.26
Silty clay	0.25	0.23	0.19

The correct *K* factor is 0.10.

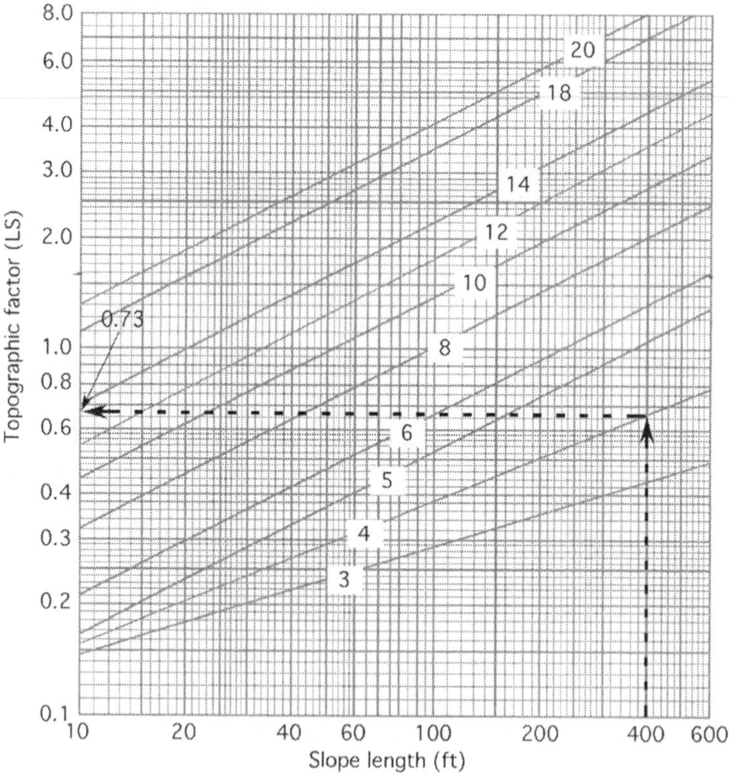

FIGURE 18.2. *LS* for sample problem.

TABLE 18.4. CP value for sample problem.

Management factors	Cropping practices: Up and down the slope	Terraces and field boundary	On the contour
Continuous small grain MRU (6/20)	0.29	0.21	0.15
Continuous small grain HRU (6/20)	0.22	0.16	0.11
Continuous small grain MRU (8/1)	0.22	0.16	0.11
Continuous small grain HRU (8/1)	0.18	0.13	0.09
Continuous small grain ROS	0.12	0.09	0.06
Continuous cotton MF no WC	0.59	0.42	0.30
RC Continuous grain sorghum (25–30 bu)	0.48	0.34	0.24
RC Continuous grain sorghum (35–45 bu)	0.42	0.30	0.21
Continuous peanuts with WC	0.43	0.30	0.22
Continuous peanuts no WC	0.54	0.38	0.27
Alfalfa 5 yr/small grain 2 yr	0.05	0.05	0.05

MRU = Moderate Residue Under; HRU = Heavy Residue Under; RC = Row Crop; ROS = Residue on Surface at Seeding Time; MF = Moderate Fertilizer; WC = Winter Cover.

18.7. Erosion Control

The best philosophy for erosion control is that it is better to prevent erosion than to try to correct it. There is no best solution for erosion control that will work in all situations. Erosion prevention and control must be developed for each site using the resources that are available.

On cultivated land, the best methods to use are those activities that will reduce the effect of wind and water on the soil particles. For the prevention of water erosion it is important to provide as much protection as possible for the surface of the soil. This includes management of tillage practices to leave residue on the surface and the use of cover crops in nonproductive seasons. Additional protection is provided by reducing the length of the continuous slope by installing terraces. To illustrate the effect of these practices, rework the sample problem with the appropriate values for terraces and residue.

Problem: Estimate the annual erosion for a field that has a rainfall factor of 220, consists mainly of loamy sand with 2% organic matter, averages a 4% slope with a slope length of 200 ft, and has been in continuous small grain with heavy residue, normally planted in June, and farmed with terraces.

Solution:

$$A = R \times K \times LS \times CP = 220 \times 0.10 \times 0.50 \times 0.21 = 2.31 \text{ or } 2.3\ \text{T/ac/yr}.$$

In the original problem the soil loss was 4.6 T/ac/yr. With the modifications the soil loss is 2.3 T/ac/yr, a reduction of 50%.

Residues on the surface also help prevent wind erosion, but reducing the length of the slope does not. The critical factors for preventing wind erosion, other than residue, are the roughness of the surface and the unobstructed distance that the wind can blow. Reducing the unobstructed distance is an effective control. This explains the prevalent use of shelterbelts in the Great Plains region. A shelterbelt will provide protection downwind for up to ten times its height.

Other methods are appropriate for the control of both water and wind on small areas for a short duration, such as construction sites. A common practice is to use commercial silt fences, bales of straw or nets to help prevent soil from leaving the site. Installing temporary wind barriers and keeping the ground covered can control wind erosion.

18.8. Metric Problems

This version of the USLE is only available in standard units and was included to illustrate the factors that affect soil erosion. Versions of the Revised Universal Soil Loss Equation (RUSLE) are available on the world wide web using metric units.

19
Irrigation

19.1. Objectives

1. Understand the purpose and use of irrigation.
2. Be able to describe the common irrigation systems.
3. Be able to calculate the required system capacity.
4. Be able to determine the irrigation interval and the depth of water to apply.

19.2. Introduction

The most limiting and most variable environmental factor affecting the productivity of plants is water. Whenever adequate water is not available, farmers have always tried to irrigate their crops. Irrigation water has always been in short supply, but it is becoming a scarce commodity in many regions. Even where it is available, pumping and/or transportation costs have increased dramatically in many locations. Today the profitability of irrigated agriculture is dependent of efficient use of water. The effective and efficient use of irrigation is dependent on four factors.

1. The effect of irrigation on plant production.
2. The best system for a given field and water supply.
3. Determining how much water to apply at peak use rate and when to apply it.
4. The quality of the water.

This chapter will discuss these aspects of irrigation.

19.3. The Effect of Irrigation

Agricultural plants need warm temperatures, sunlight, nutrients, and water to grow. In many regions of the world the required temperature and sunlight is available, but water is not. All plants have a minimum annual water requirement to survive and an optimum annual water requirement for maximum production. Historically, the availability of water has determined where crops can be grown. A high demand crop, such as rice, could not be grown in a region that has a low annual rainfall. In addition, whenever the water available to the plant is less than the optimum amount the production is reduced. The application of water through irrigation can provide the missing requirement for plant growth. The result of irrigation is that the area where some crops can be grown is expanded and the production of all crops is increased by supplementing rainfall.

19.4. Irrigation Systems

Once the decision has been made to irrigate, the next consideration is what type of system to use. Water is applied in one of three ways: from above the ground, on the ground surface, or from below the surface. The choice usually is based on cost and topography. This section will review the basic types of irrigation systems.

19.4.1. Above Ground

The most common above ground irrigation system is the sprinkler. In sprinkler systems, water is injected into the air by one of the three types of nozzles: spray, rotor, or impact. Sprinkler systems can be used on any land that can be cultivated. Sprinklers may be solid set (in a permanent position) or movable. The common means of movement are: (1) manual–lawn sprinkler or pipe system; (2) tractor-moved–skid or wheel-mounted system; (3) self-moved–side wheel roll or big gun system; and (4) self-propelled–center pivot and lateral move system. Different modifications and variations of these systems are used. The best system depends on the topography, shape and size of the field, the amount and cost of labor for movement, and the value added for the crop. For sprinkler systems, it is important that the application rate not exceed the infiltration rate of the soil.

19.4.2. Surface

In surface irrigation, water is applied by allowing it to flow over the surface by gravity or through drip valves (a process called trickle or drip irrigation). The common gravity systems are flood and furrow. Flood systems require relatively flat land and the water is distributed in level borders or contoured levees. Land for furrow irrigation can have a small slope. For level borders and contoured levees, the water usually is delivered by surface ditch. Ditches can deliver water for furrows, but a gated pipe or a siphon tube is also used. As a general rule, gravity systems require more labor than sprinkler systems to maintain the ditches and the dikes, move the gated pipe, and control the flow of the water.

In trickle irrigation, drip valves or emitters are located along a line at uniform spacing or at each large plant. These systems have high application efficiency because a very small amount of water is applied and it is applied directly to the plant.

19.4.3. Subsurface

The two most common subsurface irrigations system is a porous pipe buried in the root zone or a trickle system buried alongside the plants.

19.5. Depth of Water to Apply

Seasonal water demand and peak daily use vary considerably from crop to crop and from one field to the next. Deciding when to irrigate and how much water to apply are the two most difficult decisions to make in managing irrigation systems. Many different methods have been developed to help answer these two questions. The following discussion will explore some of the factors influencing these decisions and present one method for determining how much water to apply and the irrigation interval.

The amount of water used by plants depends on five factors:

1. The length of the growing season
2. The amount of daylight per day
3. The daily temperature
4. The speed and direction of the wind
5. The crop's stage of growth

For any given plant, the daily rate of use will increase until the plant reaches maturity, and then it will decrease, Figure 19.1. The peak water use rate occurs at the height of the growing season. Table 19.1 shows the peak use for a number of crops with both short and long growing seasons, and the root zone depth from which each crop extracts most of its moisture.

Two characteristics of soil must be considered in determining how much water should be applied: (1) the rate at which soil can absorb (store) water, or the infiltration rate; and (2) the total amount of water that can be stored. The infiltration rate is determined by the soil texture and the total amount of water that can be stored is determined by the soil depth. Loams and clays can hold more water than sands, as shown in Table 19.2. Also note that the type and the depth of subsoil make a difference in the amount of water that can be stored per irrigation. For example, soils having more compact subsoil can store a greater amount of water than the same soils having a uniform depth.

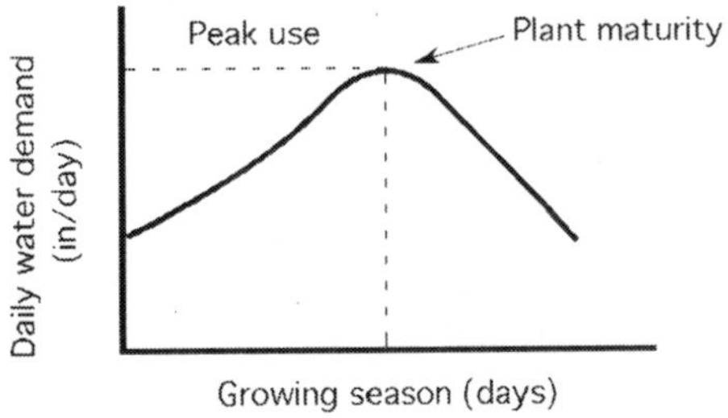

FIGURE 19.1. Typical daily water demand.

TABLE 19.1. Root zone depth and peak water use rate.

		Length of growing season	
Crop	Root zone depth of principle moisture extraction (in)	180–210 days Peak use (in/day)	210–250 days Peak use (in/day)
Alfalfa	48	0.29	0.32
Beans	24	0.20	0.20
Corn	36	0.30	0.35
Cotton	36	—	0.28
Grain sorghum	30	0.20	0.20
Melons	30	0.20	0.22
Other truck	24	0.20	0.22
Pasture	24	0.28	0.30
Peas	36	0.19	—
Potatoes	24	0.38	0.20
Small grain	30	0.20	0.22
Sugar beets	36	0.28	0.30
Tomatoes	48	0.20	0.22
Vineyards	48	0.22	0.25

The rooting depth of the crop also influences the amount of water that should be stored. Any water that infiltrates into the soil to a depth below the root zone is lost to the plant. These factors were taken into consideration during the development of Table 19.1 and Table 19.2.

To use these tables first consult Table 19.1 for a particular crop and growing season length and obtain values for (1) the root zone depth of principal moisture extraction and (2) the peak daily water use. Then Table 19.2 can be used to determine the number of inches of water to store per irrigation. Note that you must know the soil type or profile on which the irrigation is to take place and the root zone depth of principal moisture extraction of the crop.

Problem: What is the root zone depth, peak daily use, and net amount of water to store per irrigation for a crop of corn grown on a location where the season is greater than 210 days on a silt loam soil over compacted subsoil?

TABLE 19.2. Net amount of water to store per irrigation.

	Net amount of water to store (in) for various root zone depths			
Soil profile	24 in	30 in	36 in	48 in
Coarse sandy soil, uniform to 6 ft	0.85	1.10	1.30	1.75
Coarse sandy soil over compacted subsoil	1.50	1.75	2.00	2.50
Fine sandy loam uniform to 6 ft	1.75	2.20	2.60	3.00
Fine sandy loam over compacted subsoil	2.00	2.40	2.80	3.25
Silt loam uniform to 6 ft	2.25	2.75	3.00	4.00
Silt loam over compacted subsoil	2.50	3.00	3.25	4.25
Heavy clay or clay loam soil	2.00	2.40	2.85	3.85

Solution: Using Table 19.1 and Table 19.2:

Crop	Root zone depth of principle moisture extraction (in)	Length of growing season 180–210 days Peak use (in/day)	210–250 days Peak use (in/day)
Alfalfa	48	0.29	0.32
Beans	24	0.20	0.20
Corn	36	0.30	0.35
Cotton	36	—	0.28
Grain sorghum	30	0.20	0.20
Melons	30	0.20	0.22
Other truck	24	0.20	0.22
Pasture	24	0.28	0.30
Peas	36	0.19	—
Potatoes	24	0.38	0.20
Small grain	30	0.20	0.22
Sugar beets	36	0.28	0.30
Tomatoes	48	0.20	0.22
Vineyards	48	0.22	0.25

Root zone depth = 36 in.
Peak daily use = 0.35 in/day.

These figures show that the net amount of water to store per irrigation is 3.25 inches, Table 19.3.

Next determine the amount of water to apply. The amount of water to apply must be greater than the amount to store because irrigations systems do not operate at 100% efficiency. Some of the water will evaporate, some runs off, and some percolates below the crop root zone. The term application efficiency is used to describe these losses. It is defined as the ratio of the depth of water stored to the depth of water applied, expressed as a percent. The application efficiency of a well-designed irrigation system will be between 60 and 80%. Application efficiency can

TABLE 19.3. Net amount of water to store for sample problem.

Soil profile	Net amount of water to store (in) for various root zone depths 24 in	30 in	36 in	48 in
Coarse sandy soil, uniform to 6 ft	0.85	1.10	1.30	1.75
Coarse sandy soil over compacted subsoil	1.50	1.75	2.00	2.50
Fine sandy loam uniform to 6 ft	1.75	2.20	2.60	3.00
Fine sandy loam over compacted subsoil	2.00	2.40	2.80	3.25
Silt loam uniform to 6 ft	2.25	2.75	3.00	4.00
Silt loam over compacted subsoil	2.50	3.00	3.25	4.25
Heavy clay or clay loam soil	2.00	2.40	2.85	3.85

be used to determine the amount of water to apply in the following manner:

$$\text{Depth of water to apply} = \frac{\text{Depth of water stored (in)}}{\text{Application efficiency}}$$

or:

$$\text{DWA} = \frac{\text{DWS}}{\text{AE}}$$

where DWA = Depth of water to apply (in); DWS = Depth of water to store (in); AE = Application efficiency (as a decimal).

Problem: How much water should be applied for the corn crop in the previous problem if the application efficiency is 70%?

Solution:

$$\text{DWA} = \frac{\text{DWS}}{\text{AE}} = \frac{3.25\text{ in}}{0.70} = 4.642\ldots \text{ or } 4.64\text{ in}$$

It is also necessary to determine an irrigation interval, which is the number of days it takes the crop to use up the water stored in the soil. The irrigation interval is determined by dividing the amount of water stored in the soil by the plants daily use. For peak use the irrigation interval is:

$$\text{IRI} = \frac{\text{DWS}}{\text{PDU}}$$

where IRI = Irrigation interval (days); DWS = Depth of water to store (in); PDU = Peak water use (in/day).

Problem: What is the irrigation interval for the corn crop in the previous problem?

Solution:

$$\text{IRI} = \frac{\text{DWS}}{\text{PDU}} = \frac{3.25\text{ in}}{0.35\ \frac{\text{in}}{\text{day}}} = 9.285\ldots \text{ or } 9.3\text{ day}$$

For this corn crop, if 3.25 inches of water is stored in the soil per irrigation, at the time of peak use it must be irrigated every 9.3 days.

If there is rainfall during the irrigation period, the irrigation interval should be adjusted accordingly. For example, if there were 1.25 inches of rain on the corn crop in the previous problem, then we would divide the amount of rain (in) by the water needs of the crop (in/day) and extend the interval the corresponding number of days. For the peak demand of the corn crop, $1.25 \div 0.35 = 3.6$ days. Instead of irrigating again in 9.3 days, the irrigation interval could be extended to 12.9 or 13 days.

19.6. System Capacity

System capacity is the maximum amount of water that an irrigation system can deliver on a continuous basis. Different units are used to describe system capacity. These are acre-feet (ac-ft, that is, the amount of water it will take to cover one acre, one foot deep), acre-inch (ac-in), gallons per minute (gal/min or gpm), and cubic feet per second (ft^3/sec or cfs). The required pumping capacity of an irrigation system depends on the area to be irrigated (ac), the depth of water to apply (in), and the length of time that the irrigation system is operated (hr). Length of operation time refers to pumping time, not clock time. Pumping time is only the time water is flowing.

The amount of time per day that an irrigation system can operate depends on the type of system and the amount of maintenance it requires. A self-propelled unit may be able to run several days without stopping, whereas manual-move, tractor towed, and self-moved systems must be shut down at regular intervals. For systems other than center pivots and lateral moves, only a portion of a field is irrigated at one time and time is required to move the system from one portion of the field "set" to another "set." The term irrigation period is used to designate the number of days that a system can apply the water for one irrigation to a given area. Note that it is necessary for the irrigation period to be equal to or less than the irrigation interval. The required capacity of a system, in gallons per minute, can be determined by the following equation:

$$\mathrm{RSC} = \frac{453 \times A \times \mathrm{DWA}}{\mathrm{IRP} \times \mathrm{HPD}}$$

where RSC = Required system capacity (gal/min); 450 = Units conversion constant; A = Area irrigated (ac); DWA = Depth of water to apply per irrigation (in); IRP = Irrigation period (day); HPD = Time operating (hr/day).

Problem: Determine the required system capacity (gal/min) for the corn crop in the previous problem when the field area is 200 acres, and the system can operate for 18.0 hr per day for 7.7 days.

Solution:

$$\mathrm{RSC} = \frac{453 \times \mathrm{A} \times \mathrm{DWA}}{\mathrm{IRP} \times \mathrm{HPD}} = \frac{453 \times 200\ \mathrm{ac} \times 4.65\ \mathrm{in}}{7.7\ \mathrm{day} \times \frac{18.0\ \mathrm{hr}}{\mathrm{day}}}$$

$$= \frac{424290}{138.6} = 3039.61 \text{ or } 3{,}000\ \frac{\mathrm{gal}}{\mathrm{min}}$$

For 200 acres of long-season corn grown on silt loam soil over compacted subsoil, irrigated with a system that is 70% efficient and limited to operating 18 hours per day for 7.7 days, the system must be able to deliver 3,040 gallons of water per minute.

Note: This is an example of an equation with a units conversion constant. The same problem can be solved using the units cancellation method.

$$\frac{\mathrm{gal}}{\mathrm{min}} = \frac{1\ \mathrm{hr}}{60\ \mathrm{min}} \times \frac{1\ \mathrm{day}}{18\ \mathrm{hr}} \times \frac{1}{7.7\ \mathrm{day}} \times \frac{1\ \mathrm{gal}}{231\ \mathrm{in}^3} \times \frac{4.65\ \mathrm{in}}{1}$$

$$\times \frac{144\ \mathrm{in}^2}{1\ \mathrm{ft}^2} \times \frac{43560\ \mathrm{ft}^2}{1\ \mathrm{ac}} \times \frac{200\ \mathrm{ac}}{1} = \frac{5833555200}{1920996}$$

$$= 3036.7 \text{ or } 3{,}000\ \frac{\mathrm{gal}}{\mathrm{min}}$$

In some situations it might be necessary to use units of capacity other than gallons per minute. For example, water supplied from a large reservoir is often measured in acre-feet. In these cases, units cancellation and/or the appropriate conversion factors (Appendix I or II) can be used to convert the units.

Problem: What will the system capacity need to be in units of acre-feet/min?

Solution: Using units cancellation and system capacity from the previous problem:

$$\frac{\text{ac-ft}}{\text{min}} = 3{,}040\,\frac{\text{gal}}{\text{min}} \times \frac{1\ \text{ft}^3}{4.48\ \text{gal}} \times \frac{1\ \text{ac}}{43{,}560\ \text{ft}^2}$$

$$= 0.009319\ldots \text{or } 9.3\ \text{E} - 3\ \frac{\text{ac-ft}}{\text{min}}$$

As noted earlier, system capacity is a function of four variables: area (ac); water flow rate (gal/min, ft^3/min, ac-ft/min, etc.); depth of water applied or peak use (in); and time (min, hr, or days). This relationship is expressed mathematically as:

$$D \times A = Q \times T$$

where D = Depth of water, either applied or peak use (in); A = Area irrigated (ac); Q = Water flow rate (cfs); T = Length of time water is applied (hr).

When any three of the variables are known, the remaining one can be calculated by rearranging the equation and substituting the values of the known variables. You must enter flow rate (Q) in cubic feet per second, depth in inches, and time in hours. The following discussion will illustrate several uses of this equation.

In the previous problem we determined the system capacity using units cancellation. If it is necessary to know how much water has been applied, the peak use does not accurately describe what we are solving for. When we want to know the depth of water that has been applied, D becomes the depth of water applied (D_{WA}). This will work because the unit of measure is the same for both peak use and D_{WA} (inches).

Problem: A producer spends 120 hr irrigating 90.0 acres. The pump discharges 1,350 gallons per minute. What average depth of water (in) is applied?

Solution: Because we want to know the amount of water applied, not the amount available to the plants, the efficiency factor is not used. Also, Q must be converted from gal/min to ft^3/sec. Rearranging the equation, substituting depth of water to apply (D_{WA}) for the depth (D), and including the conversion factor

1 ft³/sec = 2.25×10^{-3} gal/min[1]:

$$D_{WA} \times A \cong Q \times T$$

$$D_{WA} \cong \frac{Q \times T}{A}$$

$$\cong \frac{\left(1{,}350\ \frac{\text{gal}}{\text{min}} \times \frac{2.25\ \text{E}{-3}\ \frac{\text{ft}^3}{\text{sec}}}{1\ \frac{\text{gal}}{\text{min}}}\right) \times 120\ \text{hr}}{90\ \text{ac}}$$

$$\cong \frac{364.5}{90}$$

$$\cong 4.05\ \text{in}$$

An examination of this problem shows that the units do not cancel. However, when we enter the values of the variables with the units listed above we can obtain an answer very close to the true value. The symbol ≅ means approximately equal. The exact solution using unit conversion/cancellation is:

$$D_{WA} \times A = Q \times T$$

$$D_{WA} = \frac{Q \times T}{A}$$

$$\cong \frac{\left(1{,}350\ \frac{\text{gal}}{\text{min}} \times 231\ \frac{\text{in}^3}{\text{gal}} \times \frac{1\ \text{ft}^3}{1{,}728\ \text{in}^3}\right) \times 120\ \text{hr} \times 60\ \frac{\text{min}}{\text{hr}}\ 12\ \frac{\text{in}}{\text{ft}}}{90\ \text{ac} \times 43{,}560\ \frac{\text{ft}^2}{\text{ac}}}$$

$$= 3.98\ \text{in}$$

In this case, the error in the approximate solution is:

$$\%\ \text{Error} = \frac{4.05 - 3.98}{3.98} \times 100$$

$$= 1.76\ \%$$

Variations occur in the use of the equation for different types of irrigation systems. In situations where the limiting factor is the availability of water, the problem is to determine the maximum area that can be irrigated with the available water supply.

Problem: What is the largest size of lawn (ft²) that can be irrigated in 12 hr if a minimum of 0.5 inch of water is applied at each irrigation, the system is 90% efficient, and the water supply delivers 3.5 gal/min?

---

[1] If $\frac{\text{ft}^3}{\text{sec}} = \frac{1\ \text{gal}}{\text{min}} \times \frac{1\ \text{min}}{60.0\ \text{sec}} \times \frac{1\text{ft}^3}{7.40\ \text{gal}} = 2.25 \times 10^{-3}\ \frac{\text{ft}^3}{\text{sec}}$ then $\frac{1\ \text{gal}}{\text{min}} = \frac{2.25 \times 10^{-3}\ \text{ft}^3}{1\ \text{sec}}$

Solution: Rearranging the equation, adding the efficiency factor, and converting the area to square feet:

$$D \times A = Q \times T$$

$$A\left(\text{ft}^2\right) = \left(\frac{Q \times T}{D} \times \frac{1\text{ ac}}{43{,}560\text{ ft}^2}\right) \times 0.90$$

$$= \left(\frac{\left(3.5\,\frac{\text{gal}}{\text{min}} \times \frac{60\text{ min}}{1\text{ hr}} \times \frac{1\text{ ft}^3}{7.48\text{ gal}}\right) \times 12.0\text{ hr}}{\frac{0.5\text{ in}}{12\,\frac{\text{in}}{\text{ft}}}} \times \frac{1\text{ ac}}{43{,}560\text{ ft}^2}\right) \times 0.90$$

$$= \left(\frac{168.5}{0.04167} \times \frac{1}{43{,}560}\right) \times 0.90$$

$$= 0.17\text{ ac}$$

If flood irrigation is used to water a field, assuming that the water flow rate is limited, it usually is necessary to determine the amount of time that the water should flow to cover the field at the desired depth.

Problem: How long will it take to apply 4 inches of water uniformly over 120 acres when the water is available at the rate of 20 cfs? (Assume 100% efficiency.)

Exact Solution:

$$D \times A = Q \times T$$

$$T(\text{hr}) = \frac{D \times A}{Q}$$

$$= \frac{4.0\text{ in} \times \frac{1\text{ ft}}{12\text{ in}} \times 120\text{ ac} \times 43{,}560\,\frac{\text{ft}^2}{\text{ac}}}{20.0\,\frac{\text{ft}^3}{\text{sec}} \times 60\,\frac{\text{sec}}{\text{min}} \times 60\,\frac{\text{min}}{\text{h}}}$$

$$= \frac{1{,}742{,}400}{72{,}000} = 24.2\text{ hr}$$

During furrow irrigation it is important to know how long the water must run to apply the desired amount for each set of furrows. Three values are necessary to calculate time: the water flow rate for each furrow or for the entire set, the area of the furrow or the set, and the amount of water to be applied. The area is determined from the number of rows in the set, the row spacing, and the length of the row.

Problem: How much time is required to apply 3 inches of water to sixty 32-inch rows when the rows are one half mile long, and the system capacity is 30 gal/min/row?

Solution:

$$D \times A = Q \times T$$

$$T = \frac{D \times A}{Q}$$

$$A = \frac{\text{Number} \times \text{Spacing (ft)} \times \text{Lenght (ft)}}{43{,}560\ \frac{\text{ft}^2}{\text{ac}}}$$

$$= \frac{60 \text{ rows} \times 32\ \frac{\text{in}}{\text{row}} \times \frac{1 \text{ ft}}{12 \text{ in}} \times 0.5 \text{ mile} \times 5280\ \frac{\text{ft}}{\text{mile}}}{43{,}560\ \frac{\text{ft}^2}{\text{ac}}}$$

$$= 9.696\ldots \text{ac}$$

$$Q\left(\frac{\text{ft}^3}{\text{sec}}\right) = \frac{30.0\ \frac{\text{gal}}{\text{min}}}{\text{row}} \times 60 \text{ row} \times \frac{2.25\ \text{E}{-3}\ \frac{\text{ft}^3}{\text{sec}}}{1\ \frac{\text{gal}}{\text{min}}} = 4.05\ \frac{\text{ft}^3}{\text{sec}}$$

$$T = \frac{3.0 \times 9.69\ldots}{4.05} = 7.182\ldots \text{or } 7.2 \text{ hr}$$

It will take 7.2 hr to apply 3.0 inches of water to the field.

19.7. Seasonal Need

Seasonal water demand is the amount of water (in inches) that a crop must have during one growing season for maximum production. The seasonal water demand will vary from season to season, and for each crop and region. Table 19.4 contains some typical values for three regions of the United States.

Knowledge of the seasonal water demand for a crop in a given area is useful for determining the contribution of irrigation to the cost of production and the total amount of water that will be needed. When the cost of water is known (usually expressed as dollars per acre-feet), as well as the seasonal demand of the crop, the normal rainfall during the growing season, and the number of acres, it is possible

TABLE 19.4. Typical seasonal water demand for some crops (in).

	Western region	Central region	Eastern region
Alfalfa	36.0	36.0	33.0
Corn	23.0	25.0	21.0
Cotton	31.0	20.0	19.0
Grain sorghum	20.0	22.0	21.0
Oranges	33.0	—	—
Hay	31.0	—	36.0
Sugar beets	36.0	29.0	—
Tomatoes	19.0	—	14.0

Source: Planning for an Irrigation System, American Association for Vocational Instructional Materials (AAVIM), Athens, Georgia.

to estimate the total amount of water that will be needed and what it will cost. To determine total seasonal use, the efficiency of the irrigation system also must be considered. Typical efficiency is 60–80%.

Problem: How much water (ac-ft/yr) is needed to supply 120 acres of cotton in the central region if the normal rainfall during the growing season is 5.0 inches, and the efficiency of the irrigation system is 70%?

Solution: Using units cancellation:

$$\frac{\text{ac-ft}}{\text{yr}} = \frac{120\ \text{ac}}{1} \times \frac{1\ \text{ft}}{12\ \text{in}} \times \frac{20.0\ \text{in} - 5.0\ \text{in}}{1\ \text{yr}} \times \frac{1}{0.70}$$

$$= \frac{1800}{8.4} = 214.285\ldots \text{ or } 210\ \frac{\text{ac-ft}}{\text{yr}}$$

Problem: What is the total water cost if the price is \$25.00 per acre-foot?

Solution: Using units cancellation:

$$\frac{\$}{\text{yr}} = 25.00\ \frac{\$}{\text{ac-ft}} \times 210\ \frac{\text{ac-ft}}{\text{yr}} = 5250 \text{ or } 5{,}200\ \frac{\$}{\text{yr}}$$

19.8. Metric Problems

Determining the net water to apply and store in the root zone is dependent on having the appropriate data for the tables in SI units. They are not provided in this text, but additional problems such as system capacity can be computed in SI units.

Problem: determine the pump capacity that will be required to apply 30.5 mm of water to 50.0 ha when the irrigation interval is 10 days and the pump can operate 20.0 hr per day.

Solution: Using units method

$$\text{Capacity}\left(\frac{\text{m}^3}{\text{min}}\right) = 30.5\ \text{mm} \times \frac{1\ \text{m}}{1{,}000\ \text{mm}} \times \frac{10{,}000\ \text{m}^2}{1\ \text{ha}} \times 50\ \text{ha} \times \frac{1\ \text{day}}{20.0\ \text{hr}}$$

$$\times \frac{1}{10\ \text{day}} \times \frac{1\ \text{hr}}{60\ \text{min}}$$

$$= 1.27\ \frac{\text{m}^3}{\text{min}}$$

By equation:

$$\text{RSC}\left(\frac{\text{m}^3}{\text{hr}}\right) = \frac{\text{A} \times \text{DWA}}{\text{IRP} \times \text{HPD} \times 0.1} = \frac{50.0 \times 30.5}{10 \times 20 \times 0.1} = \frac{305}{20}$$

$$= 15.25 \text{ or } 76.25\ \frac{\text{m}^3}{\text{hr}}$$

Or

$$= 1.27\ \frac{\text{m}^3}{\text{min}}$$

Problem: A producer spends 120 hr irrigating 35.0 hectares. The pump discharges 5.2 cubic meters per minute. What average depth of water (mm) is applied?

Solution:

$$\text{Depth (mm)} = \frac{1{,}000\ \text{mm}}{\text{m}} \times \frac{5.2\ \text{m}^3}{1\ \text{min}} \times \frac{1\ \text{ha}}{10{,}000\ \text{m}^2} \times \frac{1}{35.0\ \text{ha}}$$

$$\times \frac{60\ \text{min}}{1\ \text{hr}} \times 120\ \text{hr}$$

$$= \frac{3.744\ \text{E7}}{3.5\ \text{E5}} = 106.97\ldots \text{or } 110\ \text{mm}$$

Problem: What is the largest size of lawn (m^2) that can be irrigated in 6.0 hr if a minimum of 1.3 centimeters of water is applied at each irrigation, the system is 90% efficient, and the water supply delivers 13.0 Liters per minute (L/min)?

Solution:

$$\text{Area (m}^2) = \frac{13.0\ \text{L}}{\text{min}} \times \frac{1\ \text{m}^3}{1{,}000\ \text{L}} \times \frac{60\ \text{min}}{1\ \text{hr}} \times \frac{1}{1.3\ \text{cm}} \times \frac{100\ \text{cm}}{1\ \text{m}}$$

$$= \frac{7.8\ \text{E4}}{1.3\ \text{E3}} = 60\ \text{m}^2$$

20 Handling, Moisture Management, and Storage of Biological Products

20.1. Objectives

1. Be able to describe the common methods of handling biological products.
2. Be able to determine the size and horsepower requirements of screw-type conveyors.
3. Be able to determine the size and horsepower requirements of pneumatic conveyors.
4. Be able to determine the amount of water to extract from or add to biological products.
5. Understand the requirements of biological product storage.

20.2. Introduction

The term biological products describe all of the food, feed, and fiber produced by agriculture. These products include everything from fruits and vegetables to grain, hay, and cotton. Although the diversity of agricultural production is too broad to be totally covered in this text, the following sections will discuss some of the principles involved in the handling, drying, and storage of these products.

20.3. Handling

Because of differences in shape, size, and consistency, each product must have a handling system capable of moving that specific product. The designer of a handling system also must consider product perishability and the desired form of the finished product. A harvester designed to harvest tomatoes for the fresh vegetable market will be different from one designed to harvest tomatoes used for catsup. Because of the prevalence of grains across the United States, they will be used to illustrate some of the basic principles of handling biological products.

Grains were one of the first products to be mechanically moved because they flow by gravity, are small, and have a relatively hard outer coat. These characteristics

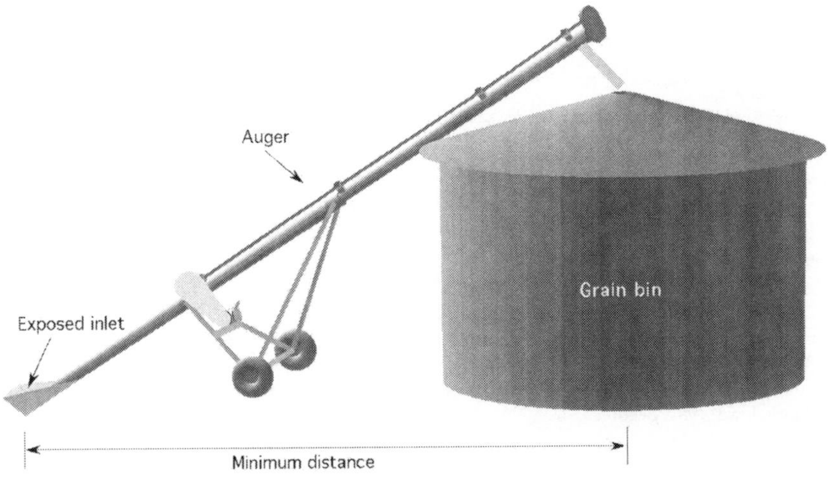

FIGURE 20.1. Auger conveyor.

allow them to be moved by different mechanical devices. Augers and pneumatic conveyors will be discussed in the following sections.

20.4. Augers

Augers are available in two types, Archimedean screw and belt. Of the two, the screw type auger is the most popular in agriculture and it is the one that will be used in the following discussion and sample problems. An auger is like a bolt, but instead of threads it uses flights that turn inside a tube. Another name for an auger is a screw conveyor. As the auger rotates, the flights move the product through the tube similarly to the way that threads move a nut on a bolt. Augers are available in several diameters and are capable of handling many different types of products. Augers have the advantage of requiring less horsepower per bushel and having fewer mechanical parts than pneumatic systems; but their disadvantages include the danger of the exposed auger at the inlet, and their requirement of more space because the inlet is some distance from the outlet, Figure 20.1.

Augers are selected on the desired capacity (bu/hr) and length requirements. The length is usually predetermined by the distance needed to move the grain, if horizontal, or by the discharge height and angle if the auger is at an angle. The selection criteria are primarily based on the auger diameter and speed because the capacity increases as the speed increases. The following discussion will illustrate how the capacity and energy requirements for augers can be determined.

Table 20.1 and Table 20.2 contain typical values for two sizes of screw augers and two different crops. This type of information can be used to make decisions in managing a grain handling system, such as determining the size of auger required to convey grain at a given rate (bu/hr).

TABLE 20.1. Screw auger capacity handling dry corn (12 in exposure).

Auger		Auger angle of elevation					
		0°		45°		90°	
Size	RPM	bu/hr	hp/10 ft	bu/hr	hp/10 ft	bu/hr	hp/10 ft
4 in	200	150	0.12	120	0.15	60	0.11
	400	290	0.29	220	0.29	130	0.24
	600	420	0.38	310	0.45	190	0.36
6 in	200	590	0.38	500	0.44	280	0.32
	400	1090	0.56	850	0.88	520	0.70
	600	1510	0.84	1160	1.28	740	1.05

Note: For total horsepower, 10% must be added for drive train losses.
Source: Structures and Environment Handbook, MWPS-1, Midwest Plan Service, Iowa State University, Ames, Iowa, 1987, Section 534.

Problem: What is the minimum size of auger that can be used to convey dry corn at the rate of 500 bushels per hour when the auger is inclined 45°?

Solution:

Auger		Auger angle of elevation					
		0°		45°		90°	
Size	RPM	bu/hr	hp/10 ft	bu/hr	hp/10 ft	bu/hr	hp/10 ft
4 in	200	150	0.12	120	0.15	60	0.11
	400	290	0.29	220	0.29	130	0.24
	600	420	0.38	310	0.45	190	0.36
6 in	200	590	0.38	**500**	**0.44**	280	0.32
	400	1090	0.56	**850**	**0.88**	520	0.70
	600	1510	0.84	**1160**	**1.28**	740	1.05

Using Table 20.1, the minimum size of auger is 6 inches.

TABLE 20.2. Screw auger capacity handling dry soybeans (12 inch exposure).

Auger		Auger angle of elevation					
		0°		45°		90°	
Size	RPM	bu/hr	hp/10 ft	bu/hr	hp/10 ft	bu/hr	hp/10 ft
4 in	200	140	1.00	125	0.17	70	0.12
	400	270	0.21	215	0.35	130	0.26
	600	390	0.34	315	0.51	180	0.40
6 in	200	500	0.40	360	0.57	220	0.40
	400	990	0.84	690	1.20	390	0.79
	600	1350	1.20	930	1.71	500	1.10

Note: For total horsepower, 10% must be added for drive train losses.
Source: Structures and Environment Handbook, MWPS-1, Midwest Plan Service, Iowa State University, Ames, Iowa, 1987, Section 534.

Another use of Table 20.1 and Table 20.2 is to determine the horsepower required to operate the auger.

Problem: How much horsepower (including drive train) is required to operate a 100-ft, 6-inch auger, installed at 45°, conveying 690 bushels of soybeans per hour?

Solution:

Auger		Auger angle of elevation					
		0°		45°		90°	
Size	RPM	bu/hr	hp/10 ft	bu/hr	hp/10 ft	bu/hr	hp/10 ft
4 in	200	140	1.00	125	0.17	70	0.12
	400	270	0.21	215	0.35	130	0.26
	600	390	0.34	315	0.51	180	0.40
6 in	200	500	0.40	360	0.57	220	0.40
	400	990	0.84	**690**	**1.20**	390	0.79
	600	1350	1.20	930	1.71	500	1.10

Using Table 20.2, the power requirement is 1.2 hp/10 ft. It is important to read the note at the bottom of Table 20.2. 10% must be added for drive train losses. Therefore:

$$\text{hp} = \left(\frac{1.2\,\text{hp}}{10\,\text{ft}} \times 100\,\text{ft}\right) \times 1.10 = 13.2 \text{ or} 13\,\text{hp}$$

The horsepower required by the auger, including the drive train, is 13 hp.

20.5. Pneumatic Conveyors

Pneumatic conveyors are used to move grain and other products using air. Pneumatic conveyors are more flexible than augers because the duct does not need to be in a straight line. They are self-cleaning, and do not have an exposed auger at the inlet. They do require more horsepower per bushel and are noisier than augers. Three types of pneumatic conveyors are used: positive pressure (push units), negative pressure (vacuum), and a combination of negative and positive pressure. In a positive pressure unit, a blower supplies the pressure, and the product enters the air stream through a rotary air lock. The material then is blown through the duct, Figure 20.2.

In a negative pressure unit, the material is vacuumed up by the inflow of air and then separated from the air in a cyclone separator. The material collects in the bottom of the separator where it can be released with a gate valve or a rotary lock. The air continues on to the pump and out through a filter into the atmosphere, Figure 20.3.

In a combination system, the material is picked up by the inflow of a negative pressure system, and then it passes through a rotary air lock into the positive pressure air stream coming from the pump.

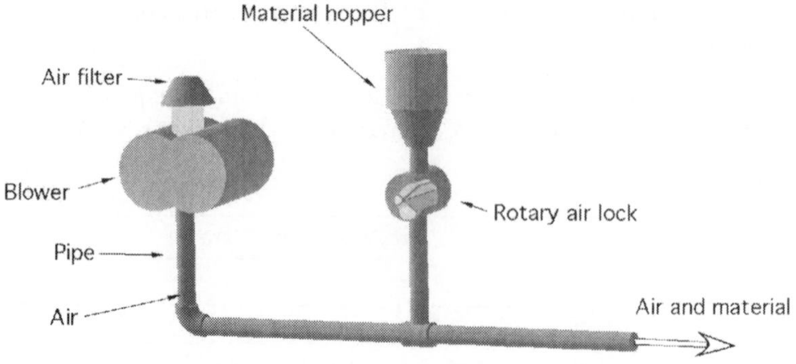

FIGURE 20.2. Positive pressure pneumatic system.

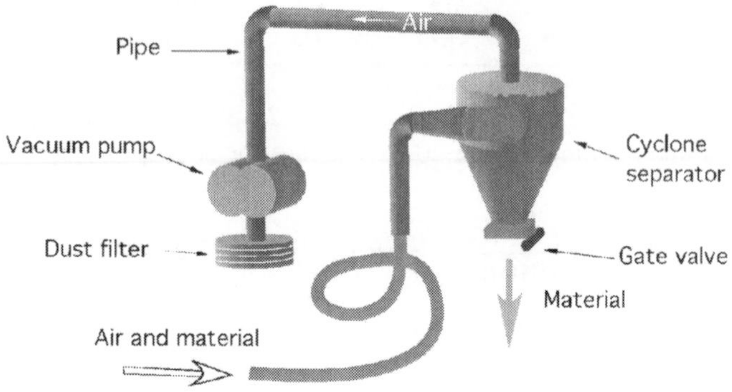

FIGURE 20.3. Negative pressure pneumatic system.

20.6. Sizing a Pneumatic System

The capacity and the horsepower requirements of pneumatic conveyors depend on eight factors:

1. The horizontal distance that the material is moved.
2. The diameter of the pipe.
3. The vertical distance that the material is moved.
4. The number of bends in the pipe.
5. The elevation above sea level.
6. The temperature of the outside air.
7. The type of material being conveyed
8. The moisture content of the material.

To size a pneumatic system the effect of each of these factors must be accounted for.

Table 20.3 is an example of the type of information that is available for sizing pneumatic systems. The data show that as the horizontal length (ft) of the pipe

TABLE 20.3. Pressure system conveyor capacities for dry corn (bu/hr).

		Equivalent Length (ft)					
Pipe size (in)	Hp	50	100	200	250	300	350
3	10	450	400	350	320	295	265
4	15	650	575	500	460	420	380
5	25	1100	1000	900	830	755	685
6	40	2100	1950	1800	1650	1510	1370
8	75	4300	3900	3500	3220	2940	2660
10	100	5800	5100	4500	4240	3980	3720

Modifications to Capacity:

1. Vertical pipe $\times 1.20 =$ Equivalent horizontal length.
2. Capacity in wheat = 90% of Corn.
3. Capacity in soybeans = 80% of Corn.
4. Add 20 ft of equivalent horizontal pipe for each 90° bend in pipe.
5. Reduce capacity by 4.0% for each 1000 ft above sea level.
6. Reduce capacity by 2.0% for each 10°F above 70°F.

Source: Beard Industries, Frankfort, Indiana 46041.

increases the capacity of the system (bu/hr) decreases and as the pipe diameter increases (in) the capacity increases.

To provide an easy way to estimate the amount of reduction caused by factors 3 and 4, the equivalent feet of pipe is determined. Equivalent feet of pipe is a procedure that uses constants to determine the amount of horizontal pipe that will cause the same reduction in capacity as the factor in question. For example, the fourth modification to capacity listed in Table 20.3 shows that each foot of vertical pipe causes the same reduction in capacity as 1.2 ft of horizontal pipe. Thus the equivalent feet for a vertical section of pipe is determined by multiplying the vertical length (ft) by 1.2.

Adjustments for factors 5 through 8 are handled differently. The capacity of the system is adjusted for the effect of factors 5, 6, and 7 by reducing the capacity and appropriate percentage. Factor 8 is accounted for by either assuming a standard moisture content, or developing a separate table for a range of moistures. Expressed as an equation:

$$\text{TEF} = (F_{\text{H}} + F_{\text{B}} + F_{\text{V}})$$

where: TEF = Total equivalent feet, F_{H} = Horizontal distance (ft), F_{B} = Equivalent feet for bends, F_{V} = Equivalent feet for vertical sections.

Problem: What is the capacity of the system in Figure 20.4 if the material to be moved is dry soybeans?

Solution: The first step is to determine the total equivalent feet for the pipe. In this example, factors 1, 2, 3, 4, and 7 are used. Note: the last section of pipe does not have a length dimension because the assumption is used that the force of gravity exceeds the resistance of the pipe. Therefore, vertical down sections of pipe are

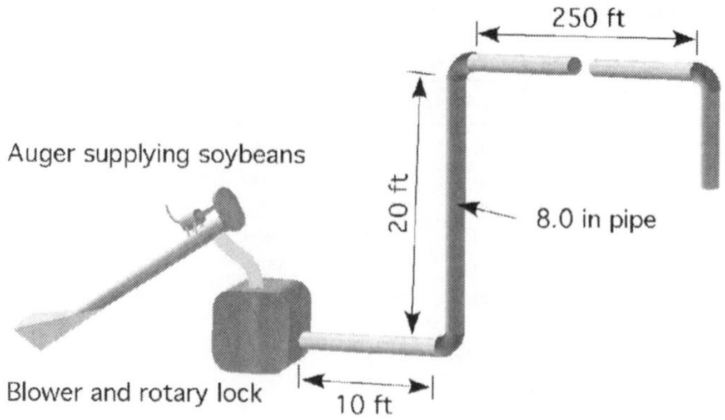

FIGURE 20.4. Pressurized conveyor for sample problem.

not included.

$$\text{TEF} = (F_H + F_B + F_V)$$
$$= \left((250\,\text{ft} + 10\,\text{ft}) + \left(3\text{ bends} \times \frac{20\,\text{ft}}{\text{bend}}\right) + (20\,\text{ft} \times 1.2)\right) = 344\,\text{ft}$$

Although the actual length of pipe in Figure 20.4 is 280 ft, because of the bends and the vertical up sections the equivalent distance is 344 ft.

The next step is to use Table 20.3 to determine the capacity of an 8.0 inch duct with an equivalent distance of 344 ft. The closest value is 2660 bu/hr. Note that Table 20.3 gives values for corn. Because the capacity of a pressurized system for soybeans is 80% of the capacity for corn (modification 3), the system capacity in Figure 20.4 is:

$$\frac{\text{bu}}{\text{hr}} = 2{,}660\,\frac{\text{bu}}{\text{hr}} \times 0.80 = 2{,}100\,\frac{\text{bu}}{\text{hr}}$$

Information to determine the capacities of auger, pneumatic, and bucket conveyor systems for other crops and situations are available from manufacturers and agricultural extension personnel.

20.7. Moisture Management

Water and its addition to or removal from agricultural products and materials is an extremely important topic in nearly all aspects of agriculture. The moisture contents of grain, feed, or hay to be bought or sold, of crops to be dried, or of meat and dairy products to be processed are but a few examples of products where moisture must be carefully managed. Moisture may be added to or removed from the product depending upon the desired final condition. Moisture is removed from products by drying. Drying usually is done to change the consistency or to extend the storage life of the product. For example, fruits and meats may be dried to change the way that they are handled, stored, and eaten. Grains and forages are dried to extend their storage life.

Some agricultural products, such as grains and forages, will dry naturally to equilibrium moisture content (the same as that of the environment) if left in the field. But with these and other products good management sometimes dictates that the crop be harvested at a wetter stage and dried artificially. Artificial drying is accomplished by causing natural or heated air to flow around and/or through the product. Artificially heated air is often used, because heating reduces the relative humidity of the air, increasing the amount of moisture each pound of air can absorb.

The management of a drying system requires the ability to be able to predict the amount of moisture that must be removed from the product. For other products, it is important to be able to predict the amount of water that must be added. The following discussion presents a method for determining the amount of moisture that must be added to or removed from biological products.

The moisture content of a given material is stated as a percent using either the wet-weight or the dry-weight basis. Because moisture content is expressed as a percent, we know that a ratio is involved. The difference between the wet-weight basis and the dry-weight basis is the value used in the denominator of the ratio. The wet-weight basis uses the weight of the product as it is received; the dry-weight basis uses the oven-dry weight (dry matter) of the product.

Study Figure 20.5. The total weight of the product is 4 pounds. When the amount of water is expressed on the dry-weight basis it is:

$$\frac{1 \text{ lb water}}{3 \text{ lb dry matter}} \times 100 = 33\%$$

Expressed on the wet-weight basis, the amount of water is:

$$\frac{1 \text{ lb water}}{4 \text{ lb total}} \times 100 = 25\%$$

These relationships are explained in the following equations:
Dry-weight basis:

$$\%\text{MDB} = \left(\frac{\text{WW} - \text{DW}}{\text{DW}}\right) \times 100$$

Wet-weight basis:

$$\%\text{MWB} = \left(\frac{\text{WW} - \text{DW}}{\text{WW}}\right) \times 100$$

where %MDB = Percent moisture, dry-weight basis; %MWB = Percent moisture, wet-weight basis; WW = Wet weight or weight of product before drying; DW = Oven-dry weight, or weight of product after drying; WW − DW = Weight of moisture removed.

It is impractical and often undesirable to remove all of the moisture from grain as well as many other products. Grain usually is considered to be dry when the moisture content is sufficiently low to discourage the growth of molds, enzymatic

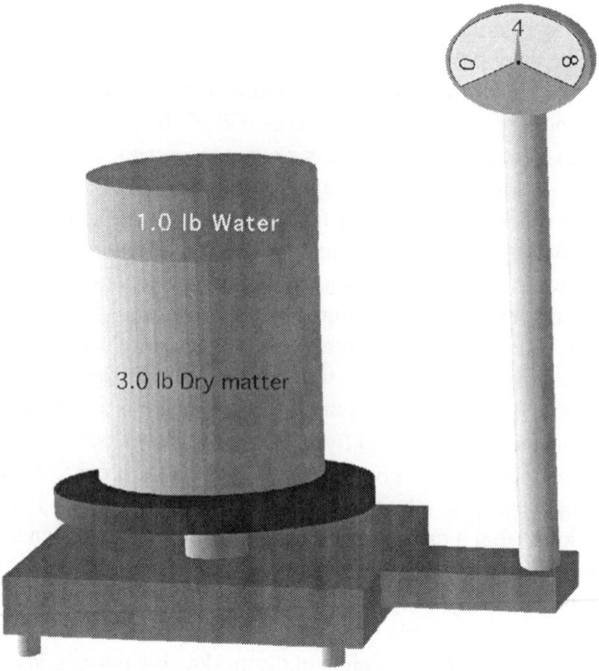

FIGURE 20.5. Dry-weight, wet-weight basis illustration.

action, and insects. This is usually about 12% moisture content, %MDB, depending upon the grain. Standards have been established to determine the heating time and temperature required to obtain an official oven dry sample of grain.

Either moisture content basis may be used with agricultural products; so to avoid confusion or misunderstanding, the basis being used should always be specified. This can be accomplished by writing the numerical value of the moisture content followed by %MDB or %MWB.

Problem: Express the moisture content on the wet-weight basis and the dry-weight basis for a product that weighs 150.0 pounds when wet, and after drying weighs 80.0 pounds.

Solution: Wet-weight basis:

$$\%MWB = \left(\frac{WW - DW}{WW}\right) \times 100 = \left(\frac{150\,lb - 80.0\,lb}{150\,lb}\right) \times 100$$

$$= 46.666\ldots \text{ or } 46.7\%$$

On the wet-weight basis, the product is 46.7% moisture.

Dry weight basis:

$$\%\text{MDB} = \left(\frac{\text{WW} - \text{DW}}{\text{DW}}\right) \times 100 = \left(\frac{150.0\,\text{lb} - 80.0\,\text{lb}}{80.0\,\text{lb}}\right) \times 100 = 87.5\%$$

On the dry-weight basis, the product is 87.5% moisture.

Notice that the only difference in the values used in the two equations is in the denominator of the ratio. For the DW basis the dry weight is used, and for the WW basis the wet weight is used.

One aspect of the DW basis is that the percentage of moisture can be greater than 100%. For example, if a product has a wet weight of 100 ounces and a dry weight of 40 ounces, the MWB is 60% and the MDB is 150%. This is why the standard moisture of many products is given on the wet-weight basis.

20.8. Adding or Removing Water

During the life of many products it may be necessary to remove or to add water. Grain must be dried for storage, and the same grain may have to be tempered (have water added) for processing. Water may be added to or removed from products such as catsup to produce the desired consistency, and so on. The dry-weight and wet-weight bases equations also can be used to determine the amount of water to put into or take out of any biological product.

The calculations for determining how much water to add or remove requires three steps, and they must be completed in the correct sequence:

1. Determine the dry weight of the material at its original state. The weight of the dry material does not change from one moisture content to another; moisture may be added or removed, but the amount of dry matter or the dry weight remains the same.
2. Using the dry weight that was determined in step one, solve for the new wet weight (the weight of the material at the new moisture content).
3. The amount of moisture to add or to remove is the difference between the weight of the product in the original state and the new wet weight.

To solve for the dry weight or the wet weight, the wet-weight basis and dry-weight basis equations are rearranged. The first two equations are used when the moisture is on the dry-weight bases and the second two equations are used when the moisture is on the wet-weight basis.

Dry-weight basis:

$$\text{DW} = \frac{\text{WW}}{\left(1 + \dfrac{\%\text{MDB}_\text{I}}{100}\right)}$$

$$\text{WW}_\text{n} = \text{DW} \times \left(1 + \frac{\%\text{MDB}_\text{E}}{100}\right)$$

Wet-weight basis:

$$DW = WW \times \left(1 - \frac{\%MWB_I}{100}\right)$$

$$WW_n = \frac{DW}{\left(1 - \frac{\%MWB_E}{100}\right)}$$

The subscripts n = new, e = ending, I = initial.
The following problems illustrate the uses of these equations.

Problem: How much water needs to be removed (lb) to dry 1,000.0 pounds of product at 70.0% MWB to 20.0% MWB?

Solution: Three step process.
Step one: Because the moisture has been measured by the wet-weight basis, the second two equations must be used. Start by solving for the dry weight of the product.

$$DW = WW \times \left(1 - \frac{\%MWB_I}{100}\right) = 1{,}000.0 \times \left(1 - \frac{70.0\%}{100}\right)$$
$$= 1{,}000.0 \times 0.30 = 300\,\text{lb}$$

Step two: Solve for the new wet-weight at the desired moisture content.

$$WW_n = \frac{DW}{1 - \frac{\%MWB_E}{100}} = \frac{300\,\text{lb}}{1 - \frac{20\%}{100}} = 375\,\text{lb}$$

Step three: The amount of water that needs to be removed is the difference between the initial wet weight (WW_I) and the new wet weight (WW_E). Thus the amount of water (lb) that must be removed is:

$$\text{lb water} = WW_I - WW_n = 1{,}000\,\text{lb} - 375\,\text{lb} = 625\,\text{lb}$$

To change the product from the original 70% MWB to 20% MWB, 625 pounds of water must be removed.

Problem: A miller needs to raise the moisture content of 1,000 bushels of wheat from the storage condition of 9.0% MDB to 16.0% MDB. How much water (lb) needs to be added?

Solution: Step one: Use the dry-weight equations and the conversion value of 1 bu = 60 lb to solve for the pounds of dry matter:

$$DW(\text{lb}) = \frac{WW}{1 + \frac{\%MDB}{100}} = \frac{1{,}000\,\text{bu} \times \frac{60\,\text{lb}}{\text{bu}}}{1 + \frac{9.0\%}{100}} = \frac{60{,}000\,\text{lb}}{1.09} = 55045.871\ldots\,\text{lb}$$

Step two: Determine the weight of the product at the new wet weight

$$WW_n = DW \times \left(1 + \frac{\%MDB}{100}\right)$$

$$= 55{,}045.87156\,\text{lb} \times \left(1 + \frac{16.0\%}{100}\right) = 63{,}853.212..\ \text{lb}$$

Step three: Determine the water to add. Because this problem is an example of adding water, the initial weight is subtracted from the new wet weight.

$$\text{lbwater} = WW_n - WW_i = 63{,}853.212101\ \text{lb} - 60{,}000 = 3{,}853.212..$$

or 3,800 lb water

The miller should add 3,800 pounds of water to the wheat to bring it up to 16% MDB.

On occasion it is useful to know how to convert from one moisture basis to the other. This can be accomplished with the following equations:

$$\%MDB = \frac{100 \times \%MWB}{100 - \%MWB}$$

$$\%MWB = \frac{100 \times \%MDB}{100 + \%MDB}$$

Problem: Determine the percent moisture, dry basis, (%MDB) of a product if the percent moisture, wet basis, (%MWB) is 50%.

Solution:

$$\%MDB = \frac{100 \times \%MWB}{100 - \%MWB} = \frac{100 \times 50\%}{100 - 50\%} = \frac{5{,}000}{50} = 100\%$$

Problem: Determine the percent moisture, wet basis, (%MWB) of a product that is 23.25% moisture, dry basis, (MDB).

Solution:

$$\%MWB = \frac{100 \times \%MDB}{100 + \%MDB} = \frac{100 \times 23.25}{100 + 23.25} = 18.864\ \ldots\ \text{or}\ 18.86\%$$

20.9. Storage of Biological Products

The term biological products has been used throughout this chapter because products such as grains, fruits, and vegetables are living organisms. Because they are alive, there are minimum requirements for moisture, temperature, and air to maintain their viability; but the life of the product can be extended if the temperature or moisture or air is modified during storage. The challenge for managers of many such products is to extend the storage life of the product without damaging its viability, color, taste, or texture. Recommendations for optimum environments for the storage of biological products can be obtained from the extension service or any department of agricultural engineering.

20.10. Metric Problems

The auger and pneumatic capacity problems are worked the same. The only difference is using the appropriate values in SI units.

Adding and removing of water from products uses the same equations, the difference is using units of mass instead of weight.

Problem: How much water (kg) must be removed from 1.2 metric tons (t) of grain to lower the moisture from 14.5 to 11.0%DB?

Solution: Three steps

Step one: Determine the dry mass.

$$\mathrm{DW(kg)} = \frac{\mathrm{WW_I}}{\left(1 + \frac{\%\mathrm{MDB_I}}{100}\right)} = \frac{1.2\ \mathrm{t}}{1 + \frac{14.5}{100}} \times \frac{1{,}000\ \mathrm{kg}}{\mathrm{t}} = \frac{1{,}200}{1.145}$$

$$= 1048.03 \ldots \ \mathrm{kg}$$

Step two: New wet mass.

$$\mathrm{WW_n} = \mathrm{DW} \times \left(1 + \frac{\%\mathrm{MDB_E}}{100}\right) = 1048.03 \cdots \times \left(1 + \frac{11.0}{100}\right)$$

$$= 1163.31 \ldots \ \mathrm{kg}$$

Step three: Subtract original and new wet mass.

$$\mathrm{Water\ (kg)} = \mathrm{WW_I} - \mathrm{WW_E} = \left(1.2\ \mathrm{t} \times \frac{1{,}000\ \mathrm{kg}}{\mathrm{t}}\right) - 1163.31 \ldots$$

$$= 36.68 \ldots \ \text{or}\ 36.7\ \mathrm{kg}$$

21
Animal Waste Management

21.1. Objectives

1. Understand the importance of animal waste management.
2. Be able to describe the methods of handling solid animal waste.
3. Be able to determine the maximum amount of solid animal waste that can be applied to the soil.
4. Be able to describe the methods of handling liquid animal waste.
5. Be able to determine the capacity of storage units for animal waste.
6. Be able to describe animal waste treatment methods.

21.2. Introduction

When animals are free ranging "dilution is the solution to pollution." Animals that range freely tend to distribute their waste over a large area and the natural processes disseminate the waste. When animals are confined the waste is concentrated in one spot and it becomes part of the management of the animals. Animal waste management is no longer an option in a livestock business; it is a requirement. A plan for the collection and disposal of animal waste must be part of any livestock facilities plan. Failure to provide adequate waste management collection and storage facilities, and disposal equipment can lead to pollution problems with legal complications, animal health problems, increased production costs, and a generally undesirable working conditions. In most states, confinement operations must have a comprehensive nutrient management plan. There are three aspects of animal waste management:

1. Waste handling
2. Waste treatment
3. Waste disposal

When the decision is made during the planning process, the producer can choose from a variety of waste handling, treatment and disposal systems to manage the animal waste. Selecting a system after the facility has been built limits the options.

21.3. Handling Solid Animal Waste

Waste containing 20% or more solids or with a moisture content of 50% MWB or less is considered to be solid waste. Proper handling of solid waste inside buildings requires solid floors that can be bedded or drained. The waste is collected and usually is not treated except by the natural processes, as it is stored. The preferred method for waste disposal is mimicking the natural system—spread it on the land. If not done correctly, disposal of solid animal waste on the land can reduce plant production, produce offensive odors, and contribute to the contamination of ground and surface water. The NRCS[1] has developed standards for the application of solid animal waste on land. These standards are based on soil type, slope, and distance from surface and ground water. The general standards are:

1. No application to frozen or snow covered soil when slope is greater than 5% unless special provisions are made to control runoff.
2. No application to cropland that exceeds soil loss tolerances.
3. No application on any cropland with slopes greater than 15%.
4. No application within 200 ft of wells, sinkholes, or surface water.
5. Liquid cannot be applied to soils with less than 10 inches of at least moderately permeable soil.
6. No application on organic soils with seasonal water table within 1 ft of the surface.
7. No application on flood plains where flooding occurs more frequent than once in 10 years.
8. Application can occur on flood plains if the injection or immediate incorporation method is used.
9. No intentional application on established waterway or any area where there many be concentrated water flow.
10. No more than 25% of the surface may be covered.

Failure to follow these standards can cause contamination of the ground or surface water and the possibility of fines or litigation.

Note: Different computer programs are available to help producers calculate the amount of livestock waste that can be applied. The manual process explained in the following sections shows the steps and types of information required to use one of the computer programs.

For accurate results, the first step in determining the amount of solid animal waste that can be applied to the soil is to analyze both the solid waste and the soil for the amounts of nitrogen and phosphorous they contain. If the waste has not been tested, Appendix VI may be used to estimate the nutritional content. These values are estimates because the animal's ration, the type and quantity of bedding used, the amount of liquid added, the type of housing and manure handling system, and the storage system all affect the nutrient content of the animal waste.

---

[1] *Waste Utilization, Code 633*, Natural Resources Conservation Service, 2002.

When the nutritional content of the soil is unknown, it must be estimated before Appendixes VI (Solid Animal Waste Production and Characteristics), VII (Nutrient Utilized by Crops), and VIII (Maximum Annual Application Rates for Phosphates Based on Soil Family) can be used to determine the amount of waste that may be applied.

The primary governing principle for the application of solid animal waste is that no more nutrients can be applied in the soil than what will be removed from the field in crops or animal weight. It is unlikely that the amount of each nutrient in the waste can be matched with the needs of the plants; so the nutrient with the most restrictive amount determines the maximum amount of waste that can be applied. *Note*: This method assumes that more than one application per year will be used to distribute the total amount of waste.

The steps in calculating the amount of waste that can be applied are:

1. Determine the amount of nutrients produced per day, Appendix VI.
2. Calculate the pounds of nutrients per pound of waste.
3. Determine the amount of nutrients used for the expected crop, Appendix VII.
4. Determine the maximum amount of phosphorous for the soil type, Appendix VIII.
5. Calculate the pounds of waste that can be applied.

Problem: Determine the amount of solid waste from 1,000 lb dairy cows that can be applied to a coarse loamy soil with a pH of 6.5 that will be used to raise 60 bushels per acre of wheat. *Note*: In this example we will assume that there are no nutrients in the soil.

Solution: In this type of a problem a table is useful for determining the answer, for example Table 21.1. First, determine the amount of nutrients in the waste and the maximum amounts of phosphorous and nitrogen that can be applied. This is accomplished by selecting the correct values from Appendixes VI, VII, and VIII, and converting them to pounds of nutrient per pound of waste. From Appendix VI the nutrients produced per day are:

$$\text{Nitrogen (N)} = 0.14\ \frac{\text{lb}}{\text{day}}$$

$$\text{Phosphorous (P)} = 0.27\ \frac{\text{lb}}{\text{day}}$$

Each cow produces 82 lb of waste per day, Appendix VI. The pounds of nutrient per pound of waste can be calculated using units cancellation:

$$\frac{\text{lb of nitrogen}}{\text{lb of waste}} = \frac{0.14\ \text{lb N}}{1\ \text{day}} \times \frac{1\ \text{day}}{82\ \text{lb waste}} = 0.1.7\ \text{E} - 3\ \frac{\text{lb N}}{\text{lb waste}}$$

$$\frac{\text{lb of phosphorus}}{\text{lb of waste}} = \frac{0.27\ \text{lb P}}{1\ \text{day}} \times \frac{1\ \text{day}}{82\ \text{lb waste}} = 0.00329 \text{ or } 3.3\ \text{E} - 3\ \frac{\text{lb P}}{\text{lb waste}}$$

TABLE 21.1. Solution to problem.

Nutrient	Soil limits (lb/ac)	Net nutrients used by crop (lb/ac)	Pounds of nutrient per pound of waste (lb/lb)	Application rate of waste (lb/ac/yr)
Nitrogen	—	125	1.7×10^{-3}	7,400*
Phosphorous	—	50	3.3×10^{-3}	15,000
Phosphorous	400	—	3.3×10^{-3}	120,000

$$^{*}\text{Application rate} = \frac{125 \text{ lb N}}{1 \text{ ac}} \times \frac{1 \text{ lb waste}}{1.7\text{ E} - 3 \text{ lb N}} = 73529.41 \text{ or } 7{,}400 \text{ lb}$$

$$^{**}\text{Application rate} = \frac{50 \text{ lb of P}}{1 \text{ ac}} \times \frac{1 \text{ lb waste}}{3.3\text{ E} - 3 \text{ lb P}} = 15151.51\ldots \text{ or } 15{,}000 \text{ lb}$$

$$^{***}\text{Application rate} = \frac{400 \text{ lb P}}{1 \text{ ac}} \times \frac{1 \text{ lb waste}}{3.3\text{ E} - 3 \text{ lb P}} = 121212.12\ldots \text{ or } 120{,}000 \text{ lb}$$

The next step is to use Appendix VII to determine the nutrients used by the crop:

$$\text{Nitrogen} = 125 \text{ lb/ac}$$
$$\text{Phosphorous} = 50 \text{ lb/ac}$$

The last information needed to find the solution is the maximum amount of phosphorous that can be applied to the soil based on soil type. Consult Appendix VIII for this information:

$$\text{Phosphorous} = 400 \text{ lb/ac}$$

The solution is found by calculating the application rate (lb/ac/yr) of waste for each nutrient. The maximum amount of waste that can be applied is determined by the smallest of three values: the nitrogen used by the plant, the phosphorous used by the plant, and the phosphorous that can be applied to the soil. The results are shown in Table 21.1.

Table 21.1 shows that nitrogen is the most restricted nutrient (7,400 lb/ac/yr). Therefore, the maximum amount of waste from the dairy cows that can be annually applied is 7,400 lb per acre per year.

In the previous sample problem, the nutrients available in the soil were not considered. The actual amount of waste than can be applied is the difference between the nutrients used by the crop and the amount of nutrients already available in the soil.

Another aspect of solid waste management is determining how many acres are required to dispose of the waste being produced.

Problem: How many acres will be required to distribute all of the waste from a 50-cow dairy? Use the cows in the previous problem.

Solution: The answer to this question can be found by using the 7,400 lb of waste per acre per year from the previous problem, additional information from Appendix VI, and units cancellation. Appendix VI shows that each 1,000 lb dairy cow produces 82.0 lb of waste per day. Then:

$$\begin{aligned} \text{ac} &= \frac{1 \text{ ac}}{7{,}400 \text{ lb waste}} \times \frac{82.0 \text{ lb waste}}{\text{cow} - \text{day}} \times 50 \text{ cows} \times 365 \text{ days} \\ &= \frac{1496500}{7{,}400} \\ &= 202.22\ldots \text{ or } 202 \text{ ac} \end{aligned}$$

For the 50-cow dairy herd, a minimum of 202 acres are required to dispose of the animal waste each year. It is important to remember that in the original problem we assumed that the soil contained no nitrogen or phosphorous. Any nitrogen or phosphorous in the soil reduces the amount of waste that may be applied and increases the number of acres required.

21.4. Handling Liquid Waste

Handling of liquid or slurry waste from buildings involves scraping or flushing the waste from where it is dropped by the animal to a pit or a storage tank. Once collected, the waste is treated and/or disposed. Open feedlot waste is handled as runoff-carried waste, with natural precipitation carrying the liquid or the slurry, flowing from the pens into a drainage and collection system for subsequent removal or treatment and disposal. Liquid waste usually is handled by pumping, and is disposed of by being spread on the surface of or injected into the soil. The maximum amount of waste that can be spread in liquid form is determined by using the same procedure as that used for solid waste.

Liquid waste is more difficult to manage than the solid because it must be stored in tanks or pits and must be pumped. In addition, because animal waste should not be spread on frozen ground, the storage unit must have the capacity to store all of the waste while the ground is frozen.

Problem: What size of above ground tank (gal) is required to store all of the waste from a 100-sow farrowing barn if the pigs are with the sows, and the waste must be stored for 6 weeks?

Solution: Using Appendix VI and units cancellation (*Note*: weights can be converted to volumes if the density is known):

$$\begin{aligned} \text{gal} &= \frac{1 \text{ gal}}{0.13368 \text{ ft}^3} \times \frac{1 \text{ ft}^3}{60.0 \text{ lb}} \times \frac{33.0 \text{ lb}}{\text{day} - \text{sow}} \times \frac{7.0 \text{ day}}{1 \text{ week}} \times \frac{6 \text{ weeks}}{1} \times \frac{100 \text{ sows}}{1} \\ &= \frac{138600}{8.0208} \\ &= 17280.07\ldots \text{ or } 17{,}300 \text{ gal} \end{aligned}$$

An additional problem that arises using liquid waste is the space required for the storage tank.

Problem: How high will a round tank be in the previous problem if the space available for the tank is 20.0 ft in diameter?

Solution: This is a good problem to use the units method.

$$\text{ft} = \frac{0.13368\ \text{ft}^3}{\text{gal}} \times \frac{17{,}300\ \text{gal}}{1} \times \frac{1}{\pi \times (20.0\ \text{ft})^2/4}$$

$$= \frac{2312.664}{314.159\ldots}$$

$$= 7.3614\ldots \text{ or } 7.36\ \text{ft}.$$

21.5. Waste Treatment

Waste treatment is an operation performed on waste that makes it more amenable to ultimate disposal. Two basic methods of biological treatment are used: aerobic and anaerobic.

21.5.1. Aerobic Treatment

Aerobic treatment occurs when there is sufficient dissolved oxygen available in the waste to allow aerobic bacteria (oxygen-using) to break down the organic matter in the waste. It is essentially an odorless process. Three methods of treatment that use aerobic bacteria are: composting, aerobic lagoons, and spray-runoff.

Composting is accomplished by piling the waste and turning it frequently to provide aeration for aerobic bacterial decomposition while maintaining a high-enough temperature in the pile to destroy pathogenic organisms and weed seeds. The volume of composted waste may be reduced to between 30% and 60% of the volume of the original waste.

An aerobic lagoon is a relatively shallow basin (3 to 5 ft deep) into which a slurry is added. The decomposition of the organic matter is accomplished by aerobic bacteria. Beating or blowing air into the liquid provides oxygen for the bacteria and accelerates the process.

In the spray-runoff treatment, an area is leveled and sloped so that water flows evenly, and then is planted to grass. The bacteria that live on the wet surface of the grass and soil accomplish decomposition of the waste.

21.5.2. Anaerobic Treatment

Anaerobic treatment is accomplished by anaerobic bacteria that consume the oxygen in the organic matter itself as the organic matter decomposes. In this process, odorous gases are produced. The most common method using this treatment is the anaerobic lagoon.

The anaerobic lagoon is a relatively deep (12 ft to 14 ft) basin that contains liquid waste and provides a climate for decomposition by anaerobic bacteria. Thus, no attempt is made to introduce oxygen into the liquid. The anaerobic lagoon is able to decompose more organic matter per unit of volume than the aerobic lagoon. However, the odor it produces may determine where it can be located.

21.6. Metric Problems

The principles of animal waste handling treatment and disposal are the same in SI units. The only differences in the problems are the units that are used.

Problem: Determine the amount of solid waste from 91 kg finishing pigs that can be applied to a coarse loamy soil with a pH of 6.5 that will be used to raise 5.0 cubic meters per hectare of wheat. A soil test reveals 35.3 kg of nitrogen and 15.0 kg of phosphorus per hectare available for the plants.

Solution: The first step is to use Appendix VI to determine the amount of nitrogen and phosphorus produced by the animals. Note: SI values in Appendix VI are direct conversions.

$$\text{Nitrogen} = 0.0305 \text{ kg/day}$$
$$\text{Phosphorus} = 0.0204 \text{ kg/day}$$

The next step is to determine the pounds of nutrients per pound of waste.

$$\frac{\text{kg of nitrogen}}{\text{kg of waste}} = \frac{0.0305 \text{ kg}_n}{1 \text{ day}} \times \frac{1 \text{ day}}{5.9 \text{ kg}_w} = 0.005169\ldots \text{ or } 0.0052 \frac{\text{kg of nitrogen}}{\text{kg of waste}}$$

$$\frac{\text{kg of phosphorous}}{\text{kg of weight}} = \frac{0.0204 \text{ kg}_p}{1 \text{ day}} \times \frac{1 \text{ day}}{5.9 \text{ kg}_w} = 0.003457\ldots \text{ or } 0.0034 \frac{\text{kg of phosphorous}}{\text{kg of waste}}$$

The next step is to determine the net amount of nutrients needed by the crop. This is determined by subtracting the nutrients available in the soil from the nutrients used by the crop, Appendix VII.

$$\frac{\text{kg N}}{\text{ha}} = \frac{140 \text{ kg N}}{\text{ha}} - \frac{35.3 \text{ kg N}}{\text{ha}} = 104.7 \frac{\text{kg N}}{\text{ha}}$$

$$\frac{\text{kg P}}{\text{ha}} = \frac{56 \text{ kg P}}{\text{ha}} - \frac{15 \text{ kg P}}{\text{ha}} = 40 \frac{\text{kg P}}{\text{ha}}$$

The last step is to determine the kilograms of waste that can be applied per hectare. This is determined by multiplying the net nutrients used by the crop by the kilograms of nutrients per kilogram of waste.

$$\text{Nitrogen waste: } \frac{\text{kg waste}}{\text{ha}} = \frac{1 \text{ kg waste}}{0.0052 \text{ kg N}} \times \frac{104.7 \text{ kg N}}{\text{ha}} = 20134.61\ldots \text{ or } 20{,}000 \frac{\text{kg waste}}{\text{ha}}$$

$$\text{Phosphorus waste: } \frac{\text{kg waste}}{\text{ha}} = \frac{1 \text{ kg waste}}{0.0034 \text{ kg P}} \times \frac{40 \text{ kg P}}{\text{ha}} = 11764.7\ldots \text{ or } 12{,}000 \frac{\text{kg waste}}{\text{ha}}$$

$$\text{Phosphorus soil limits: } \frac{\text{kg waste}}{\text{ha}} = \frac{1 \text{ kg}}{0.0034 \text{ kg P}} \times \frac{448 \text{ kg P}}{\text{ha}} = 131764.7\ldots \text{ or } 130,\ 000 \frac{\text{kg waste}}{\text{ha}}$$

The least amount (kg/ha) is 12,000 for the phosphorus in the waste. This sets the limits of 12,000 kg of waste per hectare per year that can be applied.

22
Insulation and Heat Flow

22.1. Objectives

1. Understand the principles of insulation and heat flow.
2. Understand the function of insulation.
3. Understand R-values and U-values.
4. Be able to calculate the total thermal resistance of a building component.
5. Be able to determine the amount of heat flowing through a building component.

22.2. Introduction

All living organisms have a preferred environment. Whenever the environment is not within the preferred parameters, the organism is stressed. Stress can reduce productivity, increase the incidents of illness, shorten the storage live of products and cause many other problems. In the natural environment, organism modify their environment by moving from one location to another. Once animals and other organisms are confined, they cannot roam, and the environment must be modified to meet their needs. Two ways environments are modified are heating and cooling.

Heat is a form of energy. It is transmitted in three different ways:

1. Radiation—the exchange of thermal energy, heat, between objects by electromagnetic waves.
2. Convection—the transfer of heat from or to an object by a gas or a liquid by movement of the gas or liquid.
3. Conduction—the exchange of heat between contacting bodies that are at different temperatures.

The movement of heat by radiation and convection is more difficult to calculate than conduction. Conduct heat losses will be explained in more detail to illustrate the factors that influence the amount and rate of heat movement. The amount of and rate of conduction heat movement is controlled by the area through which the heat passes, the difference in temperatures, and the thermal resistance of the materials. These concepts and others will be discussed in this chapter. Latter sections will show that the greater the area, the greater the heat flow. A similar relationship exists for temperature. The greater the temperature difference the greater the heat flow. The thermal resistance of materials is primarily dependent on the density of the material. A material with many small, trapped air spaces will have higher thermal resistance. A material with high thermal resistance is called insulation.

22.3. Insulation

Insulation is any material that reduces the rate of heat transfer by conduction. All building materials have some resistance to the movement of heat, but some, insulation, are more resistant than others. A common characteristic of insulation materials is low density (pounds/cubic foot, kilogram/cubic meter). Expanded polystyrene, which has a low density, has a higher thermal resistance than concrete, which has a very high density. Also, as the density of a material increases, for example, by compressing the material or adding water, the thermal resistance is reduced.

Some materials, such as wood, concrete, and certain insulation materials, are homogeneous; that is, they have a uniform consistency throughout. For these materials the insulating values are listed as per inch of thickness, Appendix IX. Other materials such as concrete blocks and insulation backed siding are not homogeneous. They may have holes in them, or they may be composed of more than one type of material. The insulating value of these materials is based on a thermal test of a sample, Appendix IX, and a different value must be used for each different thickness.

There are three common reasons for insulating houses and other structures, Figure 22.1 and Figure 22.2.

(1) Insulation reduces loss of heat inside any type of enclosure during the winter months, reducing the amount of supplemental heat required to maintain the temperature of the enclosure; (2) insulation helps reduce heat gain during the summer months, reducing the load on cooling equipment, Figure 22.1 and (3) insulation is used to manage the condensation point so that it does not occur within the

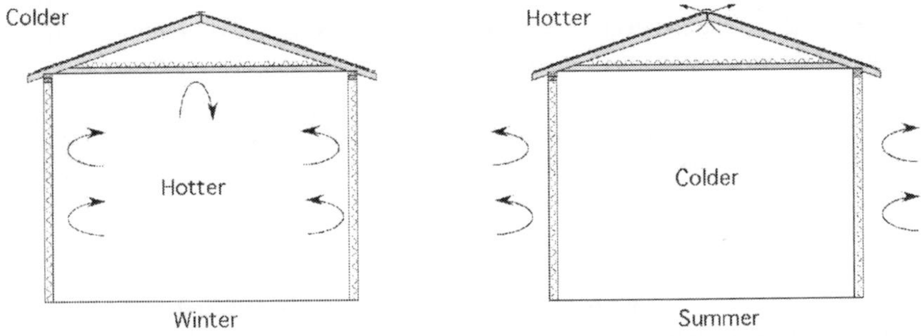

FIGURE 22.1. Advantages of insulation for reduction of the loss of heat in the winter or grain of heat in the summer.

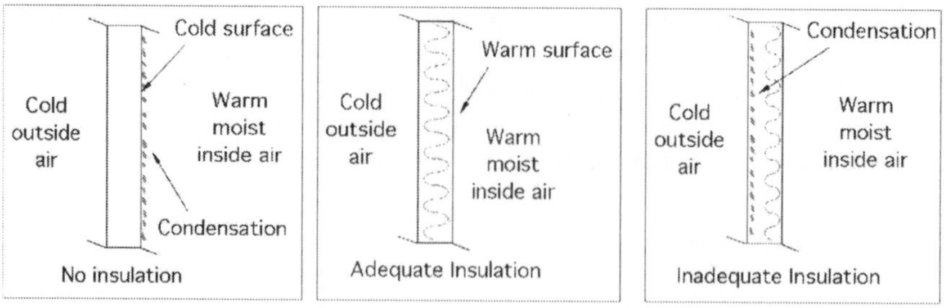

FIGURE 22.2. Adequate insulation prevents moisture condensation during the winter.

wall or on the inside surface, Figure 22.2 Condensation is the transformation of water vapor to a liquid, which occurs when warm moist air is cooled or comes in contact with a cool surface. Condensation is explained in more detail in the next chapter on psychrometrics. Condensation may occur on the outside surface of a wall, Figure 22.2, or on any surface inside the wall if no vapor barrier is used and the insulation is inadequate for the temperature difference.

22.4. *R*-values and *U*-values

Two different numbers are used to describe the insulating properties of construction materials: *R*-values and *U*-values. The *R*-value is a number that represents the thermal resistance of a material. The *U*-value is a measurement of the thermal conductivity of a material. In the customary units, a *U*-value of 1 means that 1 BTU/hr of heat will flow per square foot of area per degree difference in temperature on each side of the material. The term BTU stands for British Thermal Unit, and is the amount of heat required to raise the temperature of 1 lb of water 1 °F. In the SI unit system, the preferred unit for heat is kilojoules. A joule is equal to 9.478 E−4 BTU and is the amount of heat required to raise the temperature of 1 g of water by 1°C.

R-values are the inverse of the *U*-value ($R = 1/U$). Therefore, the insulating ability of a material increases as the *R*-value increases and as the *U*-value decreases. The *R*-value of a nonhomogeneous building component is the sum of the *R*-values of the materials that make up the component. The nonhomogeneous material may be a construction component such as a concrete block, or a part of the building such as a wall or ceiling. The units for U are $\frac{\text{BTU}}{°\text{F} \times A_u \times h}$ or $\frac{\text{J}}{°C \times A_u \times h}$, where A_u is the unit area of the material through which the heat passes (generally in ft^2 or m^2). Since the units of *R* are the inverse of the units of *U*, the units of *R* are $\frac{°\text{F} \times A_u \times h}{\text{BTU}}$ or $\frac{°\text{C} \times A_u \times h}{\text{J}}$. Appendix IX contains *R*-values for some common homogeneous and nonhomogeneous building materials.

The type of construction and the type and amount of insulating material used will determine the rate of heat flow (British Thermal Units/hr, customary, or Joules/hr, SI) for a particular building or building component. Common building components include walls, doors, windows, floors, ceilings and roofs. Knowledge of the relative insulating values of different construction materials and of how to put them together

FIGURE 22.3. Typical wood frame wall construction viewed looking down.

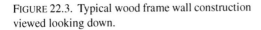

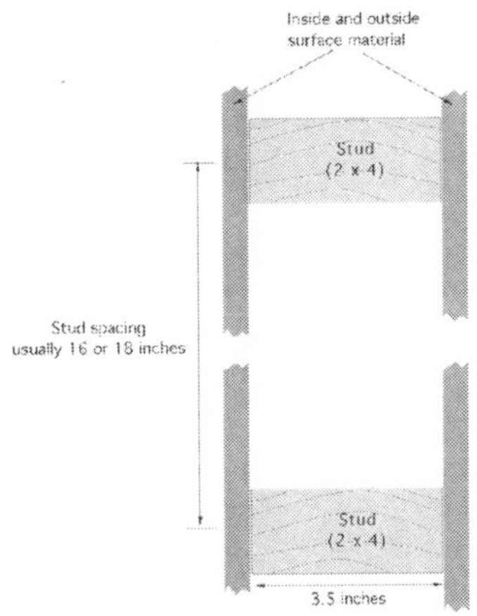

to estimate the overall insulating value of a building or a building component is a great help in selecting appropriate insulation materials for a particular application.

To determine the total thermal resistance of a wall, simply list and add together the thermal resistance of the individual parts. For homogeneous materials, the total thermal resistance value is determined by the thickness (inches) of the material used times the *R*-value per inch of thickness. For nonhomogeneous materials, select the thermal resistance value for the thickness of material specified.

Two additional factors are important in determining the total thermal resistance of a wall. The first is the thermal resistance associated with the layer of still air next to any surface, as well as any air space within the wall, floor, or ceiling. Still air is a good insulating material, and a thin air film clings to exterior and interior surfaces providing a measurable thermal resistance. Typical *R*-values for the inside and outside air film are found in Appendix IX. The second factor is an understanding of common construction methods. In wood frame construction, the walls usually are made from 2×4 or 2×6 inch lumber. Lumber is sold using nominal sizes; the actual size is less. A 2×4 is actually 1 and 1/2 inches by 3 and 1/2 inches, and a 2×6 is actually 1 and 1/2 inches by 5 and 1/2 inches. Frame walls usually are constructed with the long dimension of the board perpendicular to the wall, with different materials attached to both sides to form the wall. The cavity between the surfaces may be filled or partially filled with insulation, see Figure 22.3.

Problem: Determine the total thermal resistance (R_T) of a 2×4 wood frame wall. The outside is covered with 1/2 inch medium-density particle board, vapor barrier is permeable felt and 2.5 inches of 100 lb per cubic foot fired brick. The inside surface is 1/2 inch plasterboard. The wall is filled with medium-density, loose-fill cellulose.

Solution: The *R*-value of the wall is the sum of the *R*-value of each component. For this type of problem, a diagram of the wall and a table of the information

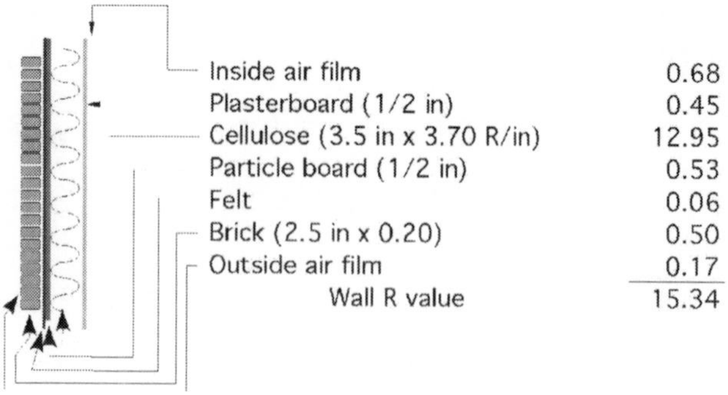

Inside air film 0.68
Plasterboard (1/2 in) 0.45
Cellulose (3.5 in x 3.70 R/in) 12.95
Particle board (1/2 in) 0.53
Felt 0.06
Brick (2.5 in x 0.20) 0.50
Outside air film 0.17
Wall R value 15.34

FIGURE 22.4. Solution to total thermal resistance problem.

will help prevent mistakes. Study Figure 22.4 for the solution. The R-values were obtained from Appendix IX.

For this particular wall the total thermal resistance (R_T) is 15.34. *Note*: Because loose fill insulation is used, the thickness of the insulation is equal to the cavity in the wall, 3.5 inches. When a batt type of insulation is used, and it is not as thick as the wall, the wall will also have an air cavity or void, the R-value for two inside surfaces is used if the air cavity is less than 3/4 inch. For example, if the insulation in the wall in the pervious problem is 3.0 inch of low density batt insulation, then the R_T is 12.71, Table 22.1.

The type of same table and procedures can be used to determine the total R-value for any building component. Simply sum the R-values for each type of material and the inside and outside air films. *Note*: some sources of R-values include the air film values with the material.

Problem: What is the total R-value (R_T) for a ceiling constructed of 1/2-inch plywood for the inside surface with 6.0 inches of low density batt insulation on top of the plywood?

Solution: This problem presents a different situation because neither surface is exposed to outside conditions. In buildings with very low attic ventilation the

TABLE 22.1. Solution for insulation problem.

Inside air film	0.68
Plaster board	0.45
Batt (3.0 R/in × 3.0 inch)	9.00
Cavity (0.68 × 2)	1.36
Particleboard	0.53
Felt	0.06
Brick	0.50
Outside air film	0.17
Total	12.75

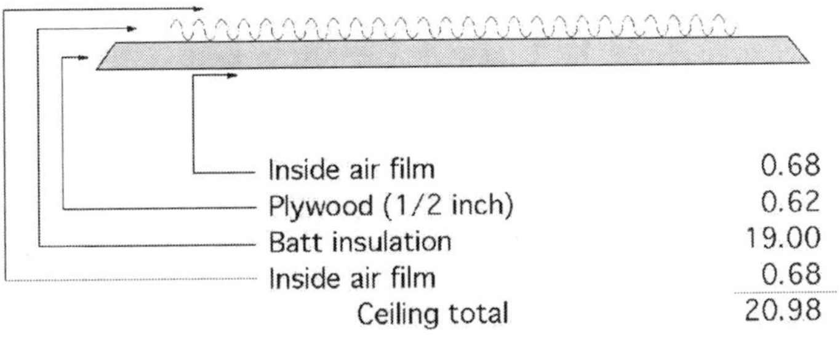

FIGURE 22.5. Solution to ceiling total resistance problem.

R-value for inside air film should be used on both sides of the ceiling. If the attic is well ventilated, an outside air film R-value would be appropriate. For this problem the assumption is that the attic is not well ventilated.

In this example the total R-value of the ceiling is 20.98, see Figure 22.5. One area of a building that is more difficult to determine a total thermal resistance is the floor. A concrete slab floor on the ground does not have a uniform temperature gradient across the floor. The insulating properties of the ground will cause the temperatures between the inside space and the ground at the center of the floor to be different from the edges. The values for insulated and un-insulated floors in Appendix IX are estimates.

22.5. Heat Flow

The rate of heat flow (BTU/h) through a building component depends upon the thermal resistance of the wall, the temperature difference on both sides of the wall, and the area of the component. Expressed as an equation:

$$Q = \frac{A \times \Delta T}{R}$$

where Q = Heat flow rate (BTU/h); A = Area of building component (ft^2); ΔT = Environmental temperature difference on each side of structure (°F); R = Total thermal resistance of the building component $\left(\frac{°F \times A_u \times h}{BTU}\right)$.

Heat flows from the area of higher temperature to the area of lower temperature. With the inside temperature greater than the outside temperature, the heat flow will be from inside to outside.

Problem: Determine the heat flow for an 8.0×10.0 ft wall with a total R-value of 1.22 and is subject to an inside temperature of 60°F and an outside temperature of 10°F.

Solution:

$$Q = \frac{A \times DT}{R} = \frac{(8.0 \text{ ft} \times 10.0 \text{ ft}) \times (60°\text{F} - 10°\text{F})}{1.22 \frac{\text{ft}^2 \times °\text{F} \times \text{h}}{\text{BTU}}}$$

$$= 3278.68 \ldots \text{ or } 3{,}300 \frac{\text{BTU}}{\text{h}}$$

Heat will move through this wall at the rate of 3,300 BTU/h.

The heat flow rate for all building components except floors can be calculated by using this method. If the floor is a concrete slab, the difference in temperature (ΔT) across the floor will change as one moves in from the outside edge of the building. The effective *R*-values for this type of floor or a concrete slab floor with insulation beneath it can be found in Appendix IX. When either of these two values is used, the perimeter of the building (ft) is substituted for the area (ft^2) in the heat flow equation. If a floor has a crawl space underneath it, the total *R*-value can be calculated by using the same procedures used for a ceiling. Floors over basements are much more complex and will not be discussed in this text.

With the information presented in the previous sections the total conduction heat flow for a building can be determined. The process for completing a total heat flow calculation is included in a later chapter because total building heat flow must also include heat lost or gained through ventilation.

22.6. SI Problems

Problem: Determine the heat flow for an 2.5 × 3 m wall with a total R-value of 6.93 $\frac{\text{kJ}}{°\text{C} \times \text{m}^2 \times \text{h}}$ and is subject to an inside temperature of 16°C and an outside temperature of 12°C.

Solution:

$$Q = \frac{A \times \Delta T}{R} = \frac{(2.5 \text{ m} \times 3.0 \text{ m}) \times (16°\text{C} - 12°\text{C})}{6.93 \frac{\text{m}^2 \times ° \text{C} \times \text{h}}{\text{kJ}}} = 30.30 \ldots \text{ or } 30.3 \frac{\text{kJ}}{\text{h}}$$

Heat will move through this wall at the rate of 30 kJ/h.

23 Heating, Ventilation, and Air-Conditioning

23.1. Objectives

1. Understand the principles of heating.
2. Be able to determine a heat balance for a building.
3. Be able to list and define the seven physical properties of air.
4. Given two properties of air, be able to find the other properties on a psychometric chart.
5. Be able to calculate a ventilation rate.
6. Understand the principles of air-conditioning.

23.2. Introduction

Air is a complex mixture of gases, water vapor, and heat. All living organisms require these three components, but not necessarily in the same proportions. Each organism has an optimum range of all three conditions, and any time the conditions are less than optimum, the organism is stressed. Management of these components may be critical within agricultural structures. Heating, cooling, and ventilation are used to modify the natural environment to reduce the environmental stress of animals and biological products.

The ability to manage building environments is based on an understanding of psychrometrics. Psychrometrics is the study of moist air. The constituents of air and their manipulation are easier to understand using a psychometric chart. The following sections explain the psychometric chart and provide examples of how it is used for heating and ventilation.

23.3. Psychometric Chart

A psychometric chart is a graphical representation of the seven physical properties of air. These physical properties are defined and described as follows:

1. Dry-bulb temperature (dbt): The dry-bulb temperature is the temperature of air measured with a standard thermometer. Dry-bulb temperatures (and all other temperatures) usually are expressed in degrees Fahrenheit (°F) with customary units or degrees Celsius (°C) with SI units.
2. Wet-bulb temperature (wbt): The wet-bulb temperature is determined with a standard thermometer having the bulb surrounded by gauze or a sock and a means for keeping the sock wet. As air passes over the wet sock, it will absorb (through evaporation) some of the water from the sock, cooling the bulb of the thermometer. The drier the air is, the greater the evaporation and the greater the cooling effect. The difference between the dry-bulb temperature and the wet-bulb temperature is called wet-bulb depression. Dry-bulb and wet-bulb temperatures can be measured with an instrument called a psychrometer.
3. Relative humidity (rh): Relative humidity is a ratio of the amount of water in the air relative to the maximum amount of water the air could hold if it were fully saturated. Relative humidity is expressed as a percent and the values can range from 0% (dry air) to 100% (fully saturated).
4. Moisture content: The moisture content is a measure of the actual amount of water held in the air in the form of vapor. Moisture content is measured in terms of pounds of water per pound of air (lb water/lb air) or grains of water per pound of air (gr. of water/lb air). (*Note:* 7000 grains of water equals one pound.) The American Society of Heating and Refrigerating Engineers (ASHRA) measures moisture in pounds per pound, but some psychrometric charts use grains of water.
5. Dew point: Dew point is the temperature at which, as air is cooled, the moisture in the air begins to condense or form droplets that are too large to remain suspended in the air. Condensation occurs on any object with a surface temperature equal to or less than the dew point.
6. The total heat is the total heat energy in the air. It includes heat due to the temperature of the air, heat required to change water vapor present in the air from liquid to vapor, and heat energy in the water vapor itself. The total heat content of air is expressed as BTU per pound of dry air (customary units or Joules per kg of dry air (SI units).
7. Specific volume: Specific volume is the volume of space occupied by a pound of dry air at standard atmospheric pressure (14.7 psi), expressed in cubic feet per pound of dry air (ft^3/lb air).

23.4. Reading a Psychrometric Chart[1]

A psychrometric chart is designed so that if any two properties of the air are known, values for the other five properties can be found on the chart. Different

[1] An ASHRAE PSYCHROMETRIC CHART NO. 1 is used in this section. Obtain from American Society of Heating, Refrigerating and Air-conditioning Engineers, Inc. 1791 Tullie Circle, NE, Atlanta, Georgia, 30329. (404-636-8400). If a different chart is used the values may be different.

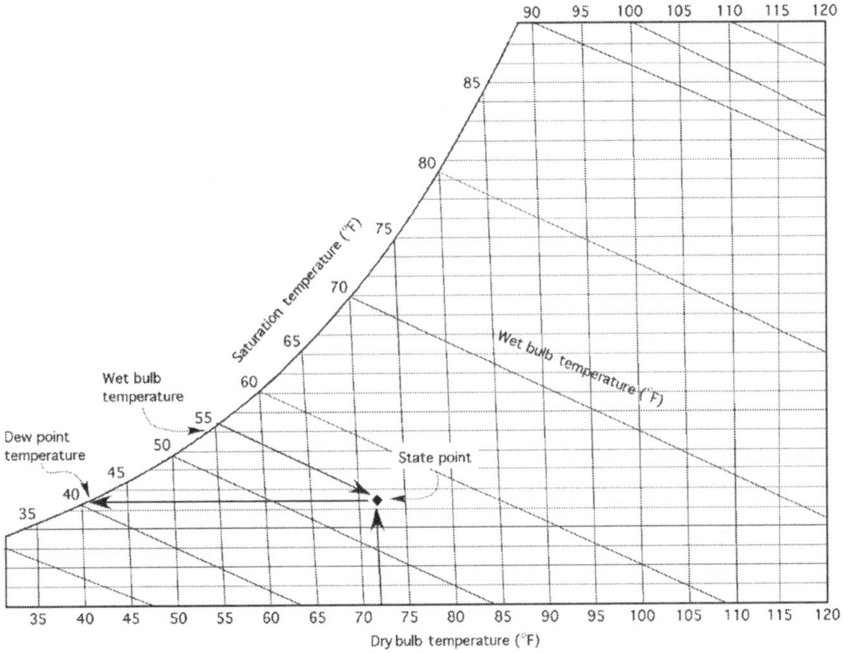

FIGURE 23.1. Dry bulb, wet bulb, and saturation temperature scales.

charts are used. It is common to see high temperature, normal temperature, and low temperature charts. In addition, charts are produced for special purposes. To complete the activities of this section you should obtain a normal temperature psychrometric chart. *Note:* the scales for each property of air are different and the number of graduations for each scale is not the same. The recommended practice is to interpolate only one decimal point beyond the scale.

A psychrometric chart has a scale and lines for each property. To find the values for the properties of air, you must be able to locate what is called the state point. The state point is the junction of any two property lines, Figure 23.1. The following section will explain how to locate the seven properties of air.

23.4.1. Dry-Bulb Temperature

The dry-bulb temperature scale is located along the bottom of the chart. Because this is a basic measurement, it usually is given or known. Locate it first, and then move vertically into the chart along the dry-bulb temperature line. In Figure 23.1 a dry bulb temperature of 72 degrees was selected.

23.4.2. Wet-Bulb Temperature

The wet-bulb scale is found along the curved side of the chart with the lines leading down and to the right. Locate the wet-bulb temperature, and then follow the line down and to the right into the chart until it meets the line of another known property of the air, for example, the vertical dry-bulb temperature line, Figure 23.1. If the temperatures of 72°Fdb and 56°Fwb are known, then the state point is located at the junction of these two lines, the "♦" shown in Figure 23.1. *Note:* the wet-bulb lines are very close to the heat lines. It is easy to confuse the two.

23.4.3. Dew Point

The dew point temperature uses the same scale as the wet-bulb temperature and the same lines as the moisture scale. To locate the dew point value, move horizontally to the left from the state point. For example, in Figure 23.1 the dew point is 41°F.

23.4.4. Total Heat

The total heat, enthalpy, scale is found by following the total heat lines up from the state point and to the left of the wet-bulb scale. The total heat lines and wet bulb lines almost coincide and this can cause errors in reading the chart, but they use different scales, Figure 23.2.

23.4.5. Specific Volume

The specific volume scale is located on vertical lines angling up and to the left from the dry-bulb temperature scale. They start at less than 12.5 ft^3/lb in the lower left corner and progress to greater than 15.0 ft^3/lb in the upper right corner of the chart, Figure 23.2. The values (scale) are contained on the lines.

23.4.6. Relative Humidity

Relative humidity lines follow the curved side of the chart. They progress from 10%, on the line closest to the dry-bulb temperature scale, up to 100%, which is the wet-bulb scale. Notice that the values (scale) are located on the lines, not at the edge of the chart, Figure 23.3.

23.4.7. Moisture Content

The moisture content scales are usually located on the right side of the chart. To locate a value for moisture, move horizontal to the right from the state point, Figure 23.3.

A psychrometric chart from the American Society of Heating, Refrigerating and Air Conditioning Engineers (ASHRAE) or similar group will contain all of seven of these scales. Figure 23.4 is a low precision chart that can be used for learning about psychrometrics and solving the sample problems, but an ASHRAE chart should be used if possible.

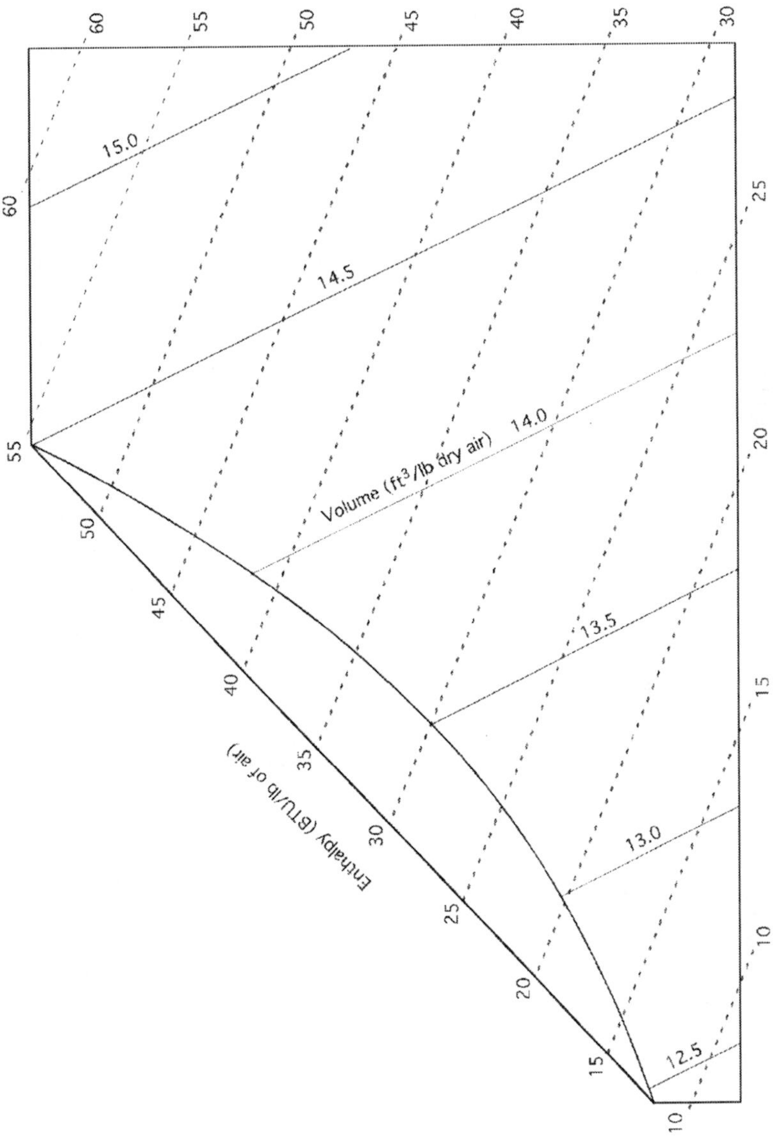

FIGURE 23.2. Enthalpy and specific volume scales.

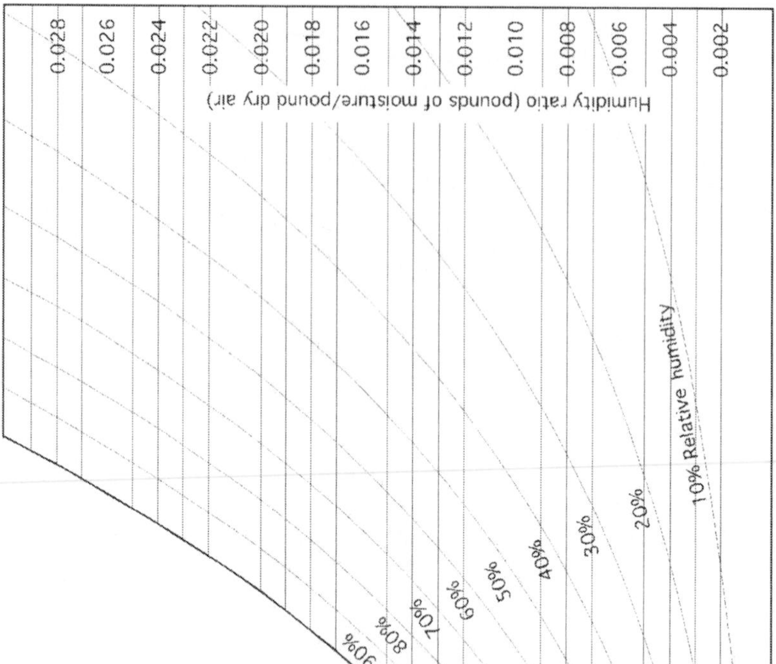

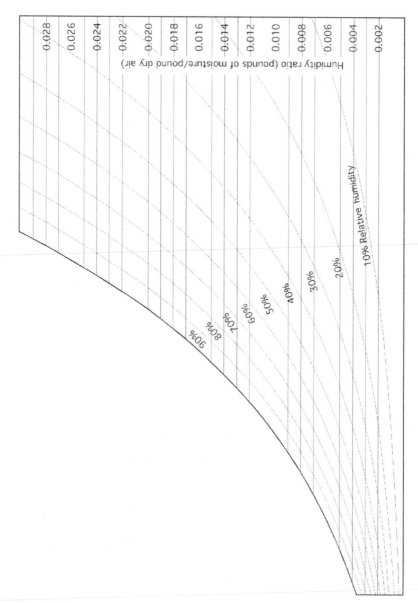

FIGURE 23.3. Relative humidity and humidity ratio scales.

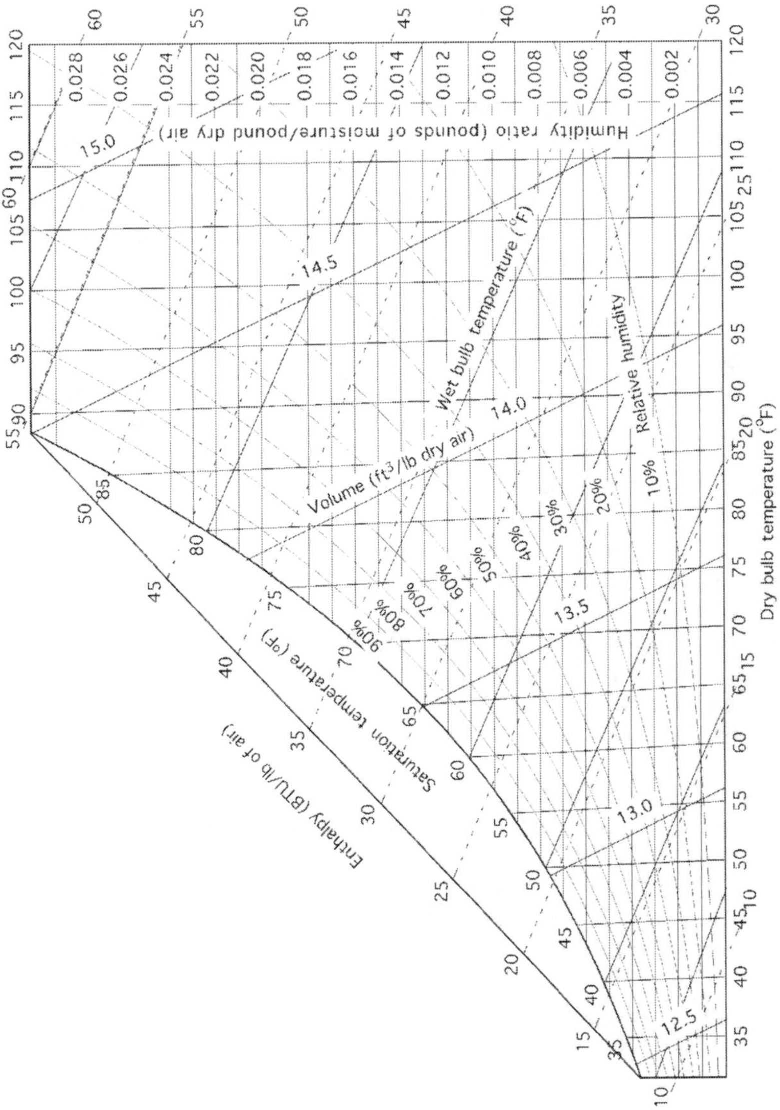

FIGURE 23.4. Complete psychrometric chart.

Problem: What are the physical properties of air if the dry-bulb temperature is 75°F, and the wet-bulb temperature is 60°F?

Solution: Using a psychrometric chart, Figure 23.5: [*Note:* not all psychrometric charts are the same. There will be small differences in the readings from different charts produced by different organizations.]

1. The first step is to find the state point. Locate the vertical line that represents the 75°F dry-bulb temperature. Then, locate the slanting wet-bulb line that represents 60°F wet-bulb. The intersection of these two lines is the state point.
2. The state point is just above the 40% rh line. Therefore the value of the relative humidity is between 40% and 50%. Because there are no other lines between 40% and 50%, the relative humidity must be estimated or interpolated. This point is estimated as 43% rh.
3. To find the dew point temperature, move horizontally to the left from the state point to the curved wet-bulb and dew point scale. The reading is slightly less than 50°F.
4. To find the moisture content move horizontally to the right from the state point to the vertical scale. You should read 0.0078 lb of water per pound of dry air.
5. The total heat is found by moving from the state point along the enthalpy line up and to the left (parallel to the dashed lines) until the total heat (BTU/lb) scale of air is reached. You should read 27 BTU/lb of air.
6. The specific volume is determined by locating and identifying the specific volume lines (ft^3/lb of dry air) that fall on either side of the state point, which in this case are 13.6 and 13.7. Estimating the state point distance between these lines gives an approximate value of 13.6 ft^3/lb.

Psychrometric charts are also used to explain the changes in the properties of air when it is cooled, heated, had water added or water removed.

Problem: What are the values for the physical properties of the air in the previous problem (75°Fdb and 60°Fwb) after the air is heated to 90°Fdb?

Solution: The first step is to determine the direction and the distance to move from the first state point to the second state point. In this problem, because heat is added and the moisture content is not changed, the second state point is located to the right of the first along the horizontal moisture line. Move right until this line intersects with the vertical 90°Fdb line, and then read the values for the characteristics at the second state point, Figure 23.6.
The values are:

Heat = 30.0 BTU/lb	Dew point = 49°F
Specific volume = 14.05 ft^3/lb	Moisture = 0.0076 lb/lb
Relative humidity = 27%	Wet-bulb = 65°F

Compare these results with those for the previous problem. When you understand the principles of air illustrated by a psychrometric chart, you should be able to

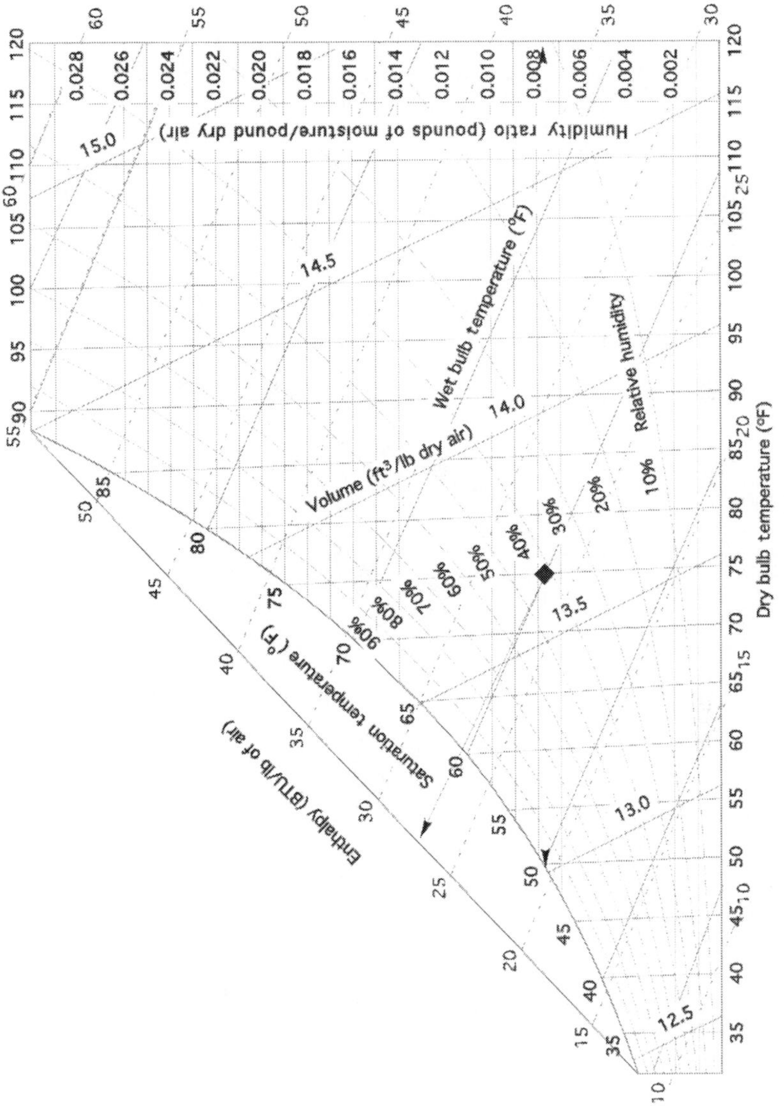

FIGURE 23.5. Psychrometric chart for problem one.

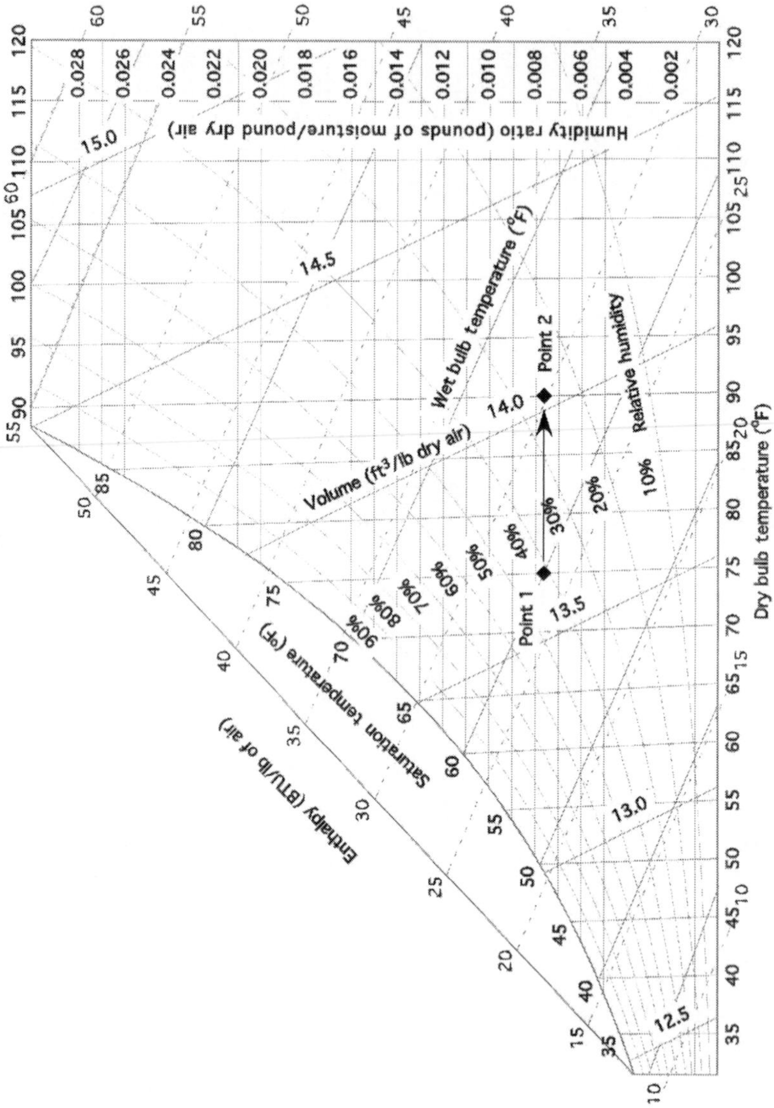

FIGURE 23.6. Psychrometric chart for problem two.

explain why the values for some of the characteristics changed, and why some did not.

23.5. Heating

Heating is used to modify an environment by raising the dry-bulb temperature, or to offset the effects of heat loss from a building. The amount of heat required depends upon the use of the building (animal housing, product storage, etc.), the heat flow out of the building, the ventilation needed, and the sources of heat within the building. In buildings used for product storage, a minimum amount of, or no ventilation may be needed, but buildings used for livestock housing must be ventilated to remove the moisture given off by the animals as they breathe and the harmful gasses that can develop in the building.

23.6. Ventilation Rate

Ventilation is the movement of air through a building. The movement may be caused by natural convection, or wind, or by forcing air movement with fans. Air is used to remove moisture, reduce the temperature if the outside air is cooler than the inside, and replace gases inside the building with outside air. The minimum ventilation rate is determined by which purpose of ventilation has the greatest requirement. For livestock buildings, if the ventilation rate is sufficient to remove excess moisture, it usually will be more than enough for the other needs.

In this section, a procedure involving the use of a psychrometric chart is used to make estimates of the amount of ventilation required to remove the excess moisture from a livestock building. For buildings used for other purposes, you must determine which aspect of the environment is critical, and base your calculations on that factor. Appendix contains values for moisture and heat released by different animals and the heat produced by mechanical equipment.

When outside air is introduced into a building, it must be uniformly distributed within the building and mixed with the building air to prevent drafts. One method commonly used is to introduce the air through small openings distributed around or throughout the building. Another method is to use heat exchangers. A heat exchanger uses the heat of the exhaust air to preheat incoming air, thus reducing the chilling effects of drafts and reducing the amount of heat that must be added. Actual airflow rates required for animal buildings will vary with the type and size of animal, management system, type of floor and floor drainage, and the number of animals in the building.

As moisture evaporates into the air it changes the physical properties of the air. The amount of change can be determined by locating the state points for both conditions. The numerical differences in the moisture content and the total heat of both conditions provide the information used to determine the amount of air and the heat required to evaporate the moisture.

TABLE 23.1. Psychrometric data for ventilation problem.

Properties	Inside	Outside	Difference
Dry bulb (°F)	55	40	
Wet Bulb (°F)	48	35	
Moisture (lb H_2O/lb of air)	0.0054	0.0028	0.0026
Total heat (BTU/lb)	19.00	13.0	6.0
Specific volume (ft^3/lb of air)	13.1	12.65	

Problem: How much air and heat will be needed to evaporate 2.0 lb of water per hour from within a building when the inside air temperature is 55°F dry bulb (db) and 48°F wet bulb (wb) and the outside air conditions are 40°Fdb and 35°Fwb?

Solution: In this problem we will use values for five of the properties of air for two different air conditions. Constructing a table, Table 23.1 is very helpful and is recommended for solving problems of this type.

The first step is to use the psychrometric chart to find the values for moisture content, total heat, and specific volume for both the inside and outside air. These values are recorded in the columns "Inside" and "Outside." *Note:* the values used are from the chart in Figure 23.4. Numbers from an ASHRAE chart will be slightly different. An ASHRAE or similar chart should be used when solving a real problem.

The next step is to determine the differences in moisture content and in total heat between the inside and the outside air. These values are recorded in the column labeled "Difference." Values for the differences in temperature and in specific volume are not needed in this problem.

The third step is to determine the amount of air (lb) required to remove the moisture. The amount of air required to evaporate the moisture is determined from the difference in the moisture content of the inside and the outside air. A difference of 0.0026 lb of water per pound of air means that every pound of outside air brought in and raised to inside conditions is capable of absorbing 0.0026 lb of moisture. If each pound of air absorbs the maximum amount of water, 0.0026 lb, and the amount of water that must be removed is 2.0 lb per hour, then the required amount of air (lb) is:

$$\frac{\text{lb air}}{\text{hr}} = \frac{1\ \text{lb air}}{0.0026\ \text{lb water}} \times \frac{2.0\ \text{lb}}{1\ \text{hr}} = 769.23\ldots \text{ or } 770\ \frac{\text{lb air}}{\text{hr}}$$

For these inside and outside conditions, 770 lb of air per hour moving through the building will be needed to remove 2.0 lb of water per hour, assuming that the air absorbs the maximum amount of water.

Before water in liquid form can be absorbed by air, it must be vaporized, and this requires heat. In addition, if the inside of the building is warmer than the outside, ventilation causes a loss of heat from the building. The total amount of heat lost is determined by using the difference in the heat of the incoming and the outgoing air. The amount of heat in the inside air is 19.0 BTU/lb, and that of the outside air is 13.0 BTU/lb—a difference of 6.0 BTU/lb. This amount of heat will be lost with each pound of air ventilated. We have already determined that 770 lb of air per hour will be needed to evaporate the water produced in the building; so the heat loss due to ventilation is:

$$\frac{\text{BTU}}{\text{hr}} = 770\,\frac{\text{lb air}}{\text{hr}} \times \frac{6.0\text{ BTU}}{\text{lb air}} = 4{,}620 \text{ or } 4{,}600\,\frac{\text{BTU}}{\text{hr}}$$

For these conditions, 4,600 BTU per hour of heat will be transported outside the building with the ventilation air.

Ventilation fans usually are sized in units of cubic feet per minute. This will require a units conversion from weight to volume. This conversion is made using the specific volume values from the psychrometric chart. When the ventilation fan is exhausting building air (the most common situation), then the inside specific volume is used. If the fan is blowing in outside air, the outside specific volume is used. Because it takes 770 lb of air per hour to evaporate the water produced in the building, the volume of air (ft^3/hr) that is required is determined by multiplying the pounds of air per hour by the specific volume. The size of fan required is:

$$\frac{\text{ft}^3}{\text{hr}} = 770\,\frac{\text{lb air}}{\text{hr}} \times 13.1\,\frac{\text{ft}^3}{\text{lb air}} = 10{,}087 \text{ or } 10{,}000\,\frac{\text{ft}^3}{\text{hr}}$$

A fan, or several fans with a combined capacity of 10,000 cubic feet per hour is required to provide enough ventilation to remove the excess moisture from the building.

The values calculated in the previous problem are accurate only as long as the inside and the outside environments do not change. This only occurs for short periods of time because the outside temperature is constantly changing, the respiration rate of the animals changes with activity, and heat exchanges also occur because of conduction, convection, and radiation. Heating, ventilating, and cooling systems must be able to react to change, and if they are required to maintain a constant inside environment, they must have the capacity for the most extreme inside and outside situations.

23.7. Building Heat Balance

The principles of heating, ventilation, and air-conditioning are all used to determine the amount of ventilation, heat, or air conditioning needed to maintain the temperature inside a building. The process of determining what changes are occurring in the environment inside a structure is call a heat balance. Expressed mathematically a heat balance is:

$$\pm\text{Heat}\left(\frac{\text{BTU}}{\text{hr}}\right) = \text{Total heat gain } (H_G) - \text{Total heat loss } (H_L)$$

Three possible answers can result from using the heat balance equation. (1) The amount of heat gain is larger than the heat loss. This will result is a positive (+) heat gain. In this situation, the temperature of the building will increase. (2) The heat gain and heat loss will be the same. When this is true, the answer from the heat balance equation is zero, and the temperature within the building will remain stable. (3) The heat loss is greater than heat gain. When this is true, the answer for the equation is negative (−) and the temperature inside the building will decrease.

The four factors needed to determine heat gain and loss are: mechanical equipment heat, animal heat, heat flow, and ventilation. Mechanical equipment, such as electric motors and heat lamps, and animals put heat into the building. Heat flows from areas of high temperature to areas of low temperature. The term heat flow is used to describe the movement of heat through the components of a building. Ventilation and its effect on the environment are discussed in a previous section.

These factors usually result in either a heat gain or a heat loss, depending on the season of the year and the temperatures. For a winter heat balance, the heat gain is the sum of the animal heat and the mechanical equipment heat, and the heat losses are due to heat flow and ventilation.

$$\pm \text{Heat}_W = (\text{Animal heat} + \text{Mechanical heat}) - (\text{Heat flow} + \text{Ventilation})$$

For a summer heat balance, the heat gain is the sum of the animal heat, mechanical equipment heat, and heat flow.

$$\pm \text{Heat}_S = (\text{Animal heat} + \text{Mechanical heat} + \text{Heat flow}) - (\text{Ventilation})$$

The following problem illustrates the procedure for calculating a winter heat balance for a building. *Note:* in this calculation the effects of radiation are ignored.

Problem: Determine the daytime heat balance (BTU/hr) for a 30.0 ft by 60.0 ft structure with 8.0 ft walls, which houses two hundred 150.0-lb feeder pigs. The building has two, 3.0-ft by 6.6-ft pine doors, 1.0 inch thick. In addition, it has ten 24.0-inch by 42.0-inch single-pane windows. The walls are 8.0-inch lightweight concrete block with the cores filled with lightweight perlite. The ceiling is 3/8-inch plywood with 6.0 inches of low-density batt insulation between the joists. The mechanical equipment includes nine, 100 W incandescent lights that operate 10 hr per day, two, 1/3 horsepower auger motors which operate 30 min per day, and ten, 1/4 horsepower ceiling fans that operate 5 hours per day. The attic space is well ventilated and is at the same temperature as the outdoors. The building is on an insulated concrete slab. The temperatures are 70°Fdb and 60°Fwb inside and 40° Fdb and 34° Fwb outside.

Solution: Two values are required to determine a heat balance—the total heat loss and the total heat gain. Begin by determining total heat loss. For winter conditions, total heat losses are heat flow and ventilation.

Heat losses: The heat flow equation used in the previous chapter is used to determine the total heat flow into or out of a building. The total heat flow is found by calculating and summing the heat flow for each component of the building having a unique *R*-value. The different *R*-values in this problem include those of the walls, windows, door, ceiling, and floor. Errors can be reduced by setting up a table of the required information. Because heat flow is a function of area, temperature difference, and thermal resistance, table columns are included for this

TABLE 23.2. Solution for building heat flow problem.

Component	Total R value	Area (ft^2)	ΔT (°F)	Q (BTU/hr)
Walls (less openings)	7.65	1330	30	5215.68...
Windows	0.91	70	30	2307.69...
Doors	1.85	40	30	648.64...
Ceiling	20.32	1800	30	2657.48...
Floor	2.22	180*	30	2432.43...
		Total heat flow (Q_T) =		13,261.93...

*The perimeter (ft) is used for slab floors, not the area (ft^2).

information, Table 23.2. In addition, it may be helpful to sketch components, as in the example in Chapter 22. A sketch was not used in this example because each building component has a small number of parts. Using the conduction heat flow equation:

$$Q = \frac{A \times \Delta T}{R}$$

where Q = heat flow (BTU/hr); A = area of structure with similar R value (ft^2); ΔT = difference in temperature between inside and outside (°F); R = thermal resistance of the building component.

With a temperature difference of 30°F, the total heat loss (flow) is 13,262 BTU/hr. These numbers were calculated as follows:

R-values:		
Wall:		
	Outside air film	0.17
	Concrete blocks, filled cores	6.8
	Inside air film	0.68
	Total *R*-value for wall	7.65
Windows:		
	Outside air film	0.17
	Single-pane glass	0.06
	Inside air film	0.68
	Total *R*-value for windows	0.91
Door:		
	Outside air film	0.17
	Wood	1.00
	Inside air film	0.68
	Total *R*-value for door	1.85
Ceiling:		
	Inside air film	0.68
	Insulation (6-in batt)	19.00
	Plywood	0.47
	Outside air film	0.17
	Total *R*-value for ceiling	20.32
Floor:		
	Insulated concrete	2.22

Areas:

$$\begin{aligned} \text{ft}^2 &= (\text{area of endwalls} + \text{area of side walls}) \\ &\quad - (\text{area of windows} + \text{area of doors}) \\ &= [(8.00\ \text{ft} \times 30.0\ \text{ft} \times 2) + (8.00 \times 60.0\ \text{ft} \times 2)] \\ &\quad - \left(70.0\ \text{ft}^2 + 40.0\ \text{ft}^2\right) \end{aligned}$$

Walls:

$$\begin{aligned} &= 1440\ \text{ft}^2 - 110\ \text{ft}^2 \\ &= 1330\ \text{ft}^2 \\ \text{ft}^2 &= \text{Width} \times \text{Height} \times \text{Number of windows} \end{aligned}$$

Windows:

$$\begin{aligned} &= [(24.0\ \text{in} \times 42\ \text{in}) \times 10] \times \frac{1\ \text{ft}^2}{144\ \text{in}^2} \\ &= 70\ \text{ft}^2 \\ \text{ft}^2 &= \text{Width} \times \text{Height} \times \text{Number of doors} \end{aligned}$$

Doors:

$$\begin{aligned} &= 3.0\ \text{ft} \times 6.6\ \text{ft} \times 2 \\ &= 40\ \text{ft}^2 \end{aligned}$$

Ceiling: $\text{ft}^2 = \text{Length} \times \text{Width} = 60.0\ \text{ft} \times 30.0\ \text{ft} = 1{,}800\ \text{ft}^2$

Perimeter:
Floor:

$$\begin{aligned} \text{ft} &= (\text{Length} \times 2) + (\text{Width} \times 2) = (60.0 \times 2) + (30.0 \times 2) \\ &= 180\ \text{ft} \end{aligned}$$

Building heat flow (Q_T):

$$\begin{aligned} Q_T &= Q_{\text{wall}} + Q_{\text{window}} + Q_{\text{door}} + Q_{\text{ceiling}} + Q_{\text{floor}} \\ &= \frac{1330 \times 30.0}{7.65} + \frac{70.0 \times 30.0}{0.91} + \frac{40.0 \times 30.0}{1.85} + \frac{1800 \times 30.0}{20.32} + \frac{180 \times 30.0}{2.22} \\ &= 5215.68\ldots + 2307.69\ldots + 648.64\ldots + 2657.48\ldots + 2432.43\ldots \\ &= 13261.92\ldots \text{ or } 13{,}300\ \frac{\text{BTU}}{\text{hr}} \end{aligned}$$

This is a wintertime heat balance so ventilation is a source of heat loss. The ventilation heat loss is determined by the ventilation rate required to remove the excess moisture from the building and the difference in temperature between the inside and the outside air. A table should be used to organize the psychrometric data, Table 23.3.

TABLE 23.3. Psychrometric data for problem.

Properties	Inside	Outside	Difference
Dry bulb (°F)	70.0	40.0	
Wet bulb (°F)	60.0	34.0	
Moisture (lb/lb)	0.0088	0.0026	0.0062
Total heat (BTU/lb)	26.5	12.5	13.5
Specific volume (ft^3/lb)	13.6	12.6	

In this problem, the building is used to house livestock. Therefore, the ventilation rate will be determined by the amount of water vapor produced by the animals. Appendix X contains values that can be used to estimate the amount of heat and moisture produced by various animals, based on weight and species.

To determine the ventilation rate, begin by finding the amount of water released into the building. A 150-lb feeder pig produces 0.219 lb of moisture per hour, Appendix X. The total water produced is:

$$\frac{\text{lb}}{\text{hr}} = \frac{0.219\ \frac{\text{lb H}_2\text{O}}{\text{hr}}}{\text{pig}} \times 200\ \text{pigs} = 43.8\ \frac{\text{lb H}_2\text{O}}{\text{hr}}$$

Next, determine the amount of air needed to remove 43.8 lb of water per hour. This is accomplished by multiplying the amount of water that each pound of air can absorb as it passes through the building by the amount of water that needs to be removed. If we assume that the air is 100% saturated when it leaves the building, then the amount of water that the air will remove is determined by the difference between the moisture content of the air entering and that of the air exiting the building. From Table 23.3, this value is 0.0062 lb of water per pound of dry air. The amount of air needed is:

$$\frac{\text{lb air}}{\text{hr}} = \frac{1\ \text{lb air}}{0.0062\ \text{lb water}} \times \frac{43.8\ \text{lb water}}{\text{hr}} = 7064.51\ldots \text{or } 7060\ \frac{\text{lb air}}{\text{hr}}$$

Now that the ventilation rate is known, the amount of heat loss due to ventilation can be calculated. This is accomplished by multiplying the pounds of air per hour that will be moving through the building by the difference between the amount of heat in the inside air and that in the outside air. From Table 23.3, this value is 13.5 BTU/lb of air. The heat loss due to ventilation is:

$$\frac{\text{BTU}}{\text{hr}} = 7060\ \frac{\text{lb air}}{\text{hr}} \times 13.5\ \frac{\text{BTU}}{\text{lb air}} = 95{,}310 \text{ or } 95{,}300\ \frac{\text{BTU}}{\text{hr}}$$

The total heat loss is obtained by adding the rates for heat flow and ventilation:

$$\begin{aligned}\frac{\text{BTU}}{\text{hr}} &= \text{Heat flow} + \text{Ventilation losses}\\ &= 13{,}300\ \frac{\text{BTU}}{\text{hr}} + 95{,}300\ \frac{\text{BTU}}{\text{hr}}\\ &= 108{,}600\ \frac{\text{BTU}}{\text{hr}}\end{aligned}$$

Heat gain: Next determine the total heat gain. The sources of heat are the mechanical equipment and the heat given off by the animals, Appendix X.

Mechanical equipment:

$$\text{Heat gain } \frac{\text{Btu}}{\text{hr}} = \text{Mechanical heat} + \text{animal heat}$$

$$\text{Mecahnical heat} = \text{Lights} + \text{Motors}$$

$$\text{Lights} = \left(9 \text{ lights} \times 100 \frac{\text{watt}}{\text{light}} \times 3.4 \frac{\text{Btu}}{\text{watt-hr}}\right) = 3060 \frac{\text{Btu}}{\text{hr}}$$

$$\text{Motors} = \left(2 \text{ motors} \times 0.33 \frac{\text{hp}}{\text{motor}}\right) + \left(10 \text{ motors} \times 0.25 \frac{\text{hp}}{\text{motor}}\right)$$

$$= \left(0.66 \frac{\text{hp–hr}}{\text{day}} + 2.5 \frac{\text{hp–hr}}{\text{day}}\right) \times 4{,}000 \frac{\text{Btu}}{\text{hp-hr}}$$

$$= 12{,}640 \frac{\text{Btu}}{\text{day}}$$

$$\text{Mechanical heat} = 3060 \frac{\text{Btu}}{\text{day}} + 12{,}640 \frac{\text{Btu}}{\text{day}}$$

$$= 15{,}700 \frac{\text{Btu}}{\text{day}}$$

$$\text{Animal heat } \frac{\text{BTU}}{\text{hr}} = \frac{\frac{354 \text{ BTU}}{\text{hr}}}{\text{pig}} \times 200 \text{ pigs} = 70{,}800 \frac{\text{BTU}}{\text{hr}}$$

$$\text{Heat gain} = \text{Mechanical heat} + \text{Animal heat}$$

$$= 15{,}700 \frac{\text{Btu}}{\text{hr}} + 70{,}800 \frac{\text{Btu}}{\text{hr}}$$

$$= 86{,}500 \frac{\text{Btu}}{\text{hr}}$$

Heat balance: All of the information needed to complete the heat balance is available.

$$\pm \text{Heat balance } \frac{\text{BTU}}{\text{hr}} = \text{Total heat gain} - \text{Total heat loss}$$

$$= 86{,}500 \frac{\text{BTU}}{\text{hr}} - 108{,}600 \frac{\text{BTU}}{\text{hr}}$$

$$= -22{,}100 \frac{\text{BTU}}{\text{hr}}$$

A heat balance with a negative number indicates that the amount of heat gain in the building is less than the amount of heat being lost from the building. When this occurs, the temperature inside the building will decrease unless additional heat is added. For this problem, 22,100 BTU/hr of heat is needed to maintain the inside temperature of the building. *Note:* livestock building environments are dynamic systems. The heat gain and loss is never constant. In this problem the heat balance was calculated with the assumption that the auger motors and ceiling fans were operating. The auger motors only operate 30 min a day. What happens to the heat balance when they are turned off? What is the effect of turning off some of the ceiling fans, or one of the waterers developing a leak?

23.8. Air-Conditioning

In some types of buildings, ventilation may not be adequate to maintain the optimum temperature. The amount of heat produced by the animals, the processing, outside temperatures and so on, may exceed the ability of the ventilation system to maintain the desired temperature. In these situations, air conditioners are used.

Historically many different systems were used to condition the air. Cooling with ice, pumping well water through a radiator, evaporative cooling, or drawing air through underground tubes has all been used. The effectiveness of these methods was limited. The use of ice will produce large amounts of cooling, but the resulting water must be disposed of, and a continuous supply of ice must be available. The amount of cooling available from well water is limited, and evaporative cooling is effective only if the outside air has a low relative humidity.

Today, the term air conditioning refers to conditioning air through the use of mechanical refrigeration. Refrigeration is the process of transferring heat from one substance to another; and for air conditioning, refrigeration moves heat from the air inside a structure to the air outside the structure. For this to occur, the air that is to be cooled must come in contact with a material at a temperature lower than that of the inside air. When this happens, heat from the air will flow into the colder material. Some air conditioners use a chilled metal surface to provide a cold mass, whereas others use a spray of chilled water. The heat is transferred by a substance circulating between the inside air and the outside air. In smaller units, the transfer substance will be a gas; for larger systems, the substance may be a liquid. Figure 23.7 is an illustration of the basic components of a mechanical refrigeration system.

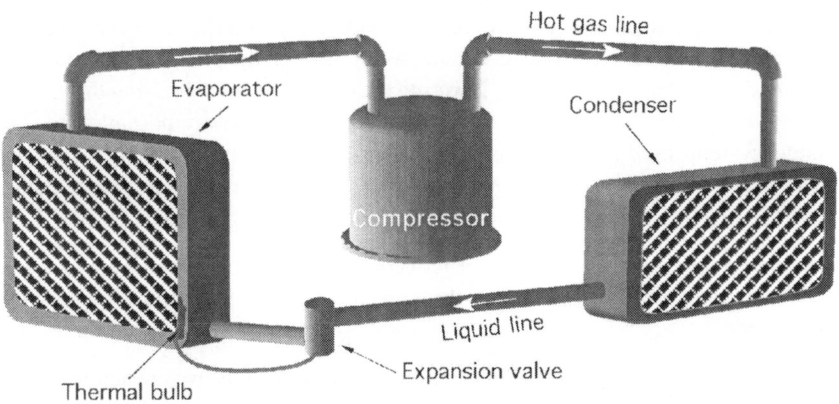

FIGURE 23.7. Mechanical refrigeration system.

In a mechanical refrigeration system, heat is moved by alternately compressing, liquefying, expanding, and evaporating a refrigerant, commonly Freon. The compressor increases the pressure and the temperature of the refrigerant as it compresses the gas. As the refrigerant passes through the condenser, the heat absorbed by the refrigerant as it passed through the evaporator, and the heat produced by compression, is transferred into the air. Thus, the condenser will be located where outside air moving across it absorbs the heat causing the refrigerant to liquefy. As the liquid refrigerant flows through the expansion valve and into the evaporator, the pressure drops and the refrigerant absorb heat from the surrounding air. The evaporator will be located inside an air duct or other location where air can pass through it and be cooled, causing the refrigerant to be vaporized. The hot, low pressure gas flows back to the compressor and the cycle begins again. The expansion valve and a thermal bulb regulate the flow of the refrigerant to produce the desired evaporator temperature.

It is common for mechanical refrigeration air conditioning to lower the temperature 20 to 30 degrees. Often this results in an air temperature that is below the dew point. This is the reason building and automobile air conditioners produce water. This can be shown on a psychrometric chart.

Problem: How many pounds of water will be removed per pound of dry air when air at 95°Fdb and 70% rh is cooled to 75°Fdb?

Solution: The solution is mapped out in Figure 23.8. Locate the initial state point on a psychrometric chart. Air conditioning lowers the dry-bulb temperature, so move left to 75 degrees. As the air is cooled, the relative humidity increases and in this example the saturation point is reached before 75 degrees. In this situation, follow the saturation line down to 75 degrees. This is the second state point. The amount of water removed is the difference in the humidity ratio between state point one and state point two.

$$\frac{\text{lb H}_2\text{O}}{\text{lb dry air}} = 0.0254\,\frac{\text{lb H}_2\text{O}}{\text{lb dry air}} - 0.019\,\frac{\text{lb H}_2\text{O}}{\text{lb dry air}} = 0.0064\,\frac{\text{lb H}_2\text{O}}{\text{lb dry air}}$$

Evaporative cooling is still useful for reducing the temperature of air in climates where the relative humidity is low. An evaporative cooler pulls outside air through a wet pad of porous material. As the air passes through the pad the temperature of the air causes water to evaporate, lowering the temperature, and raising the relative humidity. Figure 23.9 illustrates the components of a window or portable unit. The larger systems used in greenhouses and other structures have similar components. The principles of evaporative cooling can be demonstrated using a psychrometric chart.

Problem: Determine the amount of reduction in the temperature of air that is at 100°Fdb and 40% rh after it passes through an evaporative pad and becomes 90% saturated.

Solution: Figure 23.10 shows the solution mapped out on the psychrometric chart. Start by locating state point number one. In evaporative cooling, water is converted from liquid to a vapor. This requires heat. The heat for evaporation is drawn from

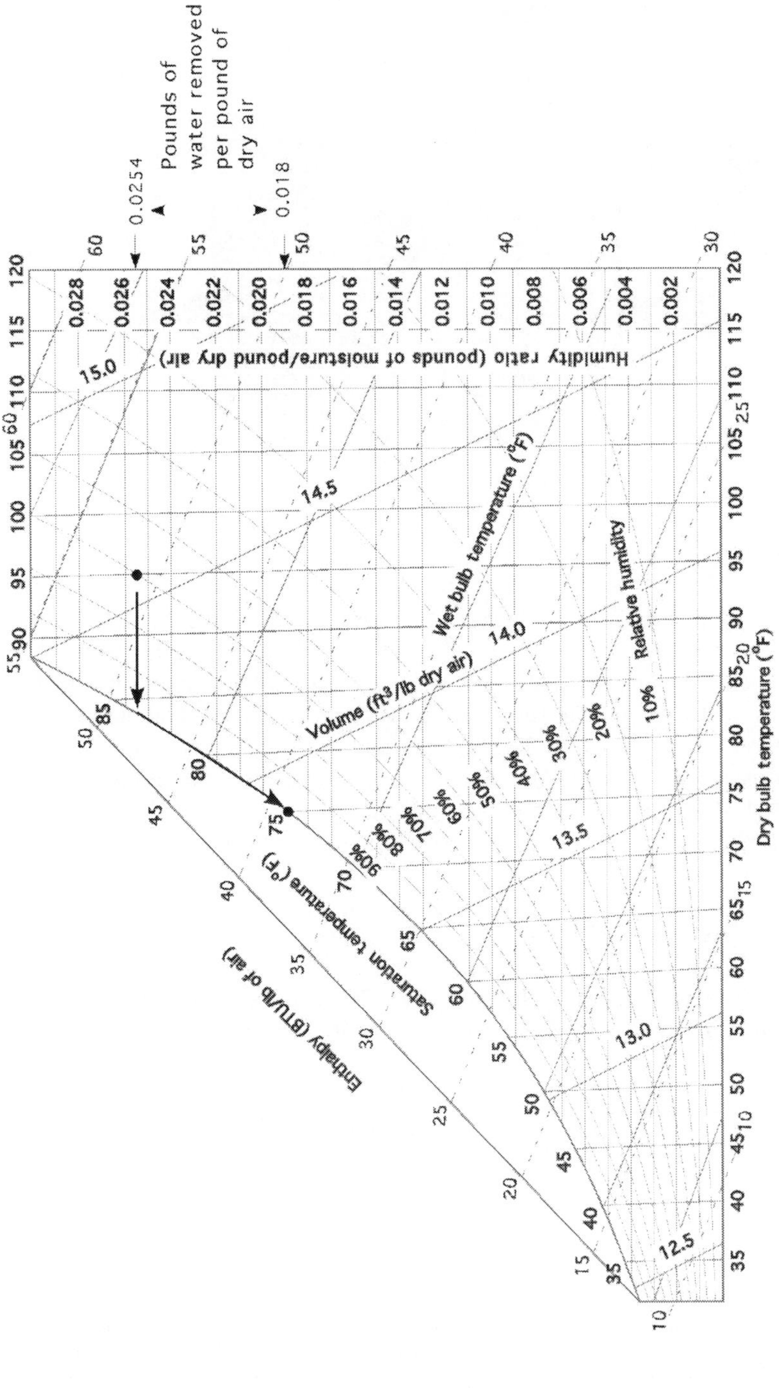

FIGURE 23.8. Solution for air conditioning problem.

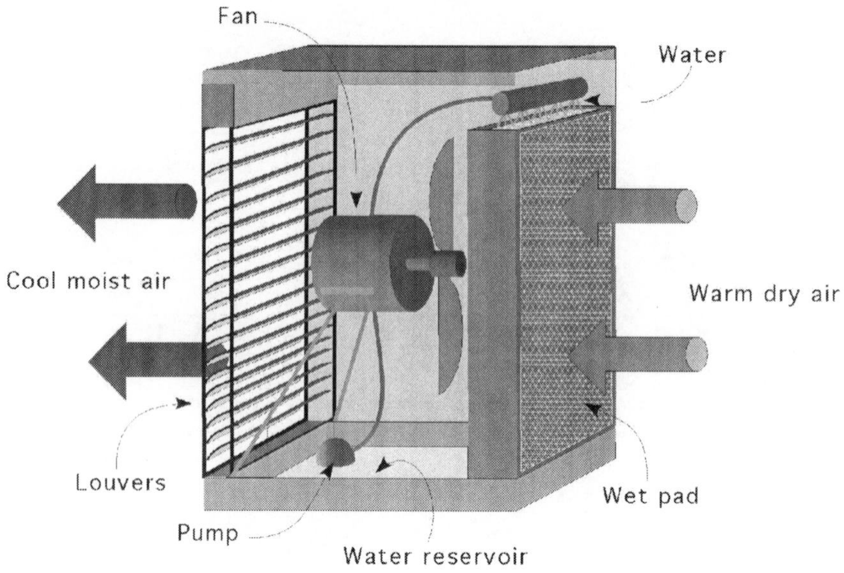

FIGURE 23.9. Evaporator cooler.

the air, which lowers the dry-bulb temperature. Lowering the dry-bulb temperature is represented by moving horizontally to the left from state point number 1. But evaporative cooling also adds water vapor to the air. Adding water vapor to the air is represented by moving vertically up the chart from state point number 2. The result is a vector problem. The best estimate of the direction to move using the chart is to follow the wet-bulb line to the left and up until the second state point is reached, 90% relative humidity. In this example, evaporative cooling lowers the temperature from 100°Fdb to 81°Fdb, a difference of 19°F.

23.9. Metric Problems

Some of the SI units for psychrometrics are the same and some are different. Refer to Table 23.4 and Figure 23.11 before attempting to complete a sample problem.

TABLE 23.4. Comparison of customary and SI psychrometric terms.

Term	Conventional	SI
Relative humidity	%	%
Volume	Ft^3/lb	M^3/kg
Dry bulb	°F	°C
Wet bulb	°F	°C
Enthalpy	Btu/lb dry air	kJ/kg dry air
Dew point	°F	°C
Moisture content	lb H_2O/lb dry air	kg H_2O/kg dry air
Time	hour	hour

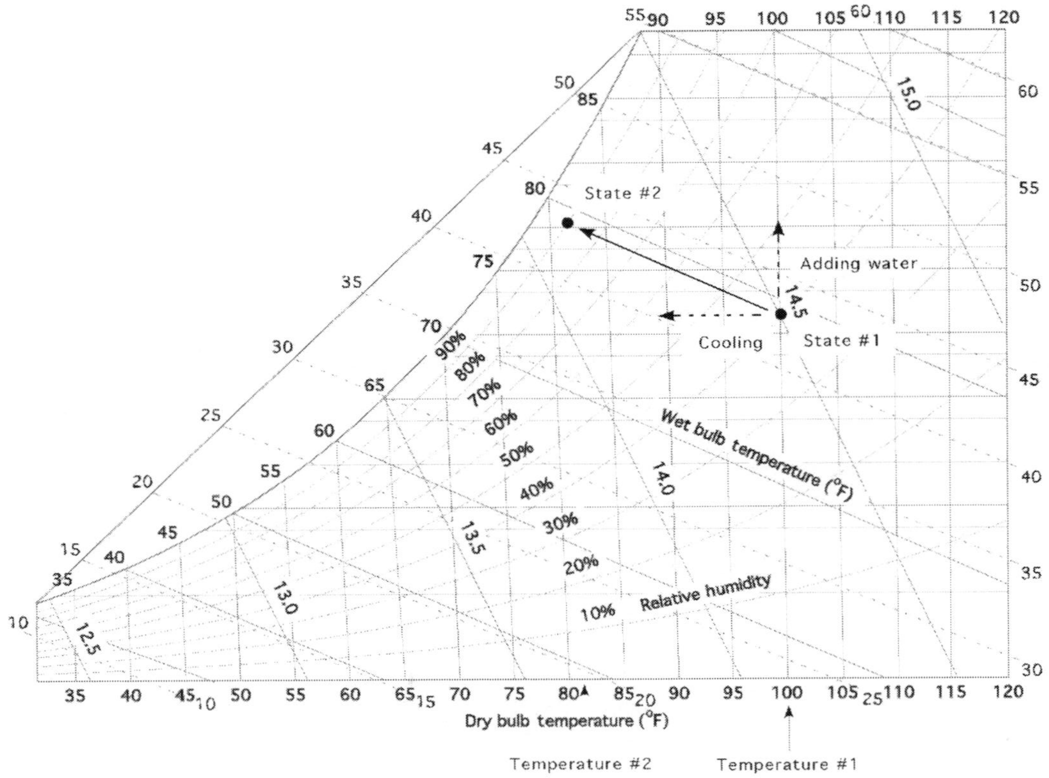

FIGURE 23.10. Solution for evaporating cooling problem.

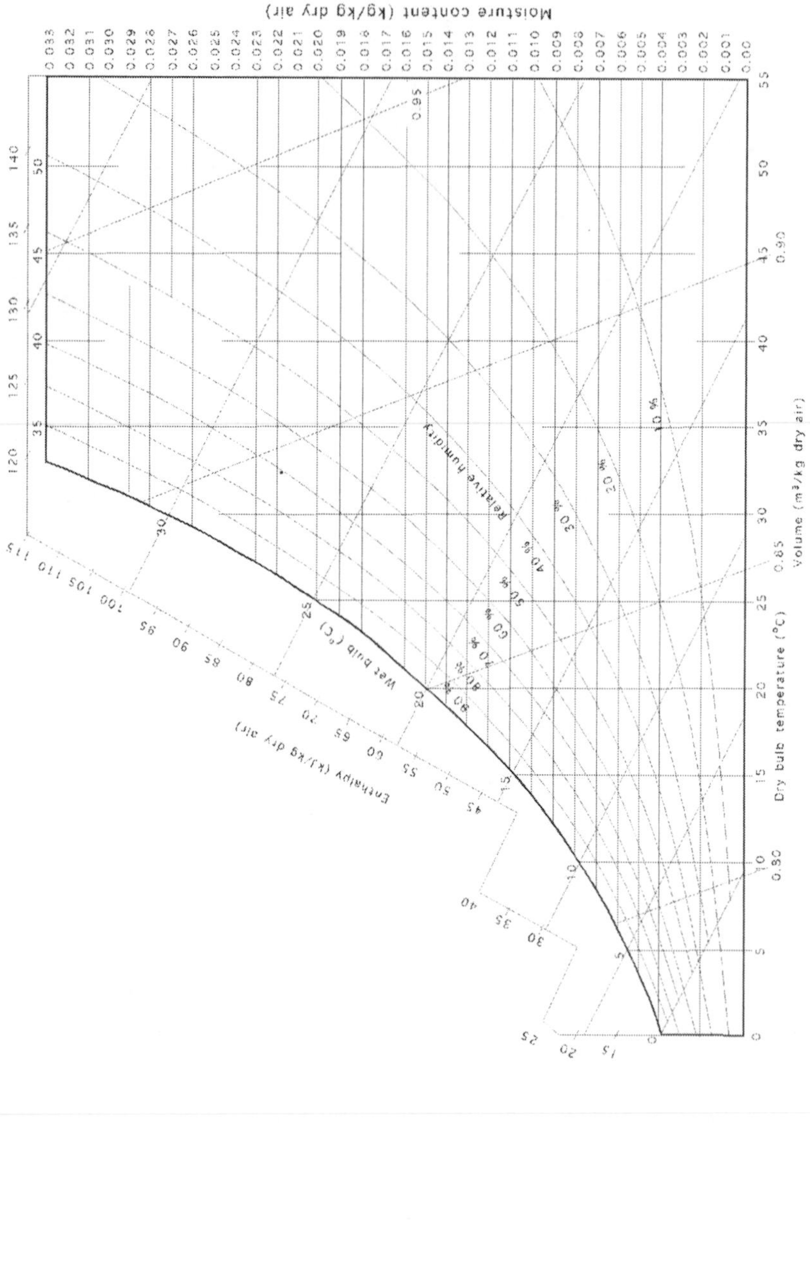

FIGURE 23.11. Metric psychrometric chart.

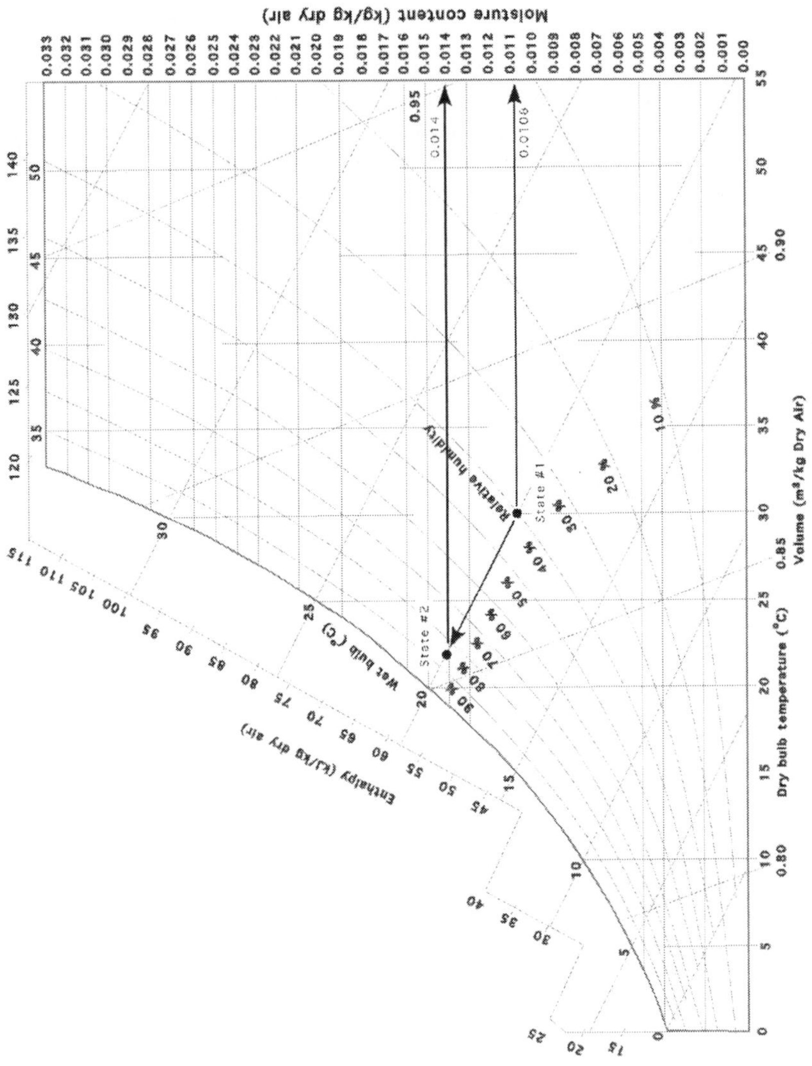

FIGURE 23.12. Psychrometric chart for drying problem.

TABLE 23.5. Information for drying problem.

Properties	Initial	Final	Difference
Dry bulb (°C)	30°Cdb		
Wet bulb (°C)	20°Cwb		
Moisture (kg H_2O/kg of air)	0.0108	0.014	0.0032
Volume (m^3/kg dry air)	0.087		

Problem: Determine the size of fan (m^3/hr) that will be required to naturally air dry 100.0 kg of product in 8.0 hr that must have 23.5 kg of water removed. The outside air conditions are 30°Cdb and 20°Cwb and the air is 85% saturated when it leaves the dryer.

Solution: Fans are sized by air moving capacity (m^3/hr). The amount of air (m^3) is determined by the amount of water that must be removed and the amount of water that can be adsorbed by each cubic meter of air. The first step is to develop a table of pertinent information, Table 23.5 and Figure 23.12.

$$\frac{\text{m}^3}{\text{hr}} = \frac{23.5\ \text{kg H}_2\text{O}}{8.0\ \text{hr}} \times \frac{1\ \text{kg air}}{0.0032\ \text{kg H}_2\text{O}} \times \frac{1\ \text{m}^3}{0.087\ \text{kg dry air}}$$

$$= \frac{23.5}{0.0022\ldots} = 10551.36\ldots \text{ or } 10{,}000\ \frac{\text{m}^3}{\text{hr}}$$

24
Selection of Structural Members

24.1. Objectives

1. Understand the importance of beam size.
2. Be able to define simple and cantilever beams.
3. Be able to calculate the maximum load that can be carried by a beam.
4. Be able to calculate the size of beam needed to support a load.

24.2. Introduction

A beam is a horizontal member used to support a load. The design of structural members for a particular load involves analysis of the forces imposed on the member by loading and selection of appropriate materials, shapes, and sizes to accommodate the loads. This procedure is best accomplished by a structural engineer and is beyond the scope of this book. Failure to use a beam of adequate size can lead to beam failure with the accompanying danger of injury and/or financial loss. However, several basic concepts are presented here to help the reader understand the properties and load-carrying abilities of simple and cantilever beams made of wood.

24.3. Simple and Cantilever Beams

Simple and cantilever beams are structural members that are loaded by forces applied at right angles to the longitudinal axis. The load that a beam can carry depends on its length, the method of support, the manner of loading, its cross-sectional size and shape, and the strength of the beam material.

Many different types of beams and loads exist in buildings and structures, but only two types of beams and two types of loads are considered in this chapter. These are simple and cantilever beams, and point and uniform loads. A simple beam is supported at each end without rigid connections; it merely rests on supports.

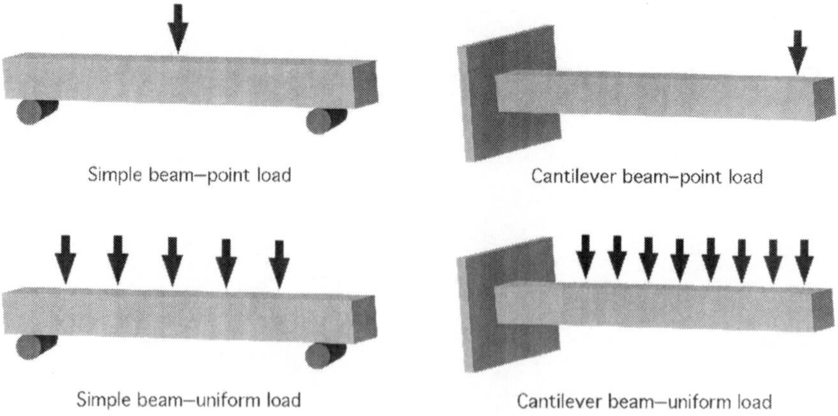

FIGURE 24.1. Two types of beams and two types of loads.

A cantilever beam is supported solidly at one end, with the other end free to move, Figure 24.1.

Point loads can be applied at any specific point along the beam, but the equations used in this chapter only apply to point loads in the middle of a simple beam and at the end of a cantilever beam. A point load is indicated by a single arrow. Uniform loads are applied along the entire length of the beam, and are indicated by a series of arrows.

24.4. Beam Loading

When a load is applied to a simple beam, the beam tends to bend. As the beam bends or flexes, the fibers along the bottom surface of the beam are stretched (put in tension), and the fibers along the top surface of the beam are pressed together (compressed). It is important to remember that the opposite is true for a cantilever beam. with a cantilever beam, the top stretches and the bottom is compressed.

As the fibers of a beam stretch, a stress is set up in the fibers. When beam loading and fiber stretching proceed to a point where the fibers yield, the maximum allowable fiber stress (*S*) has been exceeded. Estimates for maximum allowable stress are available for most materials, and values for maximum allowable fiber stress of wood can be found in Appendix XI.

The shape of a beam and the way that it is positioned to carry a load greatly affect its load-carrying ability. The common sawing method for boards results in boards that are flat sawn, the grain of the wood tends to be parallel with the long dimension, Figure 24.2. A flat sawn beam will support more weight on edge than when placed flat.

The dimensions of a beam (width and depth) are used to determine the section modulus, which provides an index of the relative stiffness or load-carrying ability of the member.

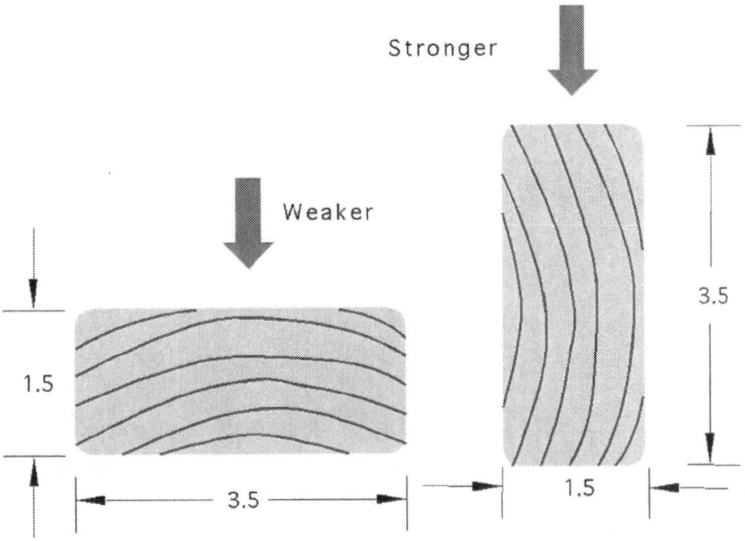

FIGURE 24.2. Relative strength of lumber oriented with the shortest dimension parallel to the load (on its side) and the shortest dimension perpendicular to the load (on its edge).

For a rectangular cross-section member, the section modulus is determined by using the following equation:

$$K = \frac{1}{6} \times a \times b^2$$

where K = Section modulus (in^3); a = Width of member, horizontal dimension (in); b = Depth of member, vertical dimension (in).

24.5. Dimensioned Lumber

The lumber industry uses two different sizes, nominal size and actual size. The nominal size is the dimensions of the lumber after it is sawed. The actual size is the size of the lumber after the surfaces have been finished. Finishing removes material, therefore the actual size is smaller than the nominal size, Table 24.1. Lumber is purchased using the nominal size, but the actual size must be used for beam calculations. An additional term may also be used and that is rough sawed or unfinished. When calculating beams using rough sawed or unfinished lumber, the actual size is the nominal size.

Problem: What is the section modulus for a 2 × 4 standing on edge?

Solution: Section modulus is calculated on actual size. Therefore:

$$K = \frac{a \times b^2}{6} = \frac{1.5 \text{ in} \times (3.5 \text{ in})^2}{6} = 3.0625 \text{ or } 3.0$$

To save time and eliminate one step in the calculations, Table 24.1 lists values for the section modulus of several common sizes of 2-inch (thick) lumber. Note that

TABLE 24.1. Section modulus of rectangular members.

Nominal size (in)	Actual size (in)	Section modulus for position (in^3) a b On edge	 a b Flat
2 × 2	1.50 × 1.50	0.56	0.56
2 × 3	1.5 × 2.50	1.56	0.94
2 × 4	1.5 × 3.50	3.06	1.31
2 × 6	1.5 × 5.50	7.56	2.06
2 × 8	1.5 × 7.25	13.14	2.72
2 × 10	1.5 × 9.25	21.39	3.47
2 × 12	1.5 × 11.25	31.64	4.22
2 × 14	1.5 × 13.25	43.89	4.97

in Table 24.1, the section modulus for a 2 × 4 on edge is 3.06. The nominal size (what one asks for at a lumber yard) and the actual size (the actual dimensions of the lumber) are shown. The section modulus is calculated for the actual size. Two columns of values are shown for the two possible positions of a member—standing on edge and lying flat. Notice that the values for the section modulus are much greater when the member is standing on edge, as it is much stiffer in this position.

The amount of weight that a beam can support is determined by the section modulus, allowable fiber stress, beam span or length, type of load, and type of support. For the two types of beams and loads being considered in this chapter, the following equations apply:

Simple beam, point load: $W = \dfrac{4SK}{L}$

Simple beam, uniform load: $W = \dfrac{8SK}{L}$

Cantilever beam, point load: $W = \dfrac{SK}{L}$

Cantilever beam, uniform load: $W = \dfrac{2SK}{L}$

where W = Maximum allowable load on the beam (lb); S = Allowable fiber stress (lb/in^2); L = Length or span of beam (in); K = Beam section modulus (in^3).

Problem: What is the maximum point load that can be supported in the middle of a 4 × 6-inch, rough sawed (dimensions are the *actual* dimensions of the board) simple beam 120 inches long when the allowable fiber stress is 1,500 lb per square inch?

Solution:

$$K = \frac{1}{6} \times a \times b^2 = \frac{1 \times 4\text{ in} \times (6.0\text{ in})^2}{6} = 24\text{ in}^3$$

$$W = \frac{4SK}{L} = \frac{4 \times 1{,}500\,\frac{\text{lb}}{\text{in}^2}\,24\text{ in}^3}{120\text{ in}} = 1{,}200\text{ lb}$$

24.6. Size of Beam

In the previous section, the method for determining the size of load that a beam will support was illustrated. In some situations, it is necessary to determine the size beam that is necessary to support a given load. The same equations are used, but they are rearranged to solve for the section modulus. Once the section modulus is known, the beam size can be determined from Table 24.1 if one dimension is 2 inches (nominal), and the orientation is known.

Problem: What is the smallest size of 2-inch, No. 1, southern pine simple beam that will support a uniform load of 2,400 lb if the beam is 100 inches long?

Solution: Using Table 24.1 and Appendix XI, and rearranging the equation:

$$W = \frac{8SK}{L} \qquad K = \frac{W \times L}{8 \times S} = \frac{2{,}400 \text{ lb} \times 100 \text{ in}}{8 \times 1{,}000 \dfrac{\text{lb}}{\text{in}^2}} = 30 \text{ in}^3$$

Referring to Table 24.1, the smallest beam size with a section modulus equal to or larger than 30 cubic inches is a beam 2 inches wide and 12 inches deep. Notice that a 2 × 14 inch beam also would carry the load, but is larger than necessary.

The procedures presented in the previous sections are not limited to boards 2 inches wide. If the section modulus and one dimension are known, the width of a beam greater than 2 inches can be determined if the equation is rearranged.

Problem: What depth of board is needed in the previous problem if a board 4 inches wide is used?

Solution: Rearranging the equation:

$$K = \frac{1}{6} \times a \times b^2 \quad b = \sqrt{\frac{K \times 6}{a}} = \sqrt{\frac{30 \text{ in}^3 \times 6}{4 \text{ in}^2}} = \sqrt{45} = 6.708\ldots \text{ or } 6.7 \text{ in}$$

This example illustrates that a beam 4 × 6.7 inches (actual size) will support the same load as a beam 2 × 12 inches (nominal size).

24.7. SI Metric

The procedure to solve SI metric is the same as that for customary units. Lumber dimensions are often given in centimeters or millimeters. Stress is in kilopascals (kPa or kN/m^2). To convert from allowable stress in Appendix XI from psi (lb/in^2) to kPa, multiply $S_{\text{Customary}}$ by 6.8947, Table 24.2.

TABLE 24.2. Metric lumber sizes.

Nominal size (cm)	Actual size (cm)
5.0 × 5.0	4.5 × 4.5
7.5 × 5.0	7.0 × 4.5
10.0 × 5.0	9.0 × 4.5
15.0 × 5.0	14.0 × 4.5
10.0 × 10.0	9.0 × 9.0
10.0 × 15.0	9.0 × 14.0

Problem: What is the maximum point load that can be supported in the middle of a 10 × 15 cm, rough sawed (dimensions are the *actual* dimensions of the board) simple beam on edge that is 3.0 m long. The board is No. 1 yellow pine?

Solution: Allowable stress from Appendix XI is 1000 lb/in^2. Convert this value to SI units

$$S_{\mathrm{SI}} = 6.8947\,\frac{\mathrm{kPa}}{\mathrm{lb}/\mathrm{in}^2} \times S_{\mathrm{Customary}}$$

$$S_{\mathrm{SI}} = 6.8947\,\frac{\mathrm{kPa}}{\mathrm{lb}/\mathrm{in}^2} \times 1{,}000\ \mathrm{lb/in}^2 = 6{,}894.7\ \mathrm{kPa}$$

$$K = \frac{1}{6} \times a \times b^2 = \frac{1 \times 10\ \mathrm{cm} \times (15.0\ \mathrm{cm})^2 \times \dfrac{1\ \mathrm{m}}{100\ \mathrm{cm}} \times \left(\dfrac{1\ \mathrm{m}^2}{10{,}000\ \mathrm{cm}^2}\right)}{6}$$

$$= \frac{0.00225}{6} = 0.000375\ \mathrm{m}^3$$

$$W = \frac{4SK}{L} = \frac{4 \times 6{,}894.7\ \dfrac{\mathrm{kN}}{\mathrm{m}} \times 0.000375\ \mathrm{m}^3}{120\ \mathrm{in} \times 0.0254\ \dfrac{\mathrm{m}}{\mathrm{in}}}$$

$$= \frac{10.34205\ \mathrm{kN\text{-}m}}{3.048\ \mathrm{m}} = 3.39306 \text{ or } 3.39\ \mathrm{kN}$$

Problem: Determine the maximum load that can be supported on the end of a finished 100 mm × 50 mm cantilever beam that is 4.0 m long. The beam is on edge and is No. 2 southern pine.

Solution: Converting allowable stress

$$S_{\mathrm{SI}} = 6.8947\,\frac{\mathrm{kN}}{\mathrm{m}^2} \times S_{\mathrm{Customary}}$$

$$= 6.8947\,\frac{\mathrm{kN}}{\mathrm{lb}/\mathrm{in}^2} \times 825\,\frac{\mathrm{lb}}{\mathrm{in}^2}$$

$$= 5{,}688.1275\ \mathrm{kN}$$

solving for K:

$$K = \frac{1}{6} \times a \times b^2$$

$$= \frac{1 \times 4.5\ \text{cm} \times (9.0\ \text{cm})^2 \times \dfrac{1\ \text{m}}{100\ \text{cm}} \times \dfrac{1\ \text{m}^2}{10{,}000\ \text{cm}^2}}{6}$$

$$= 6.075\ \text{E} - 5$$

solving for W:

$$W = \frac{SK}{L} = \frac{5688.1275\ \text{kN} \times 6.075\ \text{E} - 5}{4.0\ \text{m}}$$

$$= \frac{0.3455\ldots}{4} = 0.086388\ldots\ \text{or}\ 0.09\ \text{kN}$$

25
Principles of Electricity

25.1. Objectives

1. Be able to define electricity.
2. Be able to define basic electrical terms.
3. Understand and be able to use Ohm's law.
4. Be able to calculate electrical power.
5. Be able to calculate electrical energy use.

25.2. Introduction

Agricultural production depends heavily upon electrical energy to power machines, lights, equipment, and tools used in producing and processing agricultural products. An understanding of the principles of electricity will lead to more efficient use of electricity and reduce the risks associated with working with and around electricity.

25.3. Electricity

Electricity is the flow of electrons from one atom to another. Scientists believe that electricity was first noted when ancient people discovered the attraction between an amber rod and other materials.

An understanding of electricity begins with the atom. Electrons travel in orbit around a nucleus composed of protons and neutrons. When an atom has the same number of protons and electrons, it has no charge. When it has more electrons than protons, it has a negative charge, and if more protons than electrons, a positive charge. As electrons move from one atom to another, electric current is produced.

A flow of electrons can be produced by different methods, including friction, heat, light, pressure, chemical action, and magnetism. The last two are used most often in agriculture.

25.4. Electrical Terms

To understand electricity and how it functions, it is important to understand common electrical terms and their definitions.

Current: Current is the movement of electrons through a conductor. The amount of current moving is determined by the voltage divided by the resistance.

Alternating current (AC): One of two types of electric current, alternating current does not exhibit constant polarity or a constant voltage. The voltage builds to a maximum value in one polarity, declines to zero, builds to a maximum value in the other polarity, and declines to zero again, Figure 25.1. This sequence is called one cycle. In the United States the common current is 60 cycle; that is, 60 complete cycles occur every second.

Direct current (DC): This current flows in one direction only and at a constant voltage, Figure 25.2. The polarity will be either positive or negative.

Amperage (amp): A unit of measure for the amount of electricity flowing in a circuit. One ampere of current is equivalent to 6.28×10^{18} electrons per second.

Circuit: A continuous path for electricity from the source to the load (machine or appliance using the electricity) and back to the source.

Conductor: Any material that has a relatively low resistance to the flow of electricity. Such materials allow electrons to flow easily from one atom to another. Most metals are good conductors.

Insulator: Any material that has a relatively high resistance to the flow of electricity. Such materials do not allow easy movement of electrons from one atom to another. Glass, rubber, and many plastics are insulators.

Resistance (Ohm): The characteristic of materials that impedes the flow of electricity. All materials have varying amounts of electrical resistance. Two

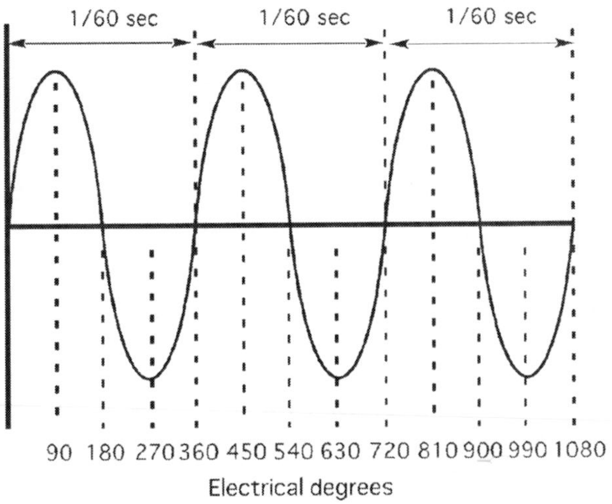

FIGURE 25.1. Three complete cycles of 60 cycle alternating current.

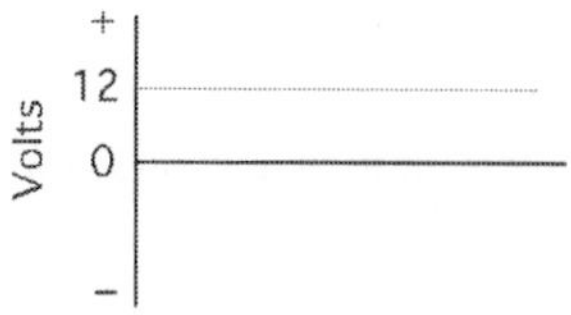

FIGURE 25.2. Direct current at positive 12 V.

characteristics of resistance are important: (1) when electricity flows through a resistant material, heat is generated; (2) when electricity flows through a resistant material, the voltage is decreased. Resistance is measured in units of ohms and designated by the Greek letter Ω (omega).

Voltage (V): The electromotive force that causes electrons to flow through a conductor. Voltage is a measure of the potential for current flow. A voltage potential may exist between objects without a flow of current. In the United States the two standard voltages are nominally 120 V and 240 V.

Watts (W): Unit of electrical power. Watts are determined by multiplying the voltage times the amperage.

Ohm's law: Ohm's law explains the relationship between current, voltage, and resistance. Ohm's law states that the voltage in a circuit is equal to the amperage times the resistance. Expressed as an equation:

$$E = IR$$

where E = Voltage (V); I = Current (amp); R = Resistance (ohms).
It is common to rearrange this equation to solve for any one of the three variables. For example:

$$I = \frac{E}{R} \text{ and } R = \frac{E}{I}$$

Problem: What is the current flow (amp) when the source supplies 120 V and the circuit has 20.0 ohms (Ω) of resistance?

Solution:

$$I = \frac{E}{R} = \frac{120 \text{ V}}{20.0\ \Omega} = 6.0 \text{ amp}$$

25.5. Electrical Power

By definition, power is the rate of doing work. In electricity, work is done by electrons moving within a conductor, and is the result of the electrical potential (V) and the flow rate of the electrons (amp). Thus, in an electrical circuit, power is the product of volts × amperes. The basic units of electrical power are the watt (W) and the kilowatt (kW). In equation form:

$$P = E \times I$$

where P = power (watts); E = voltage (V); I = current (amp).

Problem: How much power is being produced when the current is measured at 5.0 amps and 120.0 V?

Solution:

$$P = E \times I = 120\text{ V} \times 5.0\text{ amp} = 600\text{ watt}$$

Power also may be expressed in two other ways. Ohm's law states that $E = IR$, and (power) $P = E \times I$; therefore the E in the power equation can be replaced with IR from Ohm's law:

$$P = IR \times I \quad \text{or} \quad P = I^2R$$

Ohm's law also states that $I = \frac{E}{R}$; therefore:

$$P = E \times \frac{E}{R} \quad \text{or} \quad P = \frac{E^2}{R}$$

Note that the use of either of these equations requires knowledge of a value for the resistance. When the resistance and one of the other quantities are known, power can be determined.

25.6. Electrical Energy

Electrical energy is different from electrical power because it includes the element of time. The amount of electrical energy produced, or used, is determined by multiplying the electrical power (watts) by the amount of time electricity flows (hours). The result is expressed in the units for electrical energy, watt-hours (W h). For many uses of electricity, the watt-hours value is a large number; so, the electric industry has adopted a more convenient unit, the kilowatt-hour (kW h), as the basic unit of electrical energy. One kilowatt equals 1000 W. The amount of electrical energy used can be determined by:

$$EE\text{ (W h)} = P \times T$$

and:

$$EE\text{ (kW h)} = P \times T \times \frac{1\text{ kW h}}{1{,}000\text{ W h}}$$

where EE = Electrical energy (W h); P = Power (W); T = Time (hr)

Problem: How much energy (W h) is required to operate a 150-W light bulb for 3.5 h?

Solution:

$$EE = P \times T = 150\text{ W} \times 3.5\text{ hr} = 525\text{ W h}$$

Expressed in kW h:

$$\text{kW h} = 525 \text{ W h} \times \frac{1 \text{ kW}}{1{,}000 \text{ W}} = 0.525 \text{ kW h}$$

Electrical energy also is used to determine the cost of using electricity. It is common practice for an electric utility company to sell electricity on a cents (¢) per kilowatt-hour basis. When the electric rate (¢/kW h) and the amount of electricity (kW h) are known, the cost of operating any electrical appliance or motor can be determined. In the following example the units cancellation method is used.

Problem: What will it cost to operate the light bulb in the previous problem, When the electricity costs 10 ¢/kW h?

Solution: Using units cancellation:

$$\text{Cost (cents)} = 0.525 \text{ kW h} \times \frac{10 \text{ cents}}{\text{kW h}} = 5.25 \text{ or } 5.2 \text{ cents}$$

When the electricity costs 10 ¢/kW h, then it will cost 5.2 cents to operate the 150 W light bulb for 3.5 h.

25.7. SI Metric Customary Calculations

It is obvious from the examples that electrical calculations are performed in SI metric units. If the unit of power is horsepower, use the conversion constant 1 hp = 0.743 kW and work the problem in SI units.

26
Series and Parallel Circuits

26.1. Objectives

1. Be able to identify series and parallel circuits.
2. Be able to determine the total resistance of both series and parallel circuits.
3. Be able to attach a voltmeter and an ammeter correctly.
4. Be able to determine the amperage and the voltage at any point within a series or a parallel circuit.
5. Explain the importance of system and equipment grounding.

26.2. Introduction

The purpose of electrical circuits is to supply electricity for an appliance, tool, or other type of electrical device. These devices are called loads. Before the load will operate, electricity must have a complete path from the source to the load and back to the source. This path for electricity is called a circuit. Two types of circuits are commonly used to supply electrical power, series and parallel.

This chapter will explore some of the principles of series and parallel circuits and how to calculate the amperage and/or voltage of these circuits.

26.3. Series and Parallel Circuits

The following discussion will explain the differences between the two circuits and how to calculate the total resistance of the circuit. *Note:* to aid in the understanding of the following discussion of circuits, assume that the conductors in the circuit have no resistance.

26.3.1. Series Circuit

In a series circuit the electricity has no alternative paths. All of the electricity must pass through all of the components in the circuit, see Figure 26.1.

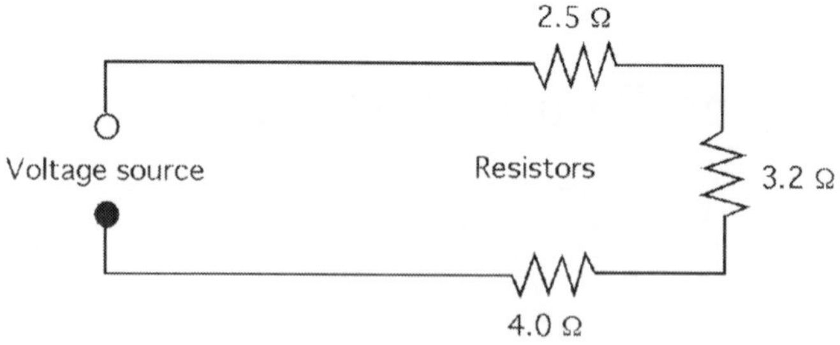

FIGURE 26.1. Series circuit.

For the electricity to leave and return to the source it must pass through all three loads, resistors. In a series circuit the total resistance of the circuit is the sum of each resistance where the unit or resistance in the Ohm (Ω).

$$R_T = R_1 + R_2 + R_3 + \cdots + R_N$$

The total resistance of the circuit in Figure 26.1 is:

$$R_T = 2.5\ \Omega + 3.2\ \Omega + 4.0\ \Omega = 9.7\ \Omega$$

The total amperage flow in the circuit can be determined by dividing the source voltage by the total resistance.

Problem: What is the current flow for the circuit in Figure 26.1 when the source voltage is 120 V?

Solution:

$$E = IR \quad I = \frac{E}{R} = \frac{120\ \text{V}}{9.7\ \Omega} = 12.3711 \cdots \text{ or } 12\ \text{A}$$

26.3.2. Parallel Circuit

In a parallel circuit the electricity has alternative paths to follow (see Figure 26.2) and consequently the total resistance of the circuit is not equal to the sum of the individual resistances. The total amperage in the circuit is determined by the total resistance, but total resistance is determined differently from series circuits. The amperage through any path is determined by the voltage at the path and the resistance of the path.

The total resistance can be determined by more than one method. In one method, the inverse of the total resistance is the sum of the inverses of each resistance in

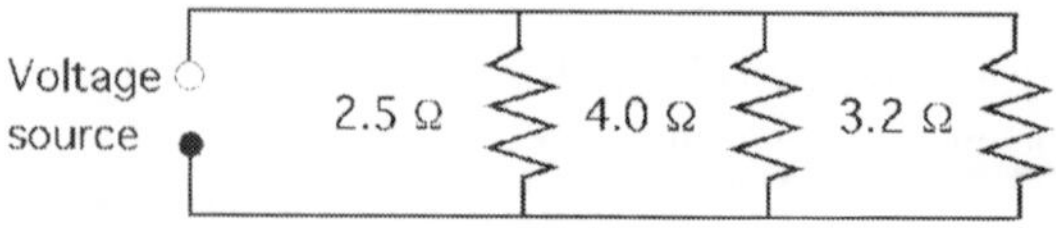

FIGURE 26.2. Parallel circuit.

the parallel circuit:

$$\frac{1}{R_T} = \frac{1}{R_1} + \frac{1}{R_2} + \cdots + \frac{1}{R_N}$$

This equation requires solving for a common denominator or reducing the fractions to a decimal. The easiest way to determine a common denominator is to multiply the denominators. When this method is used for the circuit in Figure 26.2, the total resistance is:

$$\begin{aligned}\frac{1}{R_T} &= \frac{1}{2.5\ \Omega} + \frac{1}{4.0\ \Omega} + \frac{1}{3.2\ \Omega}\\ &= \frac{12.8}{32} + \frac{8}{32} + \frac{10}{32}\\ &= \frac{30.8}{32} = 0.9625\\ R_T &= \frac{1}{0.9625} = 1.039\cdots \text{ or } 1.0\ \Omega\end{aligned}$$

Using this method, the total resistance of the circuit is 0.96 Ω.

When reducing the fractions to a decimal, it is important to remember that the result is the inverse.

Problem: Determine the total resistance of the parallel circuit in Figure 26.2 by reducing the fractions.

Solution:

$$\begin{aligned}\frac{1}{R_T} &= \frac{1}{2.5\ \mathrm{W}} + \frac{1}{4.0\ \mathrm{W}} + \frac{1}{3.2\ \mathrm{W}}\\ &= 0.4 + 0.25 + 0.3125\\ &= \frac{1}{0.9625} = 1.038\cdots \text{ or } 1.0\ \Omega\end{aligned}$$

Compare the total resistance of this circuit to the total resistance of the series circuit. With the same resistors, the total resistance is much less in a parallel circuit.

Another method is to solve for the equivalent resistance of pairs of resistors, until all of the resistance is reduced to a single resistance. This is accomplished by using the equation:

$$R_T = \frac{R_1 \times R_2}{R_1 + R_2}$$

When the circuit has more than two resistors, determine the equivalent resistance (R_E) for any two, and then combine it with the third, and so on, until only one resistance remains. The resistance of the circuit in Figure 26.2 with this method is:

$$\begin{aligned}R_E &= \frac{2.5\ \Omega \times 4.0\ \Omega}{2.5\ \Omega + 4.0\ \Omega} = \frac{10\ \Omega}{6.5\ \Omega} = 1.538\ldots \text{ or } 1.5\ \Omega\\ R_T &= \frac{1.5\ \Omega \times 3.2\ \Omega}{1.5\ \Omega + 3.2\ \Omega} = \frac{4.9\ldots}{4.738\ldots} = 1.038\ldots \text{ or } 1.0\ \Omega\end{aligned}$$

Using this method the total resistance is 1.0 Ω, which agrees with the other two methods.

26.4. Series–Parallel Circuits

A series–parallel circuit combines characteristics of both types of circuits. Some of the resistors are in series, and some are in parallel. This type of circuit is more common in electronic equipment than in the circuits used to supply electrical power for agricultural equipment.

Study Figure 26.3. In circuit A, all of the electricity must pass through the 2.5-Ω resistor, but then it has two alternative paths. Part of the current will pass through the 4.0 Ω resistor and the rest will pass through the 3.2 Ω resistor. Circuit B is the same circuit, just drawn differently so that it is easier to see the relationship between the three resistors.

To solve for the total resistance of a series–parallel circuit, start by reducing the parallel resistors. The circuit then acts as a series circuit, and the resistances can be added. It is helpful to redraw the circuit after each pair of parallel resistors is condensed. To determine the total resistance for Figure 26.3, the first step is to find the equivalent resistance of the parallel branch of the circuit. Either parallel circuit equation can be used. The equation using pairs of resistances is considered easier to use.

$$R_E = \frac{4.0\ \Omega \times 3.2\ \Omega}{4.0\ \Omega + 3.2\ \Omega} = \frac{12.8\ \Omega}{7.2\ \Omega} = 1.777\ldots\ \Omega$$

The resistance of the parallel branch of the circuit has an equivalent resistance of 1.8 Ω. The next step is to combine this equivalent resistance with the remaining resistors in the circuit.

Figure 26.4 shows that the series–parallel circuit can be reduced to a series circuit. The total resistance of this circuit is:

$$R_T = 2.5\ \Omega + 1.777\ldots\ \Omega = 4.3\ \Omega$$

The series–parallel circuit has a total resistance of 4.3 Ω.

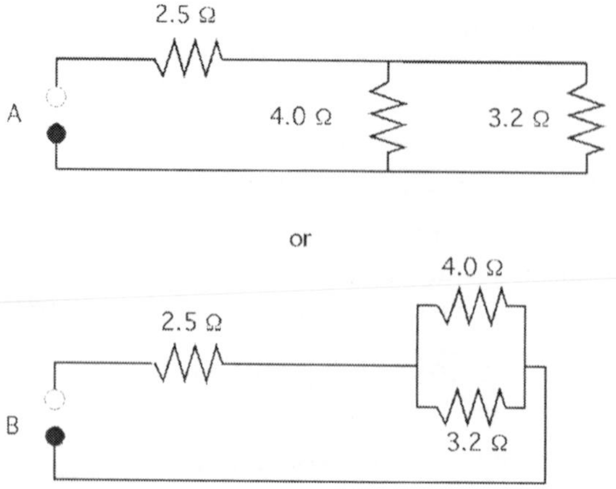

FIGURE 26.3. Two representations of a series-parallel circuit.

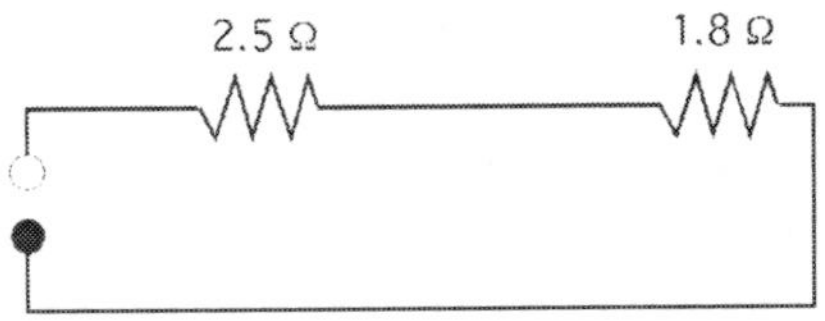

FIGURE 26.4. Equivalent series resistance of Figure 26.3.

26.5. Determining Voltage and Amperage in Circuits

In agricultural circuits the voltage is determined by the source of the electrical energy. Standard domestic service is 120 or 240 V. Agricultural equipment may have electrical systems that operate on 6, 12, 24, or 48 V.

Amperage is determined by measuring the circuit with an ammeter or by calculations using Ohm's law, $E = I \times R$. In Ohms law "E" stands for electrical motive force, in volts, "I" is intensity, in amperage or amps, and "R" is resistance, measured in Ohms. When the resistance is known, the amperage can be determined by rearranging Ohm's law. The total amperage of a circuit can be calculated by dividing the source voltage by the total resistance of the circuit. Rearranging Ohm's law for amperage:

$$I\text{ (amp)} = \frac{E\text{ (volts)}}{R\text{ (Ohms)}}$$

Problem: Determine the total amperage for the series circuit illustrated in Figure 26.5. Use 120 V for the source.

Solution: Using Ohm's Law:

$$I\text{ (amp)} = \frac{E\text{ (volts)}}{R\text{ (Ohms)}} = \frac{120\text{ V}}{2.5\ \Omega + 3.2\ \Omega + 4.0\ \Omega} = \frac{120\text{ V}}{9.7\ \Omega} = 12.371\cdots \text{ or } 12\text{ amp}$$

A circuit with 9.7 Ω of resistance and a source voltage of 120 V will have a current flow of 12 amps.

The same procedure is used to determine the amperage of a parallel circuit. The source voltage is divided by the total resistance in the circuit.

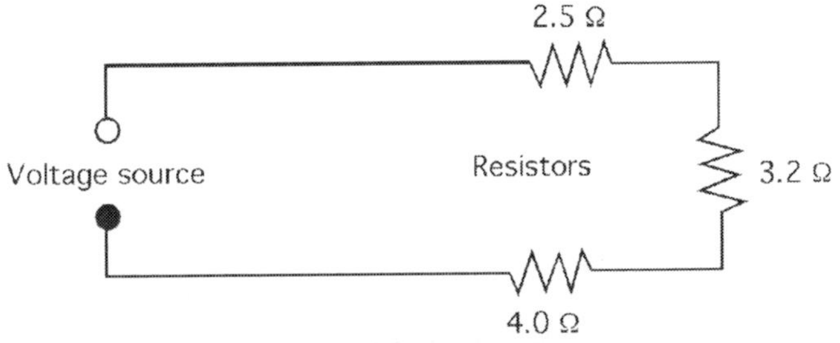

FIGURE 26.5. Series circuit for sample problem.

26.6. Using Voltmeters and Ammeters

It is helpful when troubleshooting circuits to be able to measure the voltage and amperage in the circuit. The instruments used for this purpose are called voltmeters and ammeters. When attached properly they will measure the voltage and amperage in a circuit.

26.6.1. Voltmeters

Two characteristics of circuits must be remembered in using voltmeters: (1) Voltage is the measurement of a potential between two points; so the reading on a voltmeter is the difference between the connection points. To reduce the effect of adding a voltmeter to the circuit, the meter is constructed with a very high internal resistance. (2) Anytime that electricity flows through a resistance heat is produced and the loss of energy causes the voltage to decrease. This decrease is called the voltage drop.

26.6.1.1. Voltmeters in Series Circuits

In Figure 26.6, voltmeter number one (V_1) is connected across the source; therefore, the reading on the voltmeter will be equal to the source, assuming no resistance in the conductors. Voltmeter number two (V_2) is connected across the 3.2 Ω resistor. It will measure the difference in voltage from one side of the resistor to the other—in other words, the voltage drop across the resistor.

Problem: What will voltmeter number 2 read in Figure 26.6 if the source voltage, measured by voltmeter number 1 is 120 V?

Solution: Voltmeter number 2 is measuring the voltage drop caused by the 3.2 Ω resistor. Voltage drop is caused by current passing through a resistance, therefore to calculate the voltage drop across the resistor, the total current flow in the circuit must be known. The first step is to calculate the total current flow in the circuit.

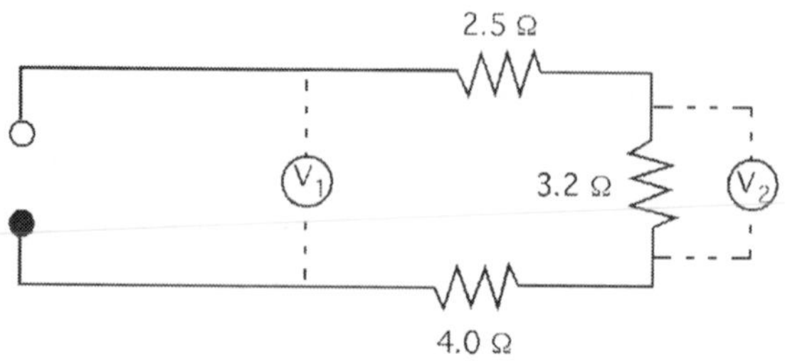

FIGURE 26.6. Voltmeters in series circuit.

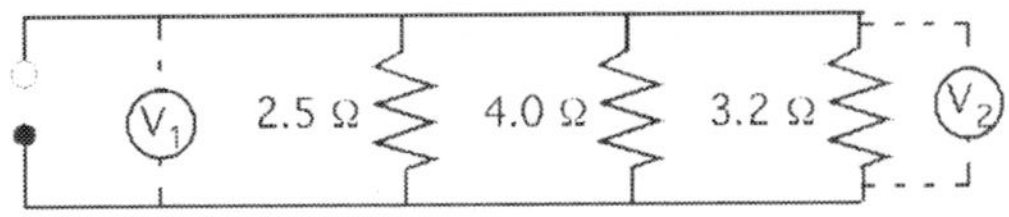

FIGURE 26.7. Voltmeters in parallel circuits.

To calculate total current flow the total resistance must be known.

$$
\begin{aligned}
R_T &= R_1 + R_2 + R_3 \\
&= 2.5\ \Omega + 3.2\ \Omega + 4.0\ \Omega \\
&= 9.7\ \Omega
\end{aligned}
$$

$$E = IR \quad I = \frac{E}{R} = \frac{120\ \text{V}}{9.7\ \Omega} = 12.37\ldots \text{ or } 12\,\text{amp}$$

The next step is to determine the voltage drop across the resistor:

$$E = IR = 12\,\text{amp} \times 3.2\ \Omega = 38.4 \text{ or } 38\ \text{V}$$

When 12 amps flow through a resistance of 3.2 Ω with a source of 120 V, there is a voltage drop of 38 V. Voltmeter 2 will have a reading of 38 V.

26.6.1.2. Voltmeters in Parallel Circuits

In a parallel circuit, voltmeters connected across the resistors are, in essence, connected across the source voltage. Therefore, assuming that there is no voltage drop from the resistance of the wire, both voltmeters in Figure 26.7 will have the same reading as the source voltage.

26.6.2. Ammeters

Ammeters are used to measure the amount of current flowing in a circuit. The laboratory type of ammeter is connected in series. To reduce the effect of the meter on the performance of the circuit, they are constructed with a very low resistance. A clamp-on type of meter is also available that measures the intensity of the electromagnetic field around a single conductor and converts the field intensity into amperage.

26.6.2.1. Ammeters in Series Circuits

When we calculated the amperage of this circuit in the previous section, we determined that with a 120 V source the current was 12 amps. In a series circuit, the electricity does not have alternative paths to follow. Because all of the amperage flows through all components in the circuit, both ammeters in Figure 26.8 will have a reading of 12 amps.

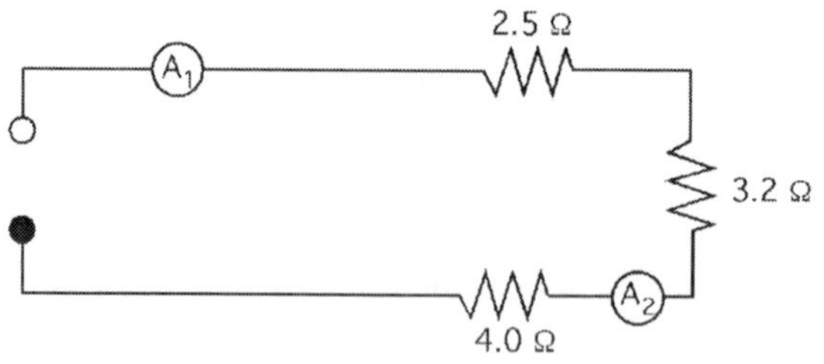

FIGURE 26.8. Ammeters in a series circuit.

26.6.2.2. Ammeters in Parallel Circuits

In parallel circuits, ammeters must also be attached so the current passes through them; but because the electricity has alternative paths, they will measure only the current in the conductor they are inserted into, see Figure 26.9.

Problem: What are the readings for ammeters number 1 and number 2 in Figure 26.9 when the source voltage is 120 V?

Solution: In this example problem, ammeter one is located between the source and the first branch on the circuit; therefore it will measure the total current flow in the circuit. Ammeter two is in the last branch of the circuit, therefore it will only measure the current flowing through that branch on the circuit. In a previous section it was explain that to measure current flow, the source voltage and the total resistance of the circuit must be known. The first step is to determine the total resistance of the circuit.

$$\begin{aligned}\frac{1}{R_T} &= \frac{1}{2.5\ \Omega} + \frac{1}{4.0\ \Omega} + \frac{1}{3.2\ \Omega}\\ &= \frac{12.8}{32} + \frac{8}{32} + \frac{10}{32}\\ &= \frac{30.8}{32} = 0.9625\\ R_T &= \frac{1}{0.9625} = 1.04 \text{ or } 1.0\ \Omega\end{aligned}$$

The next step is to calculate the total current flow in the circuit.

$$E = IR \quad I = \frac{E}{R} = \frac{120\ \text{V}}{1.0\ \Omega} = 120\ \text{amps}$$

Ammeter number 1 will have a reading of 120 amps.

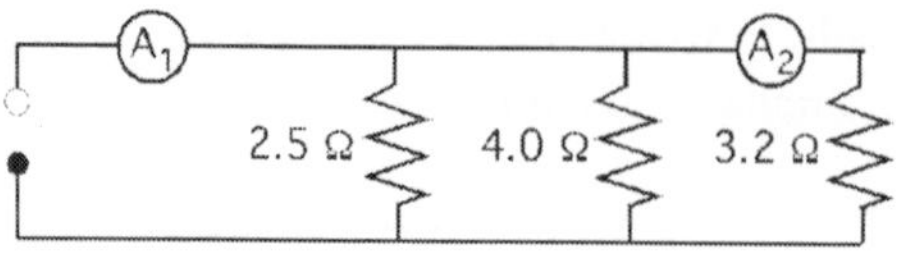

FIGURE 26.9. Ammeters in a parallel circuit.

Ammeter number 2 only will measure the current flow in that branch of the circuit. Ohm's law is used to solve this part of the problem also:

$$E = IR \quad I = \frac{E}{R} = \frac{120\,\text{V}}{3.2\,\Omega} = 37.5\text{ amps}$$

In summary, assuming no resistance in the conductors, in a series circuit the amperage remains the same, and the voltage changes as the resistance changes. In parallel circuits the voltage remains the same, and the amperage changes as the resistances changes in the circuit.

26.7. Grounding

To simplify the earlier discussion of circuits only two conductors were used, hot and neutral. To meet current wiring codes a complete wiring system uses at least three conductors; hot, neutral, and ground. The ground conductor (equipment ground) is attached to the case of metal-cased tools and it must provide a complete low resistance circuit to the earth, Figure 26.10.

This is usually accomplished by attaching it to the grounding bar in the service entrance panel. If a short occurs between the electrical components and the case or frame, the low-resistance ground circuit will permit a current flow greater than the circuit over-current protection (fuse or circuit breaker), causing the circuit to open. If a short occurs and the ground circuit is not continuous from the case or frame of the tool to the earth, the operator's body or body parts may complete the circuit resulting in a potentially fatal electric shock.

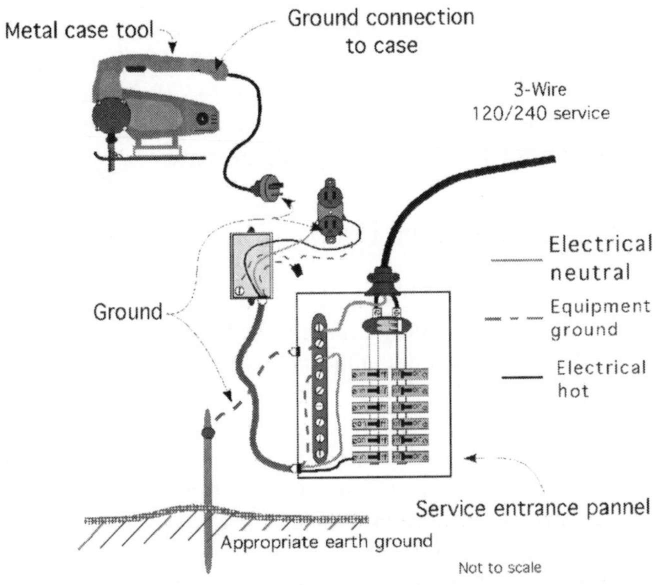

FIGURE 26.10. Case of equipment ground.

27
Sizing Conductors

27.1. Objectives

1. Understand the importance of using electrical conductors of the proper size.
2. Be able to calculate voltage drop.
3. Be able to calculate the proper size conductor.
4. Be able to select the proper size conductor from a table.
5. Understand the importance of circuit protection, and be able to identify the three common types.

27.2. Introduction

Conductors are used to provide a path for electricity. The metals used for conductors have a relatively low resistance, but they have enough resistance to cause a voltage drop when current passes through them. If the wires are overloaded, the voltage drop increases, causing a reduction in the efficiency of the circuit and an increase in the heat produced. Heating the conductors above their designed operating temperatures will cause the insulation to fail, and the heat generated can be sufficient to start a fire if flammable materials are close.

27.3. Calculating Voltage Drop

To maintain the heat produced by electricity passing through a resistance at the designed level, national standards, primarily the National Electric Code, limit the voltage drop allowed in a circuit. For most circuits this is 2 or 3%. Voltage drop is the term used to describe the reduction in voltage that occurs as electricity flows through a resistance. Voltage drop occurs because materials resist the flow of electricity; the critical issue is to ensure it does not become excessive. The amount of voltage drop is determined by the amount of current and the total resistance in the circuit. The size of a conductor required for an electrical load is determined by the allowable voltage drop for the circuit, the size of the load, and the length of wire from the source of electricity to the load. The resistance of conductors usually is listed as ohms (Ω) per 1,000 ft of conductor length. Because the electricity must pass through the entire length of a conductor, conductors perform as one continuous series resistance.

Appendix XII includes the resistance for various sizes of copper wire using AWG sizes and Appendix XIII contains resistance values for SI wire sizes. Notice that the resistance is given in ohms/1,000 ft or ohms/100 m. Bare wire is seldom used in circuits, but because the type of insulation used influences the resistance, these values are used to illustrate the principles of voltage drop. Resistance values for wires with different types of insulation can be found in the *National Electrical Code* (NEC) or other sources and these must be used when calculating wire size for a specific application.

The voltage drop is calculated using Ohm's law. For general purpose circuits, the voltage drop must be limited to 2%. The NEC allows a 3% drop for some individual circuits. In calculating voltage drop, the length of conductor usually will be measured as either the length of wire from the source to the load and back (length of wire), or the run (the distance from the source to the load).

Problem: What is the voltage drop in a wire length of 1,500.0 ft, when No. 12 wire is used, and the load is 10.0 amps?

Solution: Using Ohm's law and, from Appendix XII, using a resistance for No. 12 wire of 1.62 Ω/1,000 ft, the voltage drop is:

$$E = IR = 10.0\text{ amp} \times \frac{1.62\ \Omega}{1{,}000.0\text{ ft}} \times 1{,}500.0\text{ ft} = \frac{24{,}300}{1{,}000} = 24.3\text{ V}$$

The 1,500.0 ft of No. 12 wire has a voltage drop of 24.3 V. If the source voltage is 120 V, is this an acceptable voltage drop?

$$\% = \frac{24.3\text{ V}}{120\text{ V}} \times 100 = 20.25\%$$

The answer then is no—a 20.25% voltage drop is excessive. If No. 12 wire is used with this load, the electrical appliance will not operate correctly, the conductors will over heat and there is the potential for a fire. A larger conductor size is needed to carry a load of 10.0 amps for a distance of 1,500.0 ft.

For wires longer or shorter than 1,000 ft, the resistance is proportional to the length. The total resistance for any length is:

$$R_{\mathrm{L}} = \frac{R(\Omega)}{1{,}000\text{ ft}} \times L(\text{ft})$$

where R_{L} = Resistance in ohms (Ω) for any length (ft); R = Resistance per 1,000 ft (Ω); L = Length of wire (ft).

27.4. Calculating Conductor Size

The previous problem was used to illustrate the principle of voltage drop. In practice, because the percent of drop is fixed by electrical codes, conductors are sized by calculating the amount of resistance per 1,000 ft that will result in an acceptable voltage drop, and then selecting the appropriate size of wire from a table similar to Appendix XII.

Problem: What size of wire is needed to carry a 120-V, 15.0-amp load with a 2% drop when the run is 200.0 ft?

Solution: The first step is to determine the permissible amount of voltage drop:

$$\text{Voltage drop}(V_\text{D}) = 120\text{ V} \times 0.02 = 2.4\text{ V}$$

The second step is to determine the amount of resistance for the load that will cause a 2.4-V drop. This is accomplished by using Ohm's law:

$$E = IR \qquad R = \frac{E}{I} = \frac{2.4\text{ V}}{15.0\text{ amp}} = 0.16\ \Omega$$

For a circuit with a resistance of less than 0.16 Ω and a load of 15.0 amps the voltage drop will be less than 2%. Remember that the voltage used to calculate the resistance is the voltage drop, not the source voltage.

To select the correct size wire the calculated circuit resistance must be converted to units of Ω/1,000 ft, and then select the appropriate wire size from Appendix XII. This can be accomplished using units cancellation.

$$\frac{\Omega}{1{,}000\text{ ft}} = \frac{\Omega}{\text{Run}} \times 1{,}000\text{ ft}$$

$$= \frac{0.16\ \Omega}{2 \times 200.0\text{ ft}} \times 1{,}000\text{ ft} = 0.40\ \frac{\Omega}{1{,}000\text{ ft}}$$

Next, compare this value to the resistance values in Appendix XII. The objective is to select a size of wire with a resistance equal to or less than the calculated value. From Appendix XII, the resistance of No. 6 wire is 0.41 Ω/1,000, and the resistance of No. 4 wire is 0.26 Ω/1,000 ft. The No. 4 wire would be the best choice as the resistance is closest to the calculated resistance without being larger.

These examples illustrate the principle that voltage drop is caused by resistance and current. Another example of the importance of understanding voltage drop is in the use of extension cords. Improper use of extension cords can lead to serious consequences. Many extension cords sold in retail stores use No. 16 or No. 18 wire. Extension cords of this size are very limited in current carrying capacity. Also, when more than one extension cord is used, the resistance of the connection can be equal or greater than the resistance of the wire. This is why the connections of an extension cord that has been overloaded will be warm or hot to the touch.

Problem: What size load (amp) can a 100.0-ft, No. 18 extension cord carry on a 120 V circuit without exceeding a 2% voltage drop?

Solution: The first step is to determine the allowable voltage drop:

$$V_\text{D} = 120\text{ V} \times 0.02 = 2.4\text{ V}$$

The next step is to determine the amount of current that will cause a 2.4-V drop in the extension cord. This is accomplished by using Ohm's law and the resistance of No. 18 wire. Remember to use two times the length (100.0 ft × 2 = 200.0 ft) to determine the total feet of conductor.

$$E = IR \qquad I = \frac{E}{R} = \frac{2.4\ \text{V}}{\dfrac{6.51\ \Omega}{1{,}000\ \text{ft}} \times 200.0\ \text{ft}} = \frac{2.4\ \text{V}}{1.302\ \Omega} = 1.843\ldots \text{ or } 1.84\ \text{amp}$$

The maximum electrical load for a 100 ft, No. 18 extension cord is 1.84 amps. If the extension cord is used for a larger load (more amps), it will overheat.

An often unrecognized factor when using extension cords is the resistance of the connections. Two 50-ft extension cords may have more resistance than a 100 ft cord with the same size of conductors. To more clearly understand the potential problem with extension cords, consider the following example.

Problem: What is the maximum capacity of the No. 18 extension cord in the previous problem if two 50-ft cords are used, and the connection has a resistance of 8.0 Ω?

Solution: Using an allowable voltage drop of 2.4 V and Ohm's law:

$$E = IR \quad I = \frac{E}{R} = \frac{2.4\ \text{V}}{\left(\dfrac{6.51\ \Omega}{1{,}000\ \text{ft}} \times 200.0\ \text{ft}\right) + 8.0\ \Omega}$$

$$= \frac{2.4\ \text{V}}{9.302\ \Omega} = 0.2580\ldots \text{ or } 0.26\ \text{amp}$$

This example shows that using two 50-ft extension cords instead of one 100-ft cord reduces the capacity by seven times when the connection resistance is 8.0 ohms because of the additional resistance of the connection.

Many appliances are rated in watts. To determine the size of conductors required to supply an appliance rated in watts, the electrical power (watts) must be used first to determine the load in amps.

Problem: A 25.0-ft extension cord will be used to operate a 1,100-W, 120-V electrical iron. What size of extension cord should be selected?

Solution: The first step is to determine the amperage used by the iron.

$$P = EI \qquad I = \frac{P}{E} = \frac{1{,}100\ \text{W}}{120\ \text{V}} = 9.166\ldots \text{ or } 9.27\ \text{amp}$$

Using an allowable voltage drop of 2%, the next step is to determine total allowable resistance:

$$E = IR \qquad R = \frac{E}{I} = \frac{2.4\ \text{V}}{9.27\ \text{amp}} = 0.2618\ldots \text{ or } 0.26\ \Omega$$

The total allowable resistance in the extension cord is 0.26 Ω. The next step is to determine the ohms of resistance per 1,000 ft:

$$\frac{\Omega}{1{,}000\ \text{ft}} = \frac{0.26\ \Omega}{50.0\ \text{ft}} \times 1{,}000\ \text{ft} = 5.236\ldots \text{ or } 5.2\ \frac{\Omega}{1{,}000\ \text{ft}}$$

The resistance of a No. 18 wire is 6.51 Ω/1,000 ft, and that of a No. 16 wire is 4.09 Ω/1,000 ft. Assuming the iron and cord connectors are in good condition, a 25-ft extension cord with No. 16 wire is adequate for the 1,100-W iron.

27.5. Selecting Conductor Sizes from a Table

An alternative method for sizing conductors is to use tables provided for that purpose, such as the examples found in Appendixes XIV and XV. These tables have several important limitations. They apply only to wires with insulation types of R, T, TW, RH, RHW, and THW, and they can only be used with a 2% voltage drop and 120 or 240 V. For any other type of conductor, insulation, voltage drop, or voltage, a different table must be used.

Problem: Determine the size of wire needed to supply 120-V electricity to the pump house in Figure 27.1.

Solution: The total length of wire is two times the sum of the distance from the building to the first pole (A), the distance between the two poles, and the distance from the last pole to the pump (B). First determine lengths A and B by using Pythagorean's theorem:

$$a^2 = b^2 + c^2$$

Distance A is:

$$\begin{aligned}\text{Distance}_A &= \sqrt{25.0^2 + 20.0^2}\\ &= \sqrt{625 + 400}\\ &= 32.015\ldots \text{ or } 32.0 \text{ ft}\end{aligned}$$

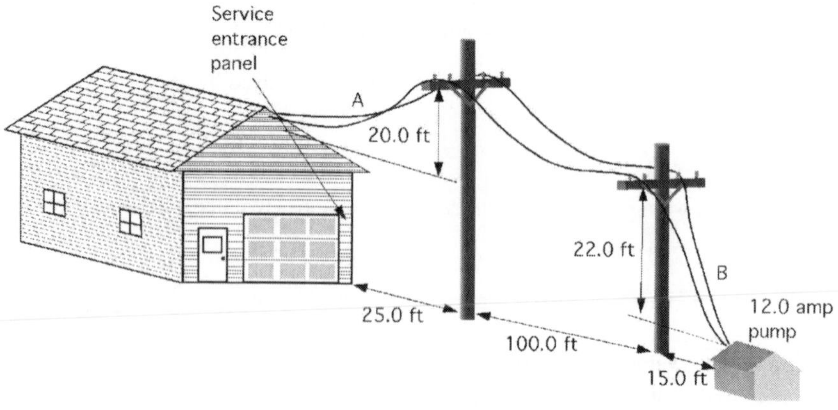

FIGURE 27.1. Illustration for sample problem.

and distance B is:

$$a = \sqrt{b^2 + c^2} = \sqrt{15.0^2 + 22.0^2} = \sqrt{225 + 484} = 26.627\ldots \text{ or } 26.6 \text{ ft}$$

The total length (L_T) of wire is:

$$L_T = (32.0 \text{ ft} + 26.0 \text{ ft} + 100.0 \text{ ft}) \times 2 = 316.0 \text{ ft}$$

Then using Appendix XIV, select the correct size of conductor. Twelve amps and 316 ft are not shown in the table; use the next larger values (15 amps and 350 ft). Then, from Appendix XIV, the required size of conductor is No. 3. Remember to compare this value with the value to the left of the vertical line and use the larger of the two sizes.

27.6. Circuit Protection

An important part of any circuit is the over current protection—fuse or circuit breaker. Over-current devices are used to prevent the conductors from overheating. Remember: when electricity passes through a resistance, heat is produced and a voltage drop occurs. As the electrical load on a circuit increases, the heat produced increases and the voltage drop increases. An electrician determines the total amperage capacity of the circuit and then installs the appropriate over-current protection. In the case of an overload or a short in the circuit, the over-current device stops the flow of electricity by opening the circuit. If the over-current device burns out or trips, the circuit has been overloaded. The load on the circuit must be reduced or the short repaired before the circuit is energized.

An additional protection device required by most codes for residential use is the Ground Fault Circuit Interrupter (GFCI). A GFCI constantly monitors the difference in the current between the hot conductor and neutral conductor. When the difference reaches 5 mA the GFCI trips and opens the circuit. This provides more protection to the user of metal cased tools and when using electrical appliances around water. Some Agricultural Engineers also recommend GFCI's be installed in circuits used around animals. This is especially true for very valuable animals.

27.7. Metric Problems

The common terms used for electricity are the same for both systems, and many European countries also use the AWG standards for sizing conductors. For those individuals that use metric sizes for wires the resistances for three common sizes are listed in Appendix XIII.

Problem: Will a 10.0 amp load on a 120 V circuit with 2.5 mm^2 conductors and a run of 24 m exceed the 2% voltage drop standard?

Solution: The first step is to determine the actual voltage drop.

$$E = IR = 10 \text{ amp} \times \frac{0.676\ \Omega}{100\text{ m}} \times (24\text{ m} \times 2) = 3.244\ldots\text{ V}$$

The second step is determining the percentage.

$$\frac{3.244\ldots\text{ V}}{120\text{ V}} \times 100 = 2.704 \text{ or } 2.7\%$$

The conclusion is that the load is not acceptable. The voltage drop is greater than 2%.

28
Electric Motors

28.1. Objectives

1. Be able to explain the advantages and disadvantages of electric motors as a power source.
2. Understand the use and performance classifications of electric motors.
3. Be able to describe the common types of motors.
4. Be able to select the correct overload protection device for the application.
5. Be able to interpret a motor nameplate.

28.2. Introduction

An electric motor is a machine that converts electrical energy to mechanical power. Modern agriculture is heavily dependent on the use of electric motors. Managers of agricultural production systems should know the common types of motors and be able to select the correct motor. Manufacturers have gone to great lengths to design motors to meet the needs of agriculture. The life of an electric motor is determined by how well the motor is matched to the job and the service environment. The following sections will discuss the characteristics and uses of 120/240-V, single phase motors.

28.3. Electric Motors

An electric motor is an energy conversion device like internal combustion engines, but instead of using the energy in a liquid fuel they use energy supplied by electricity. An electric motor requires an interaction between the rotating parts and the stationary parts. Two common types of interaction are conduction and induction. The principles of a conduction motor will be used to explain the function of electricity motors.

Electricity flowing through a conductor induces an electrical magnetic field around the conductor. This electromagnetic field will have a plus (+) and a

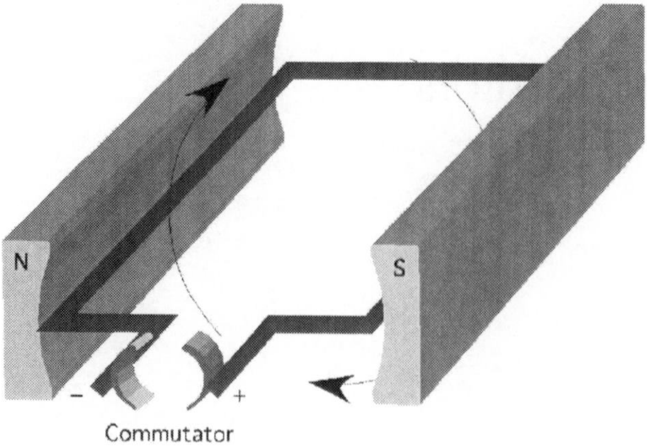

FIGURE 28.1. Simplified electric motor.

minus (−) polarity. When the electromagnetic field comes in contact with the permanent magnetic field supplied by the magnets, a force is produced, because similar polarities repel each other and opposite poles attract each other. The strength of the force is influenced by the strength of the current, the strength of the magnetic field and the angle between the conductor and the magnetic field, Figure 28.1.

In Figure 28.1, the commutator supplies electricity to the loop of wire. The electromagnetic field reacts with the permanent magnetic field supplied by the permanent magnets. This causes the wire loop to rotate. A conductive electric motor has many individual loops wrapped around the rotor and each one of them rotates a few degrees when they are energized by the commutator. Because a continuous series of loops become energized the rotor continues to rotate as long as power is being supplied to the commutator.

Many different types of motor are used and they are categorized by their type of electricity and the type of interaction they have between the rotating part and the stationary part of the motor. The following section will discuss some of the advantages and disadvantages of electric motors.

28.4. Advantages and Disadvantages

28.4.1. Advantages

Electric motors have several advantages that have made them more popular than other sources of power:

1. *Initial cost*: On a per-horsepower basis, the purchase price of an electric motor is relatively low.
2. *Design*: Electric motors have very few parts and are very easy to operate. In addition, because they are started and stopped with switches, their operation is easily automated.

3. *Operating costs*: The operating costs of electric motors are low. They have very low maintenance costs, and the cost of electricity allows them to be operated for a few cents per horsepower hour.
4. *Environmental impact*: An electric motor gives off no exhaust fumes and does not use a flammable fuel; and although it is true that the generation of electricity may affect the environment, it is easier to monitor and control this effect if it is concentrated at one location.
5. *Noise*: Electric motors operate quietly with very little vibration.
6. *Efficiency*: An electric motor is the most efficient way to produce power, operating with an efficiency range of 70–90%.

28.4.2. *Disadvantages*

1. *Portability*: Electric motors are not very portable; they must be connected to a source of electricity. Advances in battery design have reduced this disadvantage for fractional horsepower DC motors, but AC motors must be connected to a source of alternating current.
2. *Electrical hazard*: Electricity has an adverse effect on humans and other animals when they come into contact with it. With the large number of motors used in agriculture and the tendency for the environment of agricultural structures to be wetter and dustier than residences, there is a greater electrical hazard associated with the use of electrical motors in agricultural buildings.

28.5. Use and Performance Classifications

The following use and performance classifications provide a basis for comparing different types of electric motors.

28.5.1. *Type of Current*

Motors can be purchased to operate on either single phase or three phase, direct or alternating current, at several voltages. The voltage and the type of current used depend on the size of the motor and the electrical service available. The phase refers to the number of cycles of alternating current, Figure 28.2.

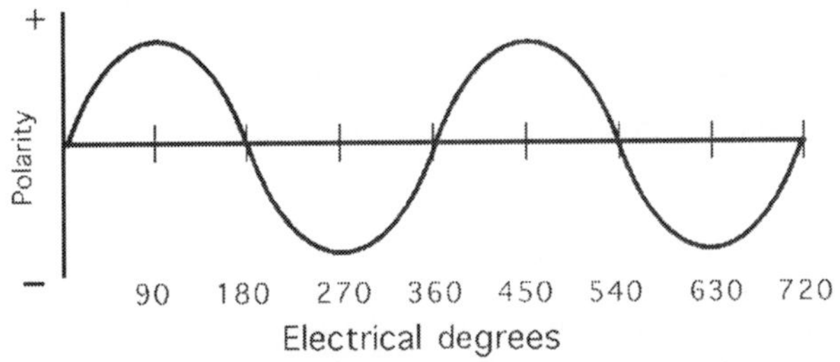

FIGURE 28.2. Two complete cycles of single phase alternating current.

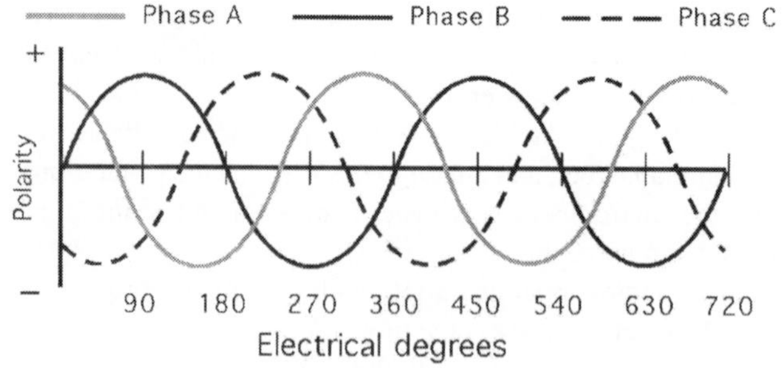

FIGURE 28.3. Two complete cycles of three phase alternating current.

In three phase alternating current service, three single-phase, 60 cycle currents are combined so the peak voltages are an equal distance apart, see Figure 28.3. Three phase current is recommended over single phase for larger loads, and is required by most electrical service companies for motors over 5–10 horsepower.

28.5.2. *Type of Enclosure*

The primary enemies of motors are liquids, dust, and heat. Motors are designed with different enclosures (cases) to operate in different environments. A drip-proof motor has ventilation vents and will operate successfully even if water occasionally drips on it. In a splash-proof motor, the vents are protected from both drips and splashes. Drip-proof and splash-proof motors usually have an internal fan that draws air through the motor to help prevent overheating. A totally enclosed motor can be used in wet, dusty environments because it has no vents; thus the air has no access to its internal parts. All of the heat generated inside the motor must be conducted to the outside surface and dissipated into the air. Some totally enclosed motors have a fan mounted outside the case to increase the air flow (cooling) around the motor enclosure.

28.5.3. *Type of Bearings*

Motors are constructed with either sleeve or ball bearings. Sleeve bearing motors are less expensive than ball bearing motors, but can withstand less force perpendicular to the shaft (belt tension, etc.) than ball bearing motors and the life of the bearings is usually shorter unless they can be lubricated.

28.5.4. *Type of Mounting Base*

Electric motor manufacturers have standardized the mounting brackets for motors. In replacing a motor, if the same base style is used, no modifications should be required in the motor mounts.

28.5.5. *Load-Starting Ability*

Some electrically driven machines offer little resistance to rotation when they are started. These loads are said to be easy to start. Other machines offer greater torsion resistance when starting, and are considered to be loads that are hard to start. The amount of torque required to start a machine depends on the type, size, and operating characteristics of the particular machine. Thus, some knowledge of the starting characteristics of a load is necessary to select the correct motor.

28.5.6. *Starting Current*

Motors require more current to start a load than they do to operate the load once it is rotating. The starting current required may be seven to eight times the operating current. The starting current of motors is classified as low, medium, or high.

28.5.7. *Reversibility*

Electric motors can be designed to operate in either a clockwise or a counter-clockwise direction of rotation, but some motors can have the direction of rotation changed by electricians. These are classified as electrically reversible or mechanically reversible.

28.5.8. *Dual Voltage Potential*

If a motor has dual voltage, it can be operated on either 120 or 240 V; the voltage used is determined by the voltage available from the source. Motors operating on 240 V are more efficient and require smaller conductors than those operating on 120 V. Motor voltage potential is considered to be dual or single.

28.6. Types of Motors

A description of each of these types of motors and information on selection of the ideal type for a specific situation is beyond the scope of this text. Each of these types of motors will have specific advantages and disadvantages for specific applications, Figure 28.4.

Three of these types of motors are more common in agricultural applications than the others. These types are listed in the following sections, with a short description of each type.

28.6.1. *Split-Phase*

Split-phase motors are inexpensive and are commonly used for easy-to-start loads. The starting current may be as high as seven times the operating current, depending on the load. They are very popular for powering fans, centrifugal pumps, and other

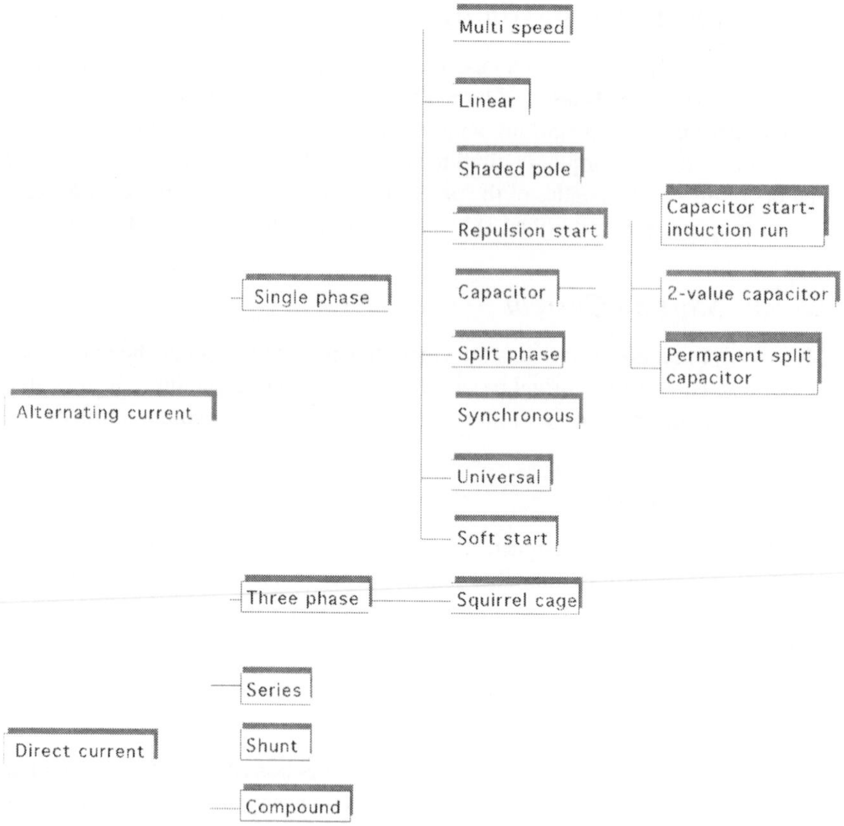

FIGURE 28.4. Common types of electric motors.

applications where the load increases as the speed increases. The common sizes are 1/6 to 3/4 horsepower.

28.6.2. Capacitor Motor

A capacitor motor is a split-phase motor with the addition of capacitors to increase the starting and/or running torque. Capacitor motors are the most widely used type in agricultural applications, particularly in the smaller sizes of less than 10 horsepower.

28.6.3. Repulsion Motors

Repulsion motors are used for hard-to-start loads. Two common variations of the basic design are used: repulsion-start induction-run, and repulsion-induction. The use of brushes and the additional construction requirements cause repulsion motors to be the most expensive type. They are available in a wide range of sizes.

28.7. Overload Protection

The over-current devices (fuses and circuit breakers) in the service entrance panel of a building are designed to protect the circuits from shorts and excessive current. The starting current of electric motors is greater than the operating current. In addition, when an electric motor is overloaded it will attempt to rotate until the electricity is disconnected or the lock-up torque is reached. Both of these situations can cause an excessive current in the motor circuit. The over current devices must be designed to withstand temporary overloads. Common breakers and fuses do not meet these requirements.

When a motor is overloaded for an extended period of time or on a frequent basis, it will overheat. Some motors include an overload protection device to protect the motor from overheating. These devices work on the principle of resistive heating. Because the current demand of a motor increases as the load increases, and because the greater the current flow through a resistance the greater the heat, engineers can predict when certain temperatures will be reached and design controlling devices using temperature as the trigger.

Three types of motor overload protection devices are used:

1. *Built-in thermal overload protection*: Many motors have this device as an integral part of the motor. It may have either an automatic reset or a manual reset. If the automatic reset type is used, the device will disconnect the motor from the circuit at the designed temperature, and when it has cooled down, will automatically reconnect the electricity. This type of device should be used where the motor is supplying power for a critical purpose, as in livestock ventilation vans and sump pumps. Automatic reset devices can be more dangerous than manual because they will attempt to restart the motor as soon as they cool down. If the power is not disconnected before anyone works on the machine, it can restart during repairs and cause an injury. When a manual reset device is used, it must be manually reset before the motor will start. Even with this device, the power should be disconnected before anyone works on the motor or the machine it is powering.
2. *Manual starting switch with overload protection*: A manual starting switch overload protection device can be added to motors without built-in thermal protection. This device is an integrated switch and thermal overload device that opens the circuit when the motor overheats. It must be reset by hand before the motor can be restarted.
3. *Magnetic starting switch with overload protection*: Magnetic starting switches are fast-acting switches for large motors and are equipped with a thermal overload (heater) device that will open the circuit if the current draw is excessive. These devices are manually reset.

28.8. Motor Nameplate Data

The National Electrical Manufacturers Association (NEMA) has developed a standardized system of identifying the important characteristics of motors. The standards do not specify a shape or location for the nameplate, and nameplates may not all contain the same information; but the information that is important for motor selection will be included, see Figure 28.5.

The following list of terms covers some of the information that can be found on motor nameplates:

Manufacturer: The company name and address are included to provide a source of further information.
Horsepower: The designed full-load horsepower rating of the motor is given.
Serial number: The serial number is included to facilitate the procurement of parts when they are needed.
Frame: Standard frame numbers are used to ensure that the motors are interchangeable. If the frame number is the same for two motors, the mounting bolt holes will have the same dimensions.
Speed: The speed is in revolutions per minute (rpm). It is the speed at which the motor will operate under a full load.
Phase: This indicates whether the motor is designed for single or three phase power.
Temperature: This is the maximum ambient temperature for the motor.
Service factor: The service factor indicates the safe overload limit of the motor. An SF of 1.20 means that the motor can be operated continuously with a 20% overload without damage.
Code letter: The code letter is used to indicate the size of the overload protection device that will be needed.
Hertz: This indicates the designed operation frequency of the electrical supply. It is shown as 60 cycles or 60 hertz.
Duty rating: Motors will be rated for either continuous or intermittent duty. If the rating is for intermittent duty, the nameplate may also indicate the maximum

Company Information	
Hp 1/3	SN 432N5-A
FR 42	RPM 1725
PH 1	TEMP 35C
SF 1.20	Code J
Hz 60	CONT
V 120/240	Amp 6/3

FIGURE 28.5. Typical motor information.

amount of time that the motor can be operated. Exceeding the duty cycle will cause the motor to overheat.

Voltage: The operating voltage for the motor will be indicated. If the nameplate contains only a single value, the motor is not dual voltage.

Amperage: This indicates the current demand at full load. If two numbers are given, it indicates the amperages for the two voltages listed. For example, in Figure 28.5, the amperage is listed as 6/3. This means that when the motor is wired for 120 V, the full load amperage is 6 amps, and it is 3 amps when the motor is wired for 240 V.

Appendices

APPENDIX I. Conventional unit conversions.

Unit	Abbrev.	Conversion values			
1 Acre	ac	43,560 ft^3	160 rods^2		
1 British Thermal Unit	BTU	778.104 ft-lb	3.93 E − 4 hp-hr		
1 Bushel	bu	2,150.42 in^3	1.24446 ft^3	32 dry qt	
1 Foot	ft	12 in	0.3333 yd	6.061 E − 2 rod*	
1 Foot, cubic	ft^3	1728 in^3	29.9 qt, liquid	7.481 gal, liquid	0.04 yd^3
1 Bushel	Bu	2,150.42 in^3	32 qt, dry	1.24446 ft^3	
1 Foot-pound	ft-lb	3.239 E − 4 BTU	5.051 E − 7 hp/hr	3.766 E − 7 kW/hr	
1 Foot per second	ft/sec	0.68182 mi/hr	0.1667 ft/min		
1 Foot, square	ft^2	144 in^2	6.9 E − 3 yd^2	9.29 E − 2 M^2	
1 Gallon, liquid	gal	231 in^3	0.13368 ft^3	4 liquid qt	
1 Gallon, dry	Gal	268.8 in^3			
1 Horsepower	hp	550 ft-lb/sec	33,000 ft-lb/min		
1 Inch	in	8.333 E − 2 ft	2.3 E − 3 yd		
1 Kilowatt	kW	737.612 ft-lb/sec	1.34111 hp	0.94796 BTU/sec	
1 Kilowatt-hour	kWh	2,655,403 ft-lb	3,412.66 BTU	1,000 W/hr	
1 Mile	mi	63,360 in	5,280 ft	1,760 yd	320 rods
1 Mile per hour	mi/hr	1.46667 ft/sec	88.0 ft/min		
1 Pound per square foot	lb/ft^2	6.944 E − 3 lb/in^2			
1 Pound per cubic foot	lb/ft^3	5.787 E − 4 lb/in^3			
1 Quart, liquid	qt	57.75 in^3	2 pt	32 oz	3.342 E − 2 ft^3
1 Ton	T	2,000 lb			
1 Watt	W	0.73761 ft-lb/sec	9.48 E − 4 BTU/sec	1.341 E − 3 Hp	
1 Yard	yd	36.0 in	3.0 ft		
1 Yard, square	Yd^3	1,296 in^2	9.0 ft^3		
1 Yard, cubic	Yd^3	46,656 in^3	27 ft^3		

E − 4 is the same as $\times 10^{-4}$.

APPENDIX II. Conventional to SI conversions.

Quantity	From	To	Multiply by
Acceleration			
Vehicle	(mile/h)/s	(km/h)/s	1.609344
General	ft/s^2	m/s^2	0.3048
Area			
General	in^2	m^2	0.00064516
	ft^2	m^2	0.09290304
Pipe	in^2	mm^2	645.16
	in^2	cm^2	6.4516
	ft^2	m^2	0.09290304
Field	acre	ha	0.40469
Area per time			
Field operations	acre/h	ha/h	0.4046873
Consumption	gal/h	L/h	3.785412
	lb/(hp-h)	g/(kW-h)	608.2774
Density (mass)			
General ag products	lb/yd^3	kg/m^3	0.5932763
	lb/ft^3	kg/m^3	16.01846
Liquid	lb/gal	kg/L	0.1198264
Efficiency, fuel			
Highway vehicles	mile/gal	km/L	0.4151437
Off highway	hp-h/gal	kW-h/L	0.1969931
Energy			
Heat	Btu	kJ	1.055056
Use	kWh	MJ	3.6
Flow, Volume			
Air	ft^3/s	m^3/s	0.02831685
	ft^3/s	m^3/min	
Liquid, general	gal/min	L/min	
	gal/sec	L/s	3.785412
Fuel	gal/h	L/h	3.785412
River, channel	ft^3/s	m^3/s	0.02831685
Force			
Pedal, lever, general	lbf	N	4.448222
Drawbar	lbf	kN	0.00448222
Length			
	mile	km	1.609344
	rod	m	5.029210
	yd	m	0.9144
	ft	m	0.3048
	in	cm	2.54
	in	mm	25.4
Mass, general	lb	kg	0.4535924
	oz	g	28.34952
	t	T	0.90719
Power			
Heating	Btu/min	W	17.58427
	Btu/h	W	0.2930711
Engine	Hp	kW	0.7456999

(continued)

APPENDIX II. (*Continued*)

Quantity	From	To	Multiply by
Pressure	lb/in^2	kPa	6.894757
	lb/ft^2	kPa	0.04788026
	inHg	kPa	3.37685
	inH_2O	kPa	0.24884
Temperature	°F	°C	$\frac{t_{°F} - 32}{1.8}$
Torque	lbf	n-m	1.355818
	lbin		0.112984
Velocity, linear	mile/h	km/h	1.609344
	ft/s	m/s	0.3048
	in/s	mm/s	25.4
Volume	yd^3	m^3	0.7645549
	ft^3	m^3	0.02831685
	bushel	L	35.23907
	in^3	L	0.01638706
	gal	L	3.78412
	gal	gal (imperial)	0.8327
	qt	L	0.9463529
	acre-ft	m^3	1233.489
	bushel	m^3	0.03523907
Volume/area	gal/acre	L/ha	9.353958

APPENDIX III. Estimating combine losses.

Crop	$Seeds/ft^2$ = Bu/ac	$Seeds/m^2$ = 50 kg/ha
Alfalfa	303	2,426
Barley	14	140
Beans-red kidney	1.4	11
Beans-white	4	32
Canola	115	1,063
Corn	2	17
Flax	1,800	15,444
Oats	11	165
Rice	30	320
Rye	22	189
Sorghum	21	180
Soybeans	4	32
Wheat	19	152

APPENDIX IV. Efficiency, speed estimated live and repair factors of common agricultural machines.

Machine	Field efficiency		Field speed				Estimated life	Repair factors	
	Range (%)	Typical (%)	Range (mi/hr)	Typical (mi/hr)	Range (km/h)	Typical (km/h)	h	RF1	RF2
Tractors									
2 wheel drive and stationary							12,000	0.007	2.0
4 wheel drive and crawler							16,000	0.003	2.0
Tillage and Planting									
Moldboard plow	70–90	85	3.0–6.0	4.5	5.0–10.0	7.0	2,000	0.29	1.8
Heavy duty disk	70–90	85	3.5–6.0	4.5	5.5–10.0	7.0	2,000	0.18	1.7
Tandem disk harrow	70–90	80	4.0–7.0	6.0	6.5–11.0	10.0	2,000	0.18	1.7
(Coulter) Chisel plow	70–90	85	4.0–6.5	5.0	6.5–10.5	8.0	2,000	0.28	1.4
Field cultivator	70–90	85	5.0–8.0	7.0	8.0–13.0	11.0	2,000	0.27	1.4
Spring tooth harrow	70–90	85	5.0–8.0	7.0	8.0–13.0	11.0	2,000	0.27	1.4
Roller packer	70–90	85	4.5–7.5	6.0	7.0–12.0	10.0	2,000	0.16	1.3
Mulcher-packer	70–90	80	4.0–7.0	5.0	6.5–11.0	8.0	2,000	0.16	1.3
Rotary hoe	70–85	80	8.0–14.0	12.0	13.0–22.5	19.0	2,000	0.23	1.4
Row crop cultivator	70–90	80	3.0–7.0	5.0	5.0–11.0	8.0	2,000	0.17	2.2
Rotary tiller	70–90	85	1.0–4.5	3.0	2.0–7.0	5.0	1,500	0.36	2.0
Row crop planter	50–75	65	4.0–7.0	5.5	6.5–11.0	9.0	1,500	0.32	2.1
Grain drill	55–80	70	4.0–7.0	5.0	6.5–11.0	8.0	1,500	0.32	2.1
Harvesting									
Corn picker sheller	60–75	65	2.0–4.0	2.5	3.0–6.5	4.0	2,000	0.14	2.3
Combine	60–75	65	2.0–5.0	3.0	3.0–6.5	5.0	2,000	0.12	2.3
Combine*	65–80	70	2.0–5.0	3.0	3.0–6.5	5.0	3,000	0.04	2.1
Mower	75–85	80	3.0–6.0	5.0	5.0–10.0	8.0	2,000	0.46	1.7
Mower (rotary)	75–90	80	5.0–12.0	7.0	8.0–19.0	11.0	2,000	0.44	2.0
Mower-conditioner	75–85	80	3.0–6.0	5.0	5.0–10.0	8.0	2,500	0.18	1.6
Mower-conditioner (rotary)	75–90	80	5.0–12.0	7.0	8.0–19.0	11.0	2,500	0.16	2.0

(continued)

APPENDIX IV. (*Continued*)

Machine	Field efficiency		Field speed				Estimated life	Repair factors	
	Range (%)	Typical (%)	Range (mi/hr)	Typical (mi/hr)	Range (km/h)	Typical (km/h)	h	RF1	RF2
Windrower (SP)*	70–85	80	3.0–8.0	5.0	5.0–13.0	8.0	3,000	0.06	2.0
Side delivery rake	70–90	80	4.0–8.0	6.0	6.5–13.0	10.0	2,500	0.17	1.4
Rectangular baler	60–85	75	2.5–6.0	4.0	4.0–10.0	6.5	2,000	0.23	1.8
Large rectangular baler	70–90	80	4.0–8.0	5.0	6.5–13.0	8.0	3,000	0.10	1.8
Large round baler	55–75	65	3.0–8.0	5.0	5.0–13.0	8.0	1,500	0.43	1.8
Forage harvester	60–85	70	1.5–5.0	3.0	2.5–8.0	5.0	2,500	0.15	1.6
Forage harvester	60–85	70	1.5–6.0	3.5	2.5–10.0	5.5	4,000	0.03	2.0
Sugar beet harvester	50–70	60	4.0–6.0	5.0	6.5–10.0	8.0	1,500	0.59	1.3
Potato harvester	55–70	60	1.5–4.0	2.5	2.5–6.5	4.0	2,500	0.19	1.4
Cotton picker (SP)	60–75	70	2.0–4.0	3.0	3.0–6.0	4.5	3,000	0.11	1.8
				Miscellaneous					
Fertilizer Spreader	60–80	70	5.0–10.0	7.0	8.0–16.0	11.0	1,200	0.63	1.3
Boom-type sprayer	50–80	65	3.0–7.0	6.5	5.0–11.5	10.5	1,500	0.41	1.3
Air-carrier sprayer	55–70	60	2.0–5.0	3.0	3.0–8.0	5.0	2,000	0.20	1.6
Bean puller-windrower	70–90	80	4.0–7.0	5.0	6.5–11.5	8.0	2,000	0.20	1.6
Beet topper/stalk chopper	70–90	80	4.0–7.0	5.0	6.5–11.5	8.0	1,200	0.28	1.4
Forage blower							1,500	0.22	1.8
Forage wagon							2,000	0.16	1.6
Wagon							3,000	0.19	1.3

*SP indicates self-propelled machine.

APPENDIX V. Draft parameters and an expected range in drafts estimated by the model parameters for tillage and seeding implements.[1]

		SI units Machine parameters				English units Machine parameters			Soil parameters			
Implement	Width (units)	C_1	C_2	C_3	Width (units)	C_1	C_2	C_3	F_1	F_2	F_3	Range ± %
Major tillage tools												
Moldboard plow	m	652	0.0	5.1	ft	113	0.0	2.3	1.0	0.70	0.45	40
Chisel plow												
Straight point	tools	91	5.4	0.0	tools	52	4.9	0.0	1.0	0.85	0.65	50
Shovel or sweep	tools	107	6.3	0.0	tools	61	5.8	0.0	1.0	0.85	0.65	50
Twisted shovel	tools	123	7.3	0.0	tools	70	6.7	0.0	1.0	0.85	0.65	50
Sweep plow												
Primary tillage	m	390	19.0	0.0	ft	68	5.2	0.0	1.0	0.85	0.65	45
Secondary tillage	m	273	13.3	0.0	ft	48	3.7	0.0	1.0	0.85	0.65	35
Disk harrow, tandem												
Primary tillage	m	3.9	16.0	0.0	ft	53	4.6	0.0	1.0	0.88	0.78	50
Secondary tillage	m	216	11.2	0.0	ft	37	3.2	0.0	1.0	0.88	0.78	30
Disk harrow, offset												
Primary tillage	m	364	18.8	0.0	ft	62	5.4	0.0	1.0	0.88	0.78	50
Secondary tillage	m	254	13.2	0.0	ft	44	3.8	0.0	1.0	0.88	0.78	30
Field cultivator												
Primary tillage	tools	46	2.8	0.0	tools	26	2.5	0.0	1.0	0.85	0.65	30
Secondary tillage	tools	32	1.9	0.0	tools	19	1.8	0.0	1.0	0.85	0.65	25
Row crop cultivator												
S-tine	rows	140	7.0	0.0	rows	80	6.4	0.0	1.0	0.85	0.65	15
C-shank	rows	260	13.0	0.0	rows	248	19.9	0.0	1.0	0.85	0.65	20
No-till	rows	435	21.8	0.0	rows	248	19.9	0.0	1.0	0.85	0.65	20

(continued)

APPENDIX V. (*Continued*)

Implement	Width (units)	SI units Machine parameters C_1	C_2	C_3	Width (units)	English units Machine parameters C_1	C_2	C_3	Soil parameters F_1	F_2	F_3	Range ± %
					Minor tillage tools							
Rotary hoe	m	600	0.0	0.0	ft	41	0.0	0.0	1.0	1.0	1.0	30
Coil tine harrow	m	250	0.0	0.0	ft	17	0.0	0.0	1.0	1.0	1.0	20
Spring tooth harrow	m	2,000	0.0	0.0	ft	135	0.0	0.0	1.0	1.0	1.0	35
Land plane	m	8,000	0.0	0.0	ft	550	0.0	0.0	1.0	1.0	1.0	45
					Seeding implements							
Row crop planter												
Mounted	rows	500	0.0	0.0	rows	110	0.0	0.0	1.0	1.0	1.0	25
Drawn	rows	900	0.0	0.0	rows	200	0.0	0.0	1.0	1.0	1.0	25
Row crop planter, no-till												
Three coulters/row	rows	3,400	0.0	0.0	rows	765	0.0	0.0	1.0	0.94	0.82	35
Grain drill w/press wheels												
<8 ft	rows	400	0.0	0.0	rows	90	0.0	0.0	1.0	1.0	1.0	25
8 ft to 12 ft	rows	300	0.0	0.0	rows	67	0.0	0.0	1.0	1.0	1.0	25
>12 ft	rows	200	0.0	0.0	rows	25	0.0	1.0	1.0	1.0	1.0	25
Grain drill, no-till	rows	720	0.0	0.0	rows	160	0.0	0.0	1.0	0.92	0.79	35
Hoe drill												
Primary tillage	m	6,100	0.0	0.0	ft	420	0.0	0.0	1.0	1.0	1.0	50
Secondary tillage	m	2,900	0.0	0.0	ft	200	0.0	0.0	1.0	1.0	1.0	50
Pneumatic drill	m	3,700	0.0	0.0	ft	250	0.0	0.0	1.0	1.0	1.0	50

[1] ASABE Standards 1997. Information for additional machines are included in the standards.

APPENDIX VI. Solid animal waste production and characteristics.

	Size		Production			Density		Nutrient Content					
								Nitrogen		Potassium		Phosphorus	
Animal	(lb)	(kg[i])	(lb/day)	(kg/day)	Water (% WB)	(lb/ft^3)	(kg/m^3)	(lb/day)	(kg/day)	(lb/day)	(kg/day)	(lb/day)	(kg/day)
Dairy cattle	150	68	12.0	5.4	87.3	62.0	993	0.06	0.0272	0.010	0.0045	0.04	0.0181
	250	113	20.0	9.1	87.3	62.0	993	0.10	0.0454	0.020	0.0091	0.07	0.0318
	500	227	41.0	18	87.3	62.0	993	0.20	0.0907	0.036	0.0163	0.14	0.0635
	1000	454	82.0	37	87.3	62.0	993	0.14	0.0635	0.073	0.0331	0.27	0.1225
	1400	635	115.0	52	87.3	62.0	993	0.57	0.2586	0.102	0.0463	0.38	0.1724
Beef cattle	500	227	30.0	14	88.4	60.0	961	0.17	0.0771	0.056	0.0254	0.12	0.0544
	750	340	45.0	20	88.4	60.0	961	0.26	0.1179	0.084	0.0381	0.19	0.0862
	1000	454	60.0	27	88.4	60.0	961	0.34	0.1542	0.110	0.0499	0.24	0.1089
	1250	567	75.0	34	88.4	60.0	961	0.43	0.1950	0.140	0.0635	0.31	0.1406
Nursery pig	35	16	2.3	1.0	90.8	60.0	961	0.016	0.0073	0.0052	0.0024	0.010	0.0045
Growing pig	65	29	4.2	1.9	90.8	60.0	961	0.029	0.0132	0.0098	0.0044	0.020	0.0091
Finishing pig	150	68	9.8	4.4	90.8	60.0	961	0.068	0.0308	0.022	0.0100	0.045	0.0204
	200	91	13.0	5.9	90.8	60.0	961	0.090	0.0408	0.030	0.0136	0.059	0.0268
Gestating sow	275	125	8.9	4.0	90.8	60.0	961	0.062	0.0281	0.021	0.0095	0.040	0.0181
Sow and litter	375	170	33.0	15	90.8	60.0	961	0.23	0.1043	0.076	0.0345	0.15	0.0680
Boar	350	159	11.0	5.0	90.8	60.0	961	0.078	0.0354	0.026	0.0118	0.051	0.0231
Sheep	100	45	4.0	1.8	75.0	65.0	1041	0.045	0.0204	0.0066	0.003	0.032	0.0145
Layers	4	1.8	0.21	0.10	74.8	60.0	961	0.0029	0.0013	0.0011	0.0005	0.0012	0.0005
Broilers	2	0.9	0.14	0.06	74.8	60.0	961	0.0024	0.0011	0.00054	0.0002	0.00075	0.0003
Horse	1000	454	45.0	20	79.5	60.0	961	0.27	0.1225	.0460	0.0209	0.1700	0.0771

[i] Direct conversion using 1 lb = 0.453924 kg.

APPENDIX VII. Nutrient utilization by crops.[a]

			N		P_2O_5		K_2O	
Crop	Yield		lb/ac	kg/ha	lb/ac	kg/ha	lb/ac	kg/ha
Corn	150 bu/ac	3.82 Mg/ha	185	207	80	90	215	241
	180 bu/ac	4.58 Mg/ha	240	269	100	112	240	269
Corn silage	32 tons	72 t/ha	200	224	80	90	245	274
Soybeans	50 bu/ac	1.36 Mg/ha	257	288	48	54	120	134
	60 bu/ac	1.64 Mg/ha	336	376	65	73	145	162
Grain sorghum	8,000 lb/ac	3.64 Mg/ha	250	280	90	100	200	224
Wheat	60 bu/ac	1.36 Mg/ha	125	140	50	56	110	123
	80 bu/ac	2.18 Mg/ha	186	208	54	60	162	181
Oats	100 bu/ac	1.45 Mg/ha	150	168	55	62	150	168
Barley	100 bu/ac	2.18 Mg/ha	150	168	55	62	150	168
Alfalfa	8 tons	18 t/ha	450	504	80	90	480	538
Orchard grass	6 tons	13 t/ha	300	336	100	112	375	420
Brome grass	5 tons	11 t/ha	166	186	66	74	254	284
Tall fescue	3.5 tons	8 t/ha	125	140	65	73	185	207
Bluegrass	3 tons	7 t/ha	200	224	55	62	180	202

Source: Imperial values reproduced with permission from: *Structures and Environment Handbook*, MWPS-1, 11th edition, revised 1987, Midwest Plan Service, Ames, IA 50011-3080.

[a] SI values are direct conversions rounded to the nearest whole unit.

APPENDIX VIII. Maximum annual application rates for phosphates based on soil family.

		PH			
Soil families	% Clay	6.0 to 7.5		<6.0 or >7.5	
Sandy	<10	300	336 kg/ha	300 lb/ac	336 kg/ha
Coarse-loamy and Coarse-silty	11–18	400 lb/ac	448 kg/ha	500 lb/ac	560 kg/ha
Fine-loamy and Fine-silty	19–35	450 lb/ac	504 kg/ha	600 lb/ac	672 kg/ha
Fine	36–60	500 lb/ac	560 kg/ha	750 lb/ac	840 kg/ha
Very fine	>60	500 lb/ac	560 kg/ha	750 lb/ac	840 kg/ha
Loamy-skeletal	15–35	150 lb/ac	168 kg/ha	200 lb/ac	224 kg/ha
Clayey-skeletal	>35	250 lb/ac	280 kg/ha	350 lb/ac	392 kg/ha

APPENDIX IX. Insulating properties of various building materials.

			Resistance	
Description	Thickness	Density (lb/ft^3)	Per inch $\left(\frac{°F ft^2 h}{Btu\ in}\right)$	Description
Building board				
Asbestos-cement		120	0.25	
Gypsum or plaster	0.375 in	50		0.32
	0.5 in	50		0.45
	0.625 in	50		0.56
Plywood (Douglas fir)		34	1.25	
Vegetable fiber board				
Sheathing, regular density	0.5 in	18		1.32
Sheathing, intermediate density	0.5 in	22		1.09
Sound deadening board	0.5 in	15		1.35
Laminated paperboard		30	2.00	
Hardboard				
Medium density		50	1.37	
High density		55	1.22	
Particleboard				
Low density		37	1.41	
Medium density		50	1.06	
High density		62	1.18	
Underlayment	0.625 in	40	1.22	
Waferboard		37	1.59	
Wood subfloor	0.75 in			0.94
Building membrane				
Vapor-permeable felt				0.06
Vapor-seal, 2 layers of mopped 15-lb. felt				0.12
Finish flooring materials				
Carpet and fibrous pad				2.08
Carpet and rubber pad				1.23
Cork tile	0.125 in			0.28
Terrazzo	1 in			0.08
Tile-asphalt, linoleum, vinyl, rubber				0.05
Insulating materials				
Blanket or batt, mineral fiber processed from rock, slag or glass				
	3–4 in	0.4–2.0		11.0
	3.5 in	0.4–2.0		13.0
	3.5 in	1.2–1.6		15.0
	5.5–6.5 in	0.4–2.0		19.0
Boards and slabs				
Cellular glass		8.0		
Glass fiber		4.0–9.0	4.0	
Expanded perlite		1.0	2.78	
Expanded rubber		4.5	4.55	
Expanded polystyrene, extruded, smooth		1.8–3.5 in	5.00	
Expanded polystyrene, molded beads		1.0	3.85	
		1.25	4.00	
		1.5	4.17	
		1.75	4.17	

(continued)

APPENDIX IX. (*Continued*)

Description	Thickness	Density (lb/ft^3)	Resistance Per inch $\left(\frac{°F ft^2 h}{Btu in}\right)$	Resistance Description
Cellular polyurethane/polyisocyanurate		1.5	6.25–5.56	
Cellular phenolic, closed cell		3.0	8.20	
Mineral Fiberboard				
Core or roof insulation		16–17	2.94	
Acoustical tile		18.0	2.86	
		21.0	2.70	
Cement fiber slabs				
Wood and Portland cement binder		25.0–27.0	2.0–1.89	
Wood and magnesia oxysulfide binder		22.0	1.75	
Lose fill				
Cellulosic		2.3–3.2	3.7–3.13	
Perlite, expanded		2.0–4.1	3.7–3.3	
		4.1–7.4	3.3–2.8	
		7.4–11.0	2.8–2.4	
Mineral fiber (rock, slag or glass)				
	3.75–5 in	0.6–2.0		11.0
	6.5–8.75 in	0.6–2.0		19.0
	7.5–10.0 in	0.6–2.0		22.0
Vermiculite, exfoliated		7.0–8.2	2.13	
		4.0–6.0	2.27	
Spray applied				
Polyurethane foam		1.5–2.5	6.25–5.56	
Ureaformaldehyde		0.7–1.6	4.53–3.57	
Cellulosic fiber		3.5–6.0	3.45–2.94	
Glass fiber		3.5–4.5	3.85–3.70	
Roofing				
Asbestos-cement shingles		120		0.21
Asphalt roll roofing		70		0.15
Asphalt shingles		70		0.44
Slate	0.5 in			0.05
Wood shingles				0.94
Plastering Materials				
Cement plaster sand aggregate		116	0.20	
Gypsum Plaster				
Lightweight aggregate	0.5 in	45		.32
Lightweight aggregate	0.625 in	45		.39
Masonry Materials				
Brick, fired clay		150	0.12–0.10	
		130	0.16–0.12	
		120	0.18–0.15	
		100	0.24–0.20	
		80	0.33–0.27	
Clay Tile, hollow				
1 cell deep	3 in			0.80
	4 in			1.11
2 cells	6 in			1.52
	10 in			2.22
3 cells	12 in			2.50

APPENDIX IX. (*Continued*)

Description	Thickness	Density (lb/ft^3)	Resistance: Per inch $\left(\frac{°F ft^2 h}{Btu in}\right)$	Resistance: Description
Concrete blocks				
Normal weight aggregate,				
8 in, 33–36 lb				1.11–0.97
With perlite filled cores				2.0
With vermiculite cores				1.92–1.37
12 in, 50 lb, 2 cores				1.23
Lightweight aggregate,				
6 in, 16–17 lb				1.93–1.65
With perlite filled cores				4.2
With vermiculite cores				3.0
8 in, 19–22 lb				3.2–1.90
With perlite filled cores				6.8–4.4
With vermiculite cores				5.3–3.9
12 in, 32–36 lb				2.6–2.3
With perlite filled cores				9.2–6.3
With vermiculite cores				5.8
Stone, lime or sand		180	0.01	
		160	0.02	
		140	0.04	
		120	0.08	
Marble and granite		180	0.03	
		160	0.05	
		140	0.06	
		120	0.09	
		100	0.13	
Woods				
Hardwoods				
Oak		41.2–46.8	0.89–0.80	
Birch		42.6–45.4	0.87–0.82	
Maple		39.8–44.0	0.92–0.84	
Ash		38.4–41.9	0.94–0.88	
Softwoods				
Southern Pine		35.6–41.2	1.00–0.89	
Douglas Fir-Larch		33.5–36.3	1.06–0.99	
Southern Cypress		31.4–32.1	1.11–1.09	
West Coast Cedar		21.7–31.4	1.48–1.11	
California Redwood		24.5–28.0	1.35–1.22	
Windows[b]				
Single glazed				0.06
With storm windows				1.15
Insulating glass, 1/4 air space				
Double pane				0.84
Triple pane				1.71
Air space, 3/4 to 4 inches				0.90

(*continued*)

APPENDIX IX. (*Continued*)

Description	Thickness	Density (lb/ft^3)	Resistance Per inch $\left(\frac{°F ft^2 h}{Btu\ in}\right)$	Resistance Description
Surface conditions				
Inside surface				0.68
Outside surface				0.17
Floor				
Concrete slab on ground				1.23
Concrete slab with insulation				2.22

[1] For conversion to SI units $R_{SI} = 5.678\ R_{Cust}$. R_{SI} is in units of $\frac{kJ}{°C\ m^2\ h}$. R_{Cust} is in units of $\frac{BTU}{°F\ ft^2\ h}$.

[b] Reprinted from Introduction to Agricultural Engineering, 2nd edition.

APPENDIX X. Moisture and heat produced.[1]

Animal	Moisture[2] (lb water/hr)	Heat* (BTU/hr)
Swine		
10–50 lb	0.065	174.0
50–100	0.177	240.0
100–200 lb	0.219	354.0
Broiler		
0.22–1.5 lb	0.0079	12.4
1.5–3.5 lb	0.0110	18.7
Turkey		
1 lb	0.0059	10.8
2 lb	0.0025	9.8
Dairy cow	1.196	1917.0
Equipment		
Incandescent lighting		3.4 Btu/hr-watt
Fluorescent lighting		4.1 Btu/hr-watt
Electric motors		4,000 Btu/hr-Hp

[1] Reproduced with permission from: *Structures and Environment Handbook*, MWPS-1, 11th edition, revised 1987, Midwest Plan Service, Ames, IA 50011-3080.

[2] The amount of water and heat released by animals changes as the air temperature changes. These are average values.

APPENDIX XI. Allowable fiber stress by species.

Use	Grade	Size		Allowable fiber stress* (lb/in^2)
		Thick	Wide	
	Douglas fir and Larch (19% moisture)			
Structural light framing				
	Select structural	2″–4″	2″–4″	1,200
	No. 1			1,050
	No. 2			850
Light framing	Construction	2″–4″	4″	625
	Standard			350
	Utility			175
	Southern Pine (19% moisture)			
Structural light framing				
	Select structural	2″–4″	2″–4″	1,150
	No. 1			1,000
	No. 2			825
	Stud			450
Light framing	Construction	2″–4″	4″	600
	Standard			350
	Utility			150

* For lumber milled and used at 15% moisture, multiply by 1.08 for allowable fiber stress.
Source: Reproduced with permission from: *Structures and Environment Handbook*, MWPS-1, 11th edition, revised 1987, Midwest Plan Service, Ames, IA 50011-3080.

APPENDIX XII. Copper wire resistance (Ohms/1,000 ft).

Wire size (AWG*)	Resistance (Ohms/1,000 ft)
22	16.46
20	10.38
18	6.51
16	4.09
14	2.58
12	1.62
10	1.02
8	0.64
6	0.41
4	0.26
2	0.16
0	0.10

APPENDIX XIII. Copper wire resistance (Ohms/100 m).

Wire size (mm^2)	Resistance (Ohms/100 m)
1.5	1.073
2.5	0.676
4.0	0.423

APPENDIX XIV. Cooper wire sizes for 120 volt, single phase, 2% voltage drop.

Minimum allowable size				Run of wire (ft)													
	In cable, conduit and earth		Overhead in air, bare and covered conductor	Compare size shown below with size shown under minimum allowable size and the used the larger size													
Load (amp)	R,T, TW	RH, RHW, THW		50	75	100	125	150	175	200	225	275	300	350	400	450	500
5	14	14	10	14	14	12	12	10	10	10	8	8	8	6	6	6	6
7	14	14	10	14	12	12	10	10	8	8	8	6	6	6	6	4	4
10	14	14	10	12	12	10	8	8	8	6	6	6	4	4	4	4	3
15	14	14	10	12	10	8	6	6	6	6	4	4	4	3	2	2	2
20	14	12	10	10	8	6	6	6	6	4	4	3	3	2	1	1	0

APPENDIX XV. Cooper wire sizes for 120 volt, single phase, 2% voltage drop.

Minimum allowable size				Run of wire (ft)													
	In cable, conduit and earth		Overhead in air, bare and covered conductor	Compare size shown below with size shown under minimum allowable size and the used the larger size													
Load (amp)	R,T, TW	RH, RHW, THW		50	75	100	125	150	175	200	225	275	300	350	400	450	500
5	14	14	10	14	14	14	14	14	12	12	12	12	10	10	8	8	6
7	14	14	10	14	14	14	14	12	12	12	10	10	10	8	8	6	6
10	14	14	10	14	14	12		2	10	10	10	10	8	6	6	6	4
15	14	14	10	14	12	12	10	10	8	8	8	8	6	4	4	4	2
20	12	12	10	12	12	10	8	8	8	6	6	6	4	4	2	2	2

Index

Printed in the United States of America

Lightning Source UK Ltd.
Milton Keynes UK
06 October 2009

144586UK00001B/19/P

9 780387 369136